Diversity and Development

Critical Contexts That Shape Our Lives and Relationships

Editor
DANA COMSTOCK
St. Mary's University

THOMSON
™
BROOKS/COLE

AUSTRALIA • CANADA • MEXICO • SINGAPORE • SPAIN
UNITED KINGDOM • UNITED STATES

*This book is dedicated to my mother, Judy Ann Bennett Blake,
whose gifts in my life are simply immeasurable.*

Executive Editor: Lisa Gebo
Acquisitions Editor: Marquita Flemming
Assistant Editor: Shelley Gesicki
Editorial Assistant: Amy Lam
Technology Project Manager: Barry Connolly
Marketing Manager: Caroline Cincilla
Marketing Assistant: Mary Ho
Advertising Project Manager: Tami Strang
Project Manager, Editorial Production:
   Mary Noel

Art Director: Vernon Boes
Print/Media Buyer: Doreen Suruki
Permissions Editor: Joohee Lee
Production Service: Stratford Publishing Services
Copy Editor: Claudia Dalton
Cover Designer: Larry Didona
Cover Image: Chris M. Rogers/Getty Images
Compositor: Stratford Publishing Services
Printer: Quebecor World/Kingsport

Library of Congress Control Number: 2004106248
ISBN   0-534-57406-8

Thomson Brooks/Cole
10 Davis Drive
Belmont, CA 94002
USA

Asia
Thomson Learning
5 Shenton Way #01-01
UIC Building
Singapore 068808

Australia/New Zealand
Thomson Learning
102 Dodds Street
Southbank, Victoria 3006
Australia

Canada
Nelson
1120 Birchmount Road
Toronto, Ontario M1K 5G4
Canada

Europe/Middle East/Africa
Thomson Learning
High Holborn House
50/51 Bedford Row
London WC1R 4LR
United Kingdom

Brief Contents

Contents

Preface

This book, *Diversity and Development: Critical Contexts That Shape Our Lives and Relationships,* is an exercise in understanding. I chose the title, in part, to honor the ways in which we are all inextricably linked. Although you may find you are reading and working to understand the context of lives very different from your own, you will ultimately come to understand yourself in a new light within the socio-cultural context in which we are all challenged to thrive.

Many of you may be helping professionals in the fields of counseling, psychology, and social work, who are committed to facilitating the growth and well-being of others. "Growth," as used in the context of this book, is not limited to individual growth, and includes the notion of assisting others to establish and maintain growth-fostering relationships necessary for psychological well-being and emotional resilience over the life span. The idea of this sounds simple enough, yet developing the type of therapeutic relationship for this to take place is often challenging and complex. Because there is no linear formula for this to happen, the process can feel overwhelming for beginning and seasoned professionals alike.

Understanding your role in this process is a lifelong effort. Our personal, professional, individual, and collective growth is not static and the ways in which we experience change in our lives take place in relationships that are always fluid. We all change and grow through the mutual impact and influence we have on each other. To say that we ever "arrive," with regard to our professional development, is dangerous—and the idea that this would ever happen is simply a myth. Therefore, the goals of this text do not strictly involve understanding the critical contexts of "others" lives; rather you, the reader, will be challenged to better understand your own life. The idea is that the better we know ourselves, the more capable we are of bringing forth our authentic experiences into all of our relationships. With regard to our professional development, "knowing" ourselves needs to include the ability to identify and dismantle our blocks to understanding a wide range of human experiences.

I wrote this text for two reasons. The first reason came from a growing concern that some of my counseling students who had completed endless hours of course work were still cautious and confused about how to develop a therapeutic alliance. The second reason was that many of them were unfamiliar and unprepared to deal with some of the more common clinical issues. In my mind's eye, I carry countless images of wide-eyed students sitting in front of me reporting that they have their first "sexual abuse client," for example, and that they have no idea about how or where to start. In spite of several things being wrong with this image, a few things are, in fact, right. On the one hand, it is great for students to be excited and if we educators have managed to instill a love of the work we have done at least part of our job.

On the other hand, the use of objectifying language, the sense of fear, and a curiosity over how clients might be impacted by the feeling that their therapist, for example, is afraid of them or their experience really began to gnaw at me. And I want to make the point that clients who come in feeling defective somehow will internalize what they sense as fear in the therapist as reflecting something about *them*, even if what they sense may be an awkwardness with respect to the process or something completely unrelated. What has become clear to me is that the fear and concern, which is so distracting for students, comes from an ideology of what they think "good" therapy *should* be, not from what it really is.

In a paper on empathy and shame, Judy Jordan (1997) makes the point that therapy is more about who we are rather than what we do. She states, "Therapy cannot be a mechanistic enterprise but must take the therapist as well as the client to deep places of vulnerability and possible shame" (p. 153). She goes on to write that "the rendering of the process in textbooks often contributes to a feeling that the books do it right, and we do it wrong" (p. 153). This idea is perpetrated by a seeming rash of new books and professional presentations on the successes of the "Master Therapist."

In Chapter 16 of this text, Linda Hartling also cites Jordan (1999), who makes the point that "'to master' is to reduce to subjection, to get the better of, to break, to tame" (pp. 1–2). As a result, "[M]astery implicit in most models of competence creates enormous conflict for many people, especially women and other marginalized groups, people who have not traditionally been 'the masters'" (p. 2).

In addition, Jordan (1997) calls for anyone writing about the therapy process to "please say what is actually happening in the therapy relationship, not what theory prescribes or what sounds smart or clever or theoretically informed" (p. 153). I think what is missing in most books on the therapeutic process is a discourse on the therapist's vulnerabilities. I also feel what is missing in most training programs is the inclusion of an emphasis on relational competence either in lieu of, or in addition to, "skill mastery," especially since many trainees feel marginalized from the whole notion of "mastery" over anything. Relational competence, as a gauge to determine preparedness for clinical work, would include the capacity to identify and name our vulnerabilities and ideologies regarding our professional development.

To facilitate this process, this text deals with the notion of "controlling images" (Collins, 2000), which are constrictive in that they define with whom and in what ways we should relate to others in any given context. In the context of the therapeutic relationship we may hold an image of a "good therapist" as someone who is skilled, competent, knowledgeable, scholarly, in control, creative, cool under pressure, confident, and maybe even aloof depending on what school of thought guides his or her practice. I think it is fair to say that most therapists (novice and seasoned alike) operate under an image much like the one constructed here. When this image is not reinforced or is threatened by some relational dynamic in the therapy process, most therapists work to restore it.

The means by which this is done can range from pathologizing the client (to clarify where and in whom the problem lies) to taking a distanced expert stance (for the purpose of reinforcing the notion of "difference") resulting in the restoration of power to the therapist who may have felt knocked off kilter. The effects on clients are too numerous to name and the therapist, who is busy trying to restore his or her image, is too distracted to notice. Any time there is a movement toward an image, there is a movement out of connection. My feeling is that novice therapists, much like the counseling interns I work with, grapple with their image of the "good therapist" (who has also "mastered" a set of skills) during this vulnerable time and in doing so they impede their ability to connect with the client.

What I have found helpful for students through the years is to have them name and dismantle their respective therapist images. Instead of a self-conscious focus on the "me," we shift our thinking to the notion of "we." In this shift, our focus turns to the therapeutic relationship and the idea that it is *mutually* constructed. And although therapists have certain responsibilities in helping define the relationship, they will be most successful in constructing it using the expertise of their clients. Relational-Cultural theorists have always encouraged the therapists to foster in themselves an attitude of "learning" and curiosity versus an attitude of "knowing" and/or "mastery." Through this approach clients sense respect rather than fear and in turn they feel safer bringing their experience into the relationship. When this happens, both the therapist and client leave the session feeling confident and energized for the work that lies ahead. My hope is that this text, with a focus on the development of relational capacities over the life span, will breathe new life into our therapeutic relationships.

The second reason I wrote this text was to mainstream many clinical issues that are often pushed to the margin in our training programs. With accreditation standards demanding so much in the way of core course work it is hard to work in an analysis of critical developmental contexts reflective of life's difficulties. The sad fact of the matter is that one in four women are sexually abused; chronic experiences of racism are traumatic (and that the combination of sexual and racial oppression has been written about as a "dual traumatization"); and the socialization of boys and men is laden with emotional and physical violence. This list goes on and is reflected in the chapter topics of this text, which are grounded in Relational-Cultural theory (RCT).

RELATIONAL-CULTURAL THEORY AND RELATIONAL DEVELOPMENT AS A CENTRAL THEME

Relational-Cultural theory was conceived after the publication of *Towards a New Psychology of Women* (Miller, 1976). After this publication, a group of scholars, namely Jean Baker Miller, Irene Stiver, Alexander Kaplan, Judy Jordan, and Janet Surrey, began a process of reconceptualizing traditional models of human development that value the ideals of individuation, separation, autonomy, and generally honor the concept of the "self" (Fedele, 1994), particularly as they applied to women's experiences. What emerged was a new developmental model termed "Self-in-Relation" theory, which suggests that we become increasingly relationally complex over the life span rather than more individuated and autonomous and that healthy psychological development occurs in the context of growth-fostering relationships.

Through the years the founding scholars began a more in-depth analysis of relationships to answer such questions as: What differentiates relationships that foster growth versus those that impede growth? What kinds of relational dynamics lead to connections and disconnections in relationships? How do our experiences of connections and disconnections in relationships contribute to our sense of agency in relationships or to experiences of chronic disconnections or condemned isolation? What does growth in connection really feel like? How do social, cultural, and political contexts play into all of this? And lastly, and probably most importantly: *How can the therapeutic relationship be constructed to foster relational competence and growth?* These are complicated and important questions that guided the emergence of what is now known as the "Relational-Cultural theory" (RCT) of psychotherapy.

Most models of individual and family development emphasize independence, autonomy, individuation, and separation as the basis for development, particularly during adolescence. Jargon such as "breaking away," or "cutting the umbilical cord" all imply a movement toward disconnection from the family, primarily from mother. When this is not achieved, our professional language describes individuals and their relationships as being "dependent," "codependent," "enmeshed," or "fused." In spite of over 20 years of research that suggest that connection, not separation, is central to female development (Gilligan, 1982; Jordan et al., 1991; Miller, 1976), our professional texts still lack a thorough discourse regarding interdependence and mutuality as central to psychological growth and emotional and physical well-being.

Bergman (1991) suggests that traditional theories of development do not address "the whole" of men's relational development, and that they "fail to describe so much of men's authentic experience." He goes on to state that "much of what is promoted in these old theories seems inaccurate, irrelevant, worn and weird" and he goes on to call for a new understanding of the psychology of men (Bergman, 1991, p. 3). Unless we address one's capacity to develop and sustain growth-fostering relationships as an essential goal of development, it seems unlikely that we can expect to be able to promote "optimum development over the life-span" for either men or women (CACREP Accreditation Standards, 2001).

As a way to examine relational development, Jordan, Kaplan, Miller, Stiver, and Surrey (1991) proposed a new construct of development, which Surrey termed "relationship-differentiation." The term "differentiation" in this regard refers to a "dynamic process that encompasses increasing levels of complexity, structure and articulation within the context of human bonds and attachment"

(p. 36). Relationship-differentiation developmental theory posits that we: (a) view critical relationships as "evolving over the life-span in a real, rather than intrapsychic form"; (b) "account for the capacity to maintain relationships with tolerance, consideration, and mutual adaptation to the growth and development of each person"; (c) "account for the ability to move closer to and further away from other people at different moments, depending on the needs of the particular individuals and the situational context"; (d) "explore the capacity for developing additional relationships based on broader, more diversified new identifications and corresponding patterns of expanding relational networks"; and (e) "examine potential problems inherent in the development of these relational capacities" (Jordan et al., 1991, pp. 38–39).

Such problems include the impact of oppression, marginalization, violence, abuse, neglect, shame, loss, gender socialization, and various developmental challenges, to name a few. All of these developmental contexts have the potential to be sources of chronic disconnections that inhibit relational growth and impede psychological well-being and are covered in depth in this text.

STYLE AND STRUCTURE

Whether or not we like to admit it, our value systems regarding what constitutes mental health and "normal" development are made up of prejudices, assumptions, and ideologies that have been taught to us, both overtly and covertly, by our families, our peers, the mass media/Hollywood (particularly advertising) and by all the institutions (religious, educational, etc.) we negotiate in our lives. Collins (2000) reminds us that "within U.S culture, racist, and sexist ideologies permeate the social structure to such a degree that they become hegemonic, namely seen as natural, normal and inevitable" (p. 5).

This process is also meant to be invisible, and as a result some individuals develop within a context of privilege while others develop within a context of marginalization. To make matters worse, we live by the "myth of meritocracy," which means that when things go well for a person in his/her life he/she has done all the right things. And when things don't go well for another, the obstacles aren't seen in any particular context, rather the problem is seen *within* the individual. This is a shame-based, complex dynamic, which is not typically addressed in most developmental texts.

The messages we all internalize as a result of our cultural ideologies contribute to our felt sense of worth and to the degree of worth we place on others and their experiences. The degree to which we hold on to oppressive ideologies inhibits our ability to take in and to understand diverse ways of being and developing in the world. This text is designed to challenge constrictive assumptions about "normal" development by encouraging a critical analysis of what we are taught, both formally and informally. This process will facilitate the integration of your personal and professional development as a parallel process, which, in turn, will increase your readiness for clinical practice.

To facilitate this process, each chapter begins with a set of reflection questions that challenges you, the reader, to think about ideas, dilemmas, and issues you may have never considered before. There may be times when you have no idea what exactly it is you are being asked to consider. If you find yourself feeling uncomfortable, confused, hurting, ashamed, angry, clueless, defensive, or perhaps you are left with more questions than answers, I simply ask that you *pay attention, hang in there,* and *go with the reading.* It may even be helpful to revisit the questions after completing the various chapters and to journal what issues come up for you during the readings.

The writing style of this text is informative, conversational, and engaging. The authors were asked to share with you much of what they have struggled with in regard to their own vulnerabilities and relational growth in their lives. My hope is that their intimate disclosures will teach you that seasoned professionals (versus "Master Therapists") are simply human, have their own struggles and make mistakes. My intention was to put forth a style that encourages and models self-reflection in a way that will serve to help you move through any obstacles that impede your own relational growth.

To get you started, Chapter 1 includes an overview and a critical analysis of traditional theories of human development including psychosexual, psychosocial, cognitive, and moral development as well as Maslow's theory of self-actualization. What is unique about this text is that we included a discussion of the sociocultural context of the respective theorists' lives, which clearly shaped their work. We also included a description of the relationships they had with each other and demonstrate the mutual impact they had on each other's work. We also included the work of Karen Horney, Lev Vygotsky, and Carol Gilligan, who led the way to a more inclusive and relational understanding of human development.

Chapter 2 is an immersion in Relational-Cultural theory. Chapters 3 to 15 focus on critical thinking; identity development and the intersection of race, gender, and ethnicity; counselor development and the interplay between one's personal and professional development; gender issues; bisexual, gay, lesbian, and transsexual issues across the life span; the impact of trauma on development; sexuality; gifted development; grief, loss, and death; marginalized family constructions and transitions; and spiritual development.

As we live in a shame-based culture, I was also interested in examining the role of shame as a secondary theme and its impact on the many aspects of our development. In particular, what is the role of shame with regard to patterns of emotional expression in men and women, our child-rearing practices, gender socialization, sexuality, one's sense of "feeling" and being different, and our sense of agency and self-worth worth, for example? The role of shame as a fallout of our cultural ideologies is also examined throughout the text.

Lastly, there is a chapter on fostering resilience across the life span. Chapter 16 highlights over 20 years of research that demonstrates how children and adolescents who have strong supportive relationships with adults (even just one) have a reduced risk for depression and many other risk factors, and enjoy overall better psychological health (Spencer, Jordan, and Sazama, 2002). As such, risk factors and protective factors are identified and delineated, along with the latest research illuminating how these protective factors operate in relationships. The text also provides guidelines for mutual engagement for parents, teachers, and mentors interested in developing and establishing growth-fostering relationships with the young people in their care. My hope is that *Diversity and Development: Critical Contexts That Shape Our Lives and Relationships* will provide for you context to the rich, rewarding, and formidable healing power of our connections, while inspiring and instilling relational hope, courage, and commitment to this crucial but all too frequently misunderstood relational process.

REFERENCES

Bergman, S. (1991) Men's psychological development: A relational perspective. *Work in Progress, No. 48*. Wellesley, MA: Stone Center Working Papers Series.

Collins, P. H. (2000). *Black feminist thought: Knowledge, consciousness and the politics of empowerment* (2nd ed.). New York: Routledge.

Council for Accreditation of Counseling and Related Educational Programs (CACREP), Accreditation Standards. (2001). Available from the American Counseling Association Web site, http://www.counseling.org.

Fedele, N. (1994). Relationships in groups: Connection, resonance and paradox. *Work in Progress, No. 69*. Wellesley, MA: Stone Center Working Papers Series.

Gilligan, C. (1982). *In a different voice*. Cambridge, MA: Harvard University Press.

Jordan, J. V. (1997). Relational development: Therapeutic implications of empathy and shame. In J. V. Jordan (Ed.), *Women's growth in diversity: More writings from the Stone Center*. New York: The Guilford Press.

Jordan, J. V. (1999). Toward connection and competence. *Work in Progress, No. 83*. Wellesley, MA: Stone Center Working Papers Series.

Jordan, J. V., Kaplan, A. G., Miller, J. B., Stiver, I. P., Surrey, J. L. (1991). *Women's growth in connection*. New York: The Guilford Press.

Miller, J. B. (1976). *Toward a new psychology of women*. Boston: Beacon Press.

Spencer, R., Jordan, J., & Sazama, J. (2002). Empowering children for life: A preliminary report. *Project Report #9*. Wellesley, MA: Wellesley Centers for Women.

Acknowledgments
in "Context"

A colleague of mine urged me to include a description of the context of my own life from which this text emerged so, for what it is worth: I undertook this project at a time in my life when things seemed somewhat settled. I had just been promoted to full professor, there had been some talk about my becoming department chair on down the line, and according to some developmental ideology I should have been looking forward to more personal and professional stability. I imagine the idea of all of that "stability" must have frightened me on some level so, at the age of 39, I decided to shake things up a bit. First, I made a decision to stop complaining about all the supplements I had to use in the development course I was teaching and decided to put together my own book. I applied for and was granted a sabbatical and was looking forward to spending a quiet fall semester writing at home.

Second, I decided I'd like to try for another addition to our family. As God and luck would have it, I conceived just as this project was getting under way. Because pregnancy complications I had anticipated began much earlier than in my previous pregnancy, I went on bed rest in April of 2003, where I remained until October, the very month the first draft of this text was due to the publisher and the same month in which my son was born. The acknowledgements in this section are made with a depth of gratitude to a multitude of individuals who kept me safe, sane, secure, and well-fed during the most intensely creative time of my life.

First, I'd like to thank my husband, Geoff Comstock, who never doubts, even for a second, that I could do just about anything and everything—all at the same time. Either I really am Superwoman or we should both reread Chapter 6! Thank you for your love, patience, and flexibility and for attributing my crankiness to my "creative process." Maybe one day we'll figure out how to do things the easy way. And if not, I am lucky to be riding these wild rides with you.

A special thanks to my daughter, Julianna, aka "Little Pea," who has been very patient in teaching me how to be a mother. You have also been a big help with your new baby brother and have given me very good advice. Thank you for showing me how to play the air guitar to "Twinkle Twinkle Little Star" (you rock, girl!), for letting John curl your hair even though you hate it, and for not crying before *every* violin lesson. All those chocolate rewards are worth it, aren't they?

Another special thanks to my new baby boy Kaleb for getting here in one piece, for looking a little like me, for thinking I'm funny, for being such good company in the middle of the night and for not getting colic. You will always be my "little sweetie cutie pie."

Thank you, Peter Kuhl, M.D., for letting me sneak out to lunch those two times, for pretending not to be as worried as you really were, and for your confidence that I could get through whatever came my way.

To my angel baby, Samantha, thank you for showing me the most tender and loving side of my family.

How do I even begin to thank my parents, Judy and Harold Blake? Let me just say there is something special about two people who can make each of their children feel like they are an only child. I can only hope my children feel as loved as I do. And thanks, Mom, for taking care of all of us during my time down. I owe you *big* time.

Thank you, Bill and Susan Comstock, for your enthusiastic support. I am lucky to have you as my family.

Thank you, Michelle, for straightening up our messes, for bathing the kids and for taking care of so many details. I know you worry about us as much as your own family.

A special thanks to Mr. Francis Farrell for taking care of the details before I've even had a chance to think about them. I hope you know how much we love and appreciate you.

And to Thelma Duffey, you truly are "Super Girlfriend!" You have given me so much confidence and support. Thanks for helping me find my wings and for insisting I use them! My work simply would not be what it is without you or our relationship.

Thank you, Jean Baker Miller, for your groundbreaking work. The Jean Baker Miller Training Institute (JBMTI) has truly had an invaluable impact on my professional development and I am so lucky to *love* my work! A huge thanks to all my JBMTI friends and mentors, including Judy Jordan, Linda Hartling, Maureen Walker, Amy Banks, Wendy Rosen, Lynn Leiberman, and the much loved and missed Irene Stiver for your brave and brilliant work.

My heartfelt gratitude to the St. Mary's University community for supporting me through my pregnancy and this project. Thanks to those of you who prayed me through what was a difficult and rewarding time. I am lucky to be able to thrive in such a loving environment.

To the authors, thank you for sharing your expertise and your lives with such courage and conviction. I learned so much from each and every one of you.

A special thanks to reviewers Priscilla Dass-Brailsford, Lesley University; C. Timothy Dickel, Creighton University; Paula R. Danzinger, William Paterson University of New Jersey; and Thomas Scofield, University of Nebraska, Kearney. Your invaluable feedback helped shape and refine this project. Thank you, Julie Martinez, Lisa Gebo, Shelley Gesicki, and Marquita Flemming, and the Brooks/Cole team for being so incredibly supportive. And thank you, Kristen Bettcher and Claudia Dalton, of Stratford Publishing, for paying such good attention to detail and for pulling the project together in record speed.

About the Editor

 Dana L. Comstock, Ph.D., LPC, received her doctorate at Mississippi State University in Counselor Education with a minor in Educational Psychology. She is a Professor of Counseling in the Department of Counseling and Human Services at St. Mary's University, San Antonio, TX. She has trained extensively at the Jean Baker Miller Training Institute (JBMTI), Wellesley College, Wellesley, MA. She recently completed the JBMTI's Practitioner's Program through the Stone Center under the supervision of Dr. Judith Jordan. Dr. Comstock integrates RCT into her teaching and writing and has published many journal articles and book chapters integrating RCT with group process, student development, and chronic illness. She is also featured in the first RCT casebook *How Connections Heal: Stories from Relational-Cultural Therapy,* published by The Guilford Press, and in the *Complete Guide to Mental Health: For Women,* published by Beacon Press. Dr. Comstock has a part-time private practice specializing in women's issues including pre- and perinatal loss and trauma in San Antonio, TX.

Contributors

Amy Elizabeth Banks, M.D.,
is Medical Director for Mental Health at the Fenway Community Health Center; on the faculty at the Jean Baker Miller Training Institute, Wellesley College; and an instructor of psychiatry at Harvard Medical School. Her private practice in Lexington, MA, specializes in the Relational–Cultural treatment of trauma survivors. Dr. Banks is also coeditor of the *Complete Guide to Mental Health: For Women,* published by Beacon Press.

Sharon Conarton, BSN, LCSW,
graduated from the University of Denver and is a Jungian psychotherapist and writer in Denver, Colorado. She founded and directed The Center for Relational Therapy and is currently writing about the midlife changes in men and women's relationships.

Thelma Duffey, Ph.D.,
is Associate Professor of Counseling and Graduate Program Director at The University of Texas at San Antonio. She is Founding President of the Association for Creativity in Counseling (American Counseling Association), Chair of the Dr. Lesley Jones Creativity in Psychotherapy Conference, and serves as Editor for the *Journal of Creativity in Mental Health*, Haworth Press, Inc., Publishing. She also maintains a private clinical practice. A graduate of Trinity University and St. Mary's University, Dr. Duffey's theoretical model is Relational–Cultural therapy (RCT). Using music, the Enneagram Personality Typology, and dream work as mediums, her research interests focus on creative interventions in relational and cultural movement.

Kimberly N. Frazier, Ph.D.,
received her degree from the University of New Orleans in Counselor Education. Her current position is the Director of Graduate Student Enrollment and Matriculation for the School of Education at Indiana University, Bloomington. Her research interests include pediatric counseling and multicultural supervision issues for counselors in training.

Linda M. Hartling, Ph.D.,
received her doctoral degree in Clinical/Community from the Union Institute and University in Cincinnati, Ohio. She is currently the Associate Director of the Jean Baker Miller Training Institute at the Stone Center, part of the Wellesley Centers for Women at Wellesley College, Massachusetts. She is a core member of an international team establishing the first Center for Human Dignity and Humiliation Studies, which is part of the Conflict Resolution Network anchored at Columbia University in New York City. She has published papers on resilience, substance abuse prevention, shame and humiliation, relational practice in the workplace, and Relational-Cultural theory. Dr. Hartling is the author of the Humiliation Inventory, a scale to assess the internal experience of derision and degradation, and the coeditor of *The Complexity of Connection: Writings from the Stone Center's Jean Baker Miller Training Institute* (2004).

Mary Howard-Hamilton, Ed.D.,
received her doctoral degree in Counselor Education, with a cognate in Psychology, from North Carolina State University. She is currently the Associate Dean of Graduate Studies and Associate Professor of Educational Leadership and Policy Studies at W. W. Wright School of Education at Indiana University, Bloomington. Her research interests include diversity issues in higher education, racial identity development theory, student development theory, as well as feminist therapy and theory. She is the coauthor of *The Convergence of Race, Ethnicity and Gender: Multiple Identities in Counseling*.

Ana M. Juárez, Ph. D.,
received her bachelor's and master's degrees from the University of Texas at Austin, and later went on to receive her doctoral degree in Anthropology from Stanford University. She is an Assistant Professor in the Department of Anthropology at Texas State University–San Marcos. She teaches courses on Mexican American Culture and Latin American Gender and Sexuality, among others. She has recent publications in *Frontiers: A Journal of Women Studies*; *The Journal of Latin American Anthropology*; *Human Organization*; *Aztlan: A Journal of Chicano Studies*; and *Latin American Indian Literatures Journal*. Juárez's research focuses on issues of race, gender, sexuality, and the effects of globalization among Latinas/os in the United States and in Latin America, especially among Mayas in Quintana Roo, Mexico.

Stella Beatriz Kerl, Ph.D.

is a Licensed Psychologist and an Associate Professor in the Professional Counseling program at Texas State University–San Marcos. She has worked in many different clinical settings, including women's centers, community mental health agencies, medical settings, and university counseling centers. Her research interests primarily involve gender, race, and cultural issues in therapy and mental health. She regularly presents at state and national conferences and has publications in several journals, including the *Journal of Multicultural Counseling and Development, Counselor Education and Supervision, Aztlan: A Journal of Chicano Studies,* and *Rehabilitation Psychology.*

Jeffrey Kottler, Ph.D.,

is Professor and Chair of the Counseling Department at California State University, Fullerton. Jeffrey has been a Fulbright Scholar in Peru, Iceland, and Thailand, and has worked extensively throughout Asia and the Pacific Rim. He is the author of over 55 books, including *On Being a Therapist; Introduction to Therapeutic Counseling; Making Changes Last; American Shaman; Bad Therapy;* and *Counselors Finding Their Way.* In addition to his counseling interests, he is an avid surfer, runner, traveler, and photographer.

Marilyn Montgomery, Ph.D., LMHC, NCC,

is Associate Professor of Psychology at Florida International University in Miami, Florida, where she teaches in the Mental Health Research and Services and the Developmental programs. She received her doctorate in Human Development at Texas Tech University. Her areas of interest include the professional development of counselors, diversity training for non-majority groups and individuals, and interventions that promote individual well-being and emotional development through relationship enhancement and personal empowerment. Her most recent book is titled, *Building Bridges to Parents*, published by Sage.

Melanie Munk, MA, LPC, NCC,

is a doctoral candidate in counselor education and supervision at St. Mary's University in San Antonio and a counselor at Family Violence Prevention Services. She received a B.S. from West Point, the United States Military Academy, and an M.A. in counseling from St. Mary's University. Her research interests include spiritual development, identity development in lesbian women, and complementary and alternative approaches to counseling.

Dongxiao Qin, Ph.D.,

received her doctoral degree in Developmental Psychology from the Lynch School of Education, Boston College, in 2000. She is currently an Assistant Professor of Psychology at Western New England College. Her research interests are in critical feminist theories and women's self-development in cross-cultural contexts. Her current research examines international women students' self-development in U.S. universities.

Stacee Reicherzer, M.A.,
is a doctoral student in Counselor Education and Supervision at St. Mary's University. She is an intern at the Gay and Lesbian Community Center of San Antonio and Waterloo Counseling Center in Austin, Texas. Her research focuses on issues relating to transgender mental health and relational orientation, and connectedness in human relationships.

JoLynne Reynolds, Ph.D.,
received her master's and doctoral degrees from the University of South Florida. She is currently an Associate Professor at Texas State University, a faculty member of the Gestalt Institute of Austin, and maintains a private psychotherapy practice in San Marcos, Texas. Her research interests include children and families, and applications of relational Gestalt therapy.

David Shepard, Ph.D.,
is Assistant Professor of Counseling in the M.S. in Counseling program, at California State University, Fullerton. He is also a private practice psychotherapist in Santa Monica. His research interests include counselor education and men's studies, with a particular emphasis on how male socialization impacts men's relational capacities. Dr. Shepard earned his Ph.D. in Counseling Psychology from the University of Southern California.

Linda Kreger Silverman, Ph.D.,
licensed psychologist, is the founder and director of the Gifted Development Center. After receiving her Ph.D. from the University of Southern California, she served on the faculty of the University of Denver in counseling psychology and gifted education. Her 300 publications include *Counseling the Gifted and Talented* and *Upside-Down Brilliance: The Visual-Spatial Learner*. She founded the juried psychological journal, *Advanced Development*. Currently, she is conducting research on profoundly gifted children, the visual–spatial learner, comparative assessment of the gifted on different instruments, and introversion.

Maureen Walker, Ph.D.,
earned her doctorate in Counseling Psychology at The Georgia State University. Dr. Walker is a licensed psychologist with an independent practice in psychotherapy and multicultural consultation in Cambridge, MA. She is a member of the faculty and Director of Program Development at the Jean Baker Miller Training Institute of the Stone Center at Wellesley College. She is a coeditor of *How Connections Heal: Stories from Relational-Cultural Therapy* and *The Complexity of Connection: Writings from the Stone Center's Jean Baker Miller Training Institute*. She is also the associate director of MBA support services at Harvard Business School.

1

Traditional Models of Development: Appreciating Context and Relationship

By Dongxiao Qin, Ph.D., and Dana L. Comstock, Ph.D.

INTRODUCTION

Dana and I (Dongxiao) met a few years ago during a Jean Baker Miller's Training Institute we had both attended to learn more about Relational-Cultural theory. It was a sunny and cool afternoon in June and a reception was being held for participants who had come from all over the world. Our conversations were passionate and our laughter was delightful. It was during this reception that Dana first shared that she was editing a new developmental textbook. She explained that the purpose of the text was to illuminate the complex contexts which we all develop that impact our lives and our relationships across the life span. She also shared that Relational-Cultural theory would be used to illuminate many of these issues when applicable. At that time, I was teaching developmental psychology at Western New England College in Springfield, MA, and felt a surge of excitement over the prospect of a text that specifically addressed contextual and relational complexities.

Over time, Dana and I have had many conversations about how we could enhance our understanding of the complexities of human development while still appreciating aspects of traditional developmental theories. Through our conversations we shared a sense of connection and discovered we also shared identities as college professors teaching developmental courses, that we were both mothers of preschoolers of similar ages, and are both fans of spicy Thai food. As much as we have in common, it is important to remember our conversations have been between women from two different backgrounds: Dana is from North America, White, and from an individualistic culture; I'm from mainland China, Han, and from a collectivist culture. Our connection, which was forged between women from different cultural and ethnic backgrounds, has allowed us to participate in a unique critical analysis of traditional developmental models in Western psychology while rethinking how we might mainstream the considerations of contextual and relational development. As we thought through much of what we wanted to accomplish, we felt it was crucial to offer something unique to our readers which included the idea that we present traditional theories in a new light by giving the reader a glimpse into the theorists' respective lives, cultures, and relationships. We decided to present this chapter first since many of you have had exposure to these theories in your undergraduate training. Thus this chapter will also serve as a review of sorts.

In this chapter we will give a philosophical overview of traditional developmental models and their relationship to Western culture. In doing so, we aim to demonstrate the cultural context out of which the respective theories emerged. And, in the spirit of context and relationship, we have put together a presentation of some of the relational aspects of the respective theorists' lives as well as insight into what their relationships were like with each other. Our feeling is that we can best appreciate traditional theories of development if we gain a perspective of the context from which they emerged. Because most of the central thinking around some of these theories focuses on the development of the *individual*, we also felt it was crucial, in an ironic sort of way, to remind our readers that most of these theorists were mutually influenced and impacted, both personally and professionally, by each other's work. We feel that by having a contextual and relational understanding of these theorists we are better able to participate in a critical analysis of both the strengths and weaknesses of these theories and to make the most of what they have to offer with regard to shifts in mental health paradigms. Porter (2002) makes the point that:

> Although in theory Western psychology is swinging back from describing human beings solely from an individualistic perspective, practice has not kept pace, especially in mental health and diagnostic areas. A developmental perspective, by definition, would attend to children within the context of their families, cultures and environments. The developmental principle of adaptation affirms that children are shaped by and adapt to the environmental contexts in which they live. Thus, these contexts do more than inform us about the child's experience, they form the child. (p. 263)

This profound statement reflects the goals of this chapter in relation to the book and of the book as a whole, which are to (a) provide you with a context from which to begin a critical analysis of traditional theories of development and (b) assist you to begin the *practice* of considering the impact of context on both individual and relational development. In this process, you will be considering your own developmental contexts and throughout your training, you will develop a *practice* of exploring how your contextual development, which has formed who you are in relation to others, intersects with the developmental contexts of others. Jenkins (1999) sums up the outcomes of this practice, which include and emphasize an understanding of:

1. The capacity to see oneself and others as individuals within a larger cultural and social frame of reference;
2. The capacity to de-center, or to accept and be open to seeing the [cultural] embeddedness of one's own perceptions, assumptions, and judgments, and to move away from seeing others through one's own culturally bound lenses;
3. The capacity to see and to know one's self is inextricably relational; and
4. To understand that self-acceptance is basic to a healthy psychological experience of difference and the process of learning to live and grow with others in relationship. (p. 8)

The practice of considering context creates relational conditions that "transcend negative perceptions of one's own social identity and differences in others . . . and facilitate and support connection through mutually empowering interactions between individuals from the same or different reference groups" (Jenkins, 1999, pp. 8–9). We recommend that a critical analysis of traditional Western developmental theories (and theories of psychotherapy in general) include a consideration of whether or not the respective theory encourages an understanding of these conditions or offers ways in which they may be facilitated.

THE PHILOSOPHICAL BASIS OF TRADITIONAL WESTERN DEVELOPMENTAL THEORIES

In reflecting on the relationship of the individual and others (or society) in the context of American culture vis-à-vis Chinese culture, we were struck with the following general contrast. In America, the ideological and philosophical stance has been that of individualism, whereas in China, the ideological and philosophical stance has been that of relationality. "That individualism has been at the heart of American thought from the beginning is hardly an issue for contention" (Kegley, 1984, p. 202). In the course of American history, the self has become ever more detached from the social and cultural contexts that embody the traditions. "Self-reliance" is a nineteenth-century term, popularized by Ralph Waldo Emerson's famous essay of that title, but it still comes easily to the tongues of many Americans today. It is clear that the image of the self-reliant, transcendent individual who believes in "being your own person" and in "paddling your own canoe" is very much a part of American thought and ideology.

In order to affirm a self-celebratory cultural belief, Western psychology has participated and encouraged the belief in a negative, and at times almost destructive, relationship between self and others. The general philosophy of traditional developmental theories concerns the common notion that development evolves through stages of ever increasing levels of separation and spheres of mastery and personal independence. These theories emphasize "the separate self": an autonomous, self-sufficient, and contained entity. Modern American theorists of early psychological development and indeed, of the entire life span from Freud, Erikson, Piaget, and Kohlberg to Maslow, tend to see all of development as a process of separating oneself out from the matrix of others by "becoming one's own man," in Levinson's words (Miller, 1991, p. 1). Development of the "self" is presumably attained via a series of painful crises by which the individual accomplishes a sequence of allegedly essential separations from others, thereby achieving an inner sense of separated individuation (Miller, 1976; Jordan, Kaplan, Miller, Stiver, Surrey, 1991). These traditional developmental theories are about a "self-in-isolation," not a "self-in-relation" (Bergman, 1991).

Markus and Kitayama (1991) suggest that "[a]chieving the cultural goal of independence requires constructing oneself as an individual whose behavior is organized and made meaningful primarily by reference to one's own internal repertoire of thoughts, feelings and actions of others" (p. 226). According to this construal of self, the person is viewed as a "container" (Sampson, 1993a, p. 54) and is "bounded, unique, more or less integrated in a motivational and cognitive universe, a dynamic center of awareness, emotion, judgment, and action organized into a distinctive whole and set contrastively both against other such wholes and against a social and natural background" (Geertz, 1975, p. 48). This construal of self derives from a belief in the "ego-pole" that is self-conscious of the wholeness and uniqueness of each person's configuration of internal attributes (Heidegger 1962; Johnson, 1985; Sampson, 1988, 1993a, 1993b; Waterman 1981).

The "American self" characterized by such a privatized self-contained entity exhibits a firm self-other boundary where "others" are not an intrinsic part of the self or "self-development." Although an independent self must also be responsive to the social environment and social situations, their responsiveness to social resources derives from the need to verify and affirm the inner core of the self (Markus and Kitayama, 1991). The degree to which the individual is related to others is exclusively skewed. This exclusive conception of the individual defines the self as a separate entity whose essence can be meaningfully abstracted from the various relationships and in-group memberships that he or she has (Sampson, 1993a). Many traditional developmental theorists are grounded in this individualistic philosophy and interestingly offer pathological perspectives when relationships are considered with regard to individual development and lack a crucial consideration of context whether it be gender, culture, or race, for example, or other pertinent factors that forge our multiple and relational identities.

TRADITIONAL DEVELOPMENTAL MODELS
IN WESTERN PSYCHOLOGY

Sigmund Freud's Psychoanalytic Theory

Sigmund Freud (1856–1939), a physician whose career started in Vienna, Austria, in the 1880s, proposed that human beings are not strictly rational in their behavior or even knowledgeable about their own motives. Freud (1923/1974) argued that much of human behavior is governed by unconscious instinctual motives involving sex/pleasure and aggression, and that human personality has its origins in early childhood relations between children and their parents. A major feature of psychoanalytic theory was Freud's belief that from an early age, boys and girls come into the world as *isolated selves,* with primary and internal drives of sex/pleasure and aggression, and that children go through stages of psychosexual development one after another "like a train crossing Austria, stopping at the right station at the right time" (Bergman, 1991, p. 2). Freud believed that over the course of childhood, sexual impulses and pleasure-seeking behaviors shift their focus from the oral to the anal and finally to the genital regions of the body.

In the first stage (Oral: birth to 1 year) the new ego directs a baby's sucking activities toward breast or bottle. If oral needs are not met appropriately, the individual may develop such habits as thumb sucking, fingernail biting, and pencil chewing in childhood and overeating and smoking later in life.

In the second stage (Anal: 1 to 3 years) toddlers and preschoolers enjoy holding and releasing urine and feces. Toilet training becomes a major issue between parent and child. If parents insist that children be trained before they are ready or make too few demands, conflicts about anal control may appear in the form of extreme orderliness and cleanliness or messiness and disorder.

During the third stage (Phallic: 3 to 6 years) unconscious id impulses transfer to the genitals, and the child finds pleasure in genital stimulation. Freud's Oedipus complex for boys and Electra complex for girls arise, and young children feel a sexual desire for the other-sex parent. To avoid punishment, they give up this desire and, instead, adopt the same-sex parent's characteristics and values. As a result, the superego is formed. The relations among id, ego, and superego established at this time determine the individual's basic personality.

In the fourth stage (Latency: 6 to 11 years), children's sexual instincts die down, and the superego develops further. The child acquires new social values from adults outside the family and from play with same-sex peers.

During the fifth stage (Genital: adolescence) puberty causes the sexual impulses of the phallic stage to reappear. If development has been successful during earlier stages, it leads to mature sexuality, marriage, and the birth and rearing of children.

With regard to gender, the important divergence between females and males supposedly took place during the phallic stage, when children's appropriate feminine or masculine identification was precipitated by the development and resolution of the Oedipus complex. This involves withdrawal from and renunciation of mother, which is primary for establishing an independent, solid, gender and sexual identity. To the small boy, the father seems a powerful and terrifying rival. He becomes fearful of what his father may do to him. His fear crystallizes around the possibility of castration. According to Freud, the boy's fear becomes so intense that he is literally forced to resolve the Oedipus complex in order to dispel it. He finds resolution by identifying with his father (accepting his father as a model and incorporating many features of his father's personality into his own) and squelching his desire for his mother.

"Identifying with father comes through competition, fear, and renunciation, not through a wish to connect. It sets the stage for hierarchy—that is, patriarchy—for dominance, entitlement, ownership of women, and men's fear of men" (Bergman, 1991, p. 2). In this process, men are taught to be fearful of women, to fight for power over women, and to blame women for the trouble "caused" by the mother. For men, self-identity comes *before* relational-intimacy. Therefore from very young ages, men themselves are taught to *disconnect* from the relationship with mother, or any intimate relationship, in the name of "becoming a man."

If the little boy is propelled into masculine identification by fear of castration, what then, becomes of the little girl? Freud argued that "penis envy" caused girls to enter a female version of the Oedipus complex referred to as the Electra complex, i.e., attraction to the father and jealousy of the mother. Girls resolve the complex gradually by realizing that they can never possess their fathers. So, eventually they reestablish their feminine identification with their mothers and concentrate on becoming attractive love objects for men.

If Freud's story of female development, with its focus on the lack of a penis, sounds somewhat improbable to us we should note that it eventually seemed improbable to him as well. Freud later believed that he had overstated the parallels between the sexes in terms of the Oedipus complex, yet he never developed another model to replace it. Regardless of his later stance, his story of male and female psychosexual development has influenced other developmental theorists and psychoanalytic therapists for years. One of the implications of his theory is that a true intimate relationship is difficult, if not impossible, if people are encouraged to develop a *"self in isolation,"* versus a *"self-in-relation"* (Bergman, 1991).

To understand how Freud developed his views of female/male development, it is important to explore not only the patriarchal Victorian society of Vienna but Freud's own personal conflicts as well. Since personal and familial dynamics as well as sociocultural forces are closely intertwined and influence one another, we will give a brief overview of how the effects of all these factors played a role in shaping Freud's personality and his theory about men and women. These factors include losses of important early childhood attachment figures; his unconscious conflicts with his mother, who is reportedly described as having been seductive (Slip, 1993), aggressive, intrusive, and exploitative; his mother's own frustration as a person and her constricted social role; and anti-Semitism, which contributed to his father's economic failure and Freud's own professional difficulties, including several moves to escape the Nazi regime.

The historical-cultural antecedents of Victorian society, which are an important cultural context to consider, had a profound influence on Freud's ideas about men and their power. The fear of and the need to control women and their sexuality are shown to have no rational basis but rather stem from the magical way of adapting to nature in ancient and primitive societies. Women and their sexuality were associated with the great mother goddess (Mother Nature) who was believed to control fertility, life, death, and rebirth. Women and their mysterious sexuality were feared and had to be controlled. This magical form of adaptation to helplessness in life and death was a major factor in the evolution of a patriarchal social structure.

In humans' historical evolution, culture has employed ways of adaptation similar to those found in each human individual's development. Individual child development is used as a template for cultural evolution. The evolution of a patriarchal society is traced to men's attempt to gain further mastery over nature by replacing the female goddess with male gods. In patriarchal Victorian society, women were deprived of an identity and needed to achieve a sense of self by identifying with the social and economic successes of their husbands. Because Freud's father was considered a failure by Victorian economic standards, his mother appears to have established a close relationship with Sigmund, her elder son. With Sigmund Freud serving as the family savior, his mother could sustain her self-esteem and identity by living vicariously through the achievement of her son instead of her husband. From a psychoanalytic perspective, this responsibility for preserving his mother's ego as well as his earlier pre-Oedipal annihilation anxiety probably prevented Freud from dealing with his unconscious ambivalence toward her.

Freud's unconscious conflicts with his mother have been shown to have affected his *biased* theory of women in part by not acknowledging the role of the *mother* while emphasizing the relationship to the father in the child's personality development. As feminist psychologist Karen Horney (1922) noted, psychoanalytic theory was created by a *male genius*, and that it was *mostly men* who elaborated on Freud's ideas. Therefore it was easier for these men to evolve a more *masculine* model than feminine model of developmental psychology.

Karen Horney's Early Feminist Voice

Karen Horney (1885–1952) was born in Germany to an affectionate mother of noble ancestry and a stern, remote Norwegian sea captain father. Her parents preferred her brother, and her father's derogatory comments about her appearance and her intelligence continued to haunt her into adulthood. Deciding that if she could not be beautiful, she could certainly be smart, Karen excelled in her studies in medical school. She later joined the Berlin Psychoanalytic Society and became one of its most prominent members. Although Horney was an early member of Freud's circle and a well-respected European psychoanalyst, she began to critique Freud's theory of female development. She felt that he had overemphasized the role of the penis in the psychology of *both* sexes, and she discounted the idea that girls automatically view their bodies as inferior (Lips, 2003). Furthermore, she noted male envy of the female womb and breasts was an important issue among her therapeutic clients (Horney, 1926/1973).

According to Horney, much of men's behavior could be explained by the envy of the female capacity for pregnancy and breast-feeding. Horney proposed that a girl's psychosexual development centered on her own anatomy, rather than on the male's. A girl's rejection of the female role could be caused by an unconscious fear of vaginal injury through penetration and not by resentment over lack of a penis. Thus Horney's critique of Freud repudiated the notion that women's development was driven in a major way by envy of the male body (Lips, 2003).

In her late work, Horney moved away from the early psychoanalytic focus on the biologically based instinctual drives of sex and aggression. Instead, she emphasized the cultural context in which children developed (Westkott, 1986). This approach provided her with a new way to conceptualize dynamics related to Freud's Oedipus complex. She theorized that the social environment in which a child was raised could create harm in two ways: *devaluation* (parents' failure to respect the child as a unique and worthwhile person) and *sexualization* (adults' taking a sexual approach to the child) (Lips, 2003). Horney argued that a child's attachment to one parent and jealousy toward the other was not a sign of Freud's Oedipus complex, but rather an expression of, and a means for coping with, the child's anxiety over being devalued and sexualized (Horney, 1939, cited in Lips).

According to Horney, in a society that values males over females, girls are especially vulnerable to the fallout of being devalued and sexualized. In addition, girls might well adopt feelings of inferiority if they are clearly valued less than their brothers. If they were treated, as many daughters were, as if their sexuality were the most important aspect of their identity, they were likely to become anxious and dependent. Although Horney constructed her story of female development within the family framework emphasized by psychoanalysis, she became one of the first members of that school to recognize the effect of cultural variables on these family dynamics (Lips, 2003).

Erik Erikson's Psychosocial Theory

Several of Freud's followers took what was useful from his theory and attempted to elaborate on and improve his theory of psychosexual development. The most important of these neo-Freudians was Erik Erikson (1902–1994). Although Erikson (1950) accepted Freud's basic psychosexual framework, he expanded the picture of development at each stage and extended the line to adulthood. According to Bergman (1991), Erickson felt it was important that "we check out who else is at the station when we stop" (p. 3). In his psychosocial theory of development, Erikson emphasized that at each stage, the "self" acquires attitudes and skills that make the individual an active, contributing member of society. A basic psychosocial conflict, which is resolved along a continuum from positive to negative, determines healthy or maladaptive outcomes at each stage. Erikson's eight stages of identity development followed Freud's stage model with Erikson adding three adult stages.

The life cycle presentation that Erikson developed began with the first stage termed "Trust vs. Mistrust" (comparable to Freud's Oral stage: birth to 1 year). During this period of development, infants gain a sense of trust or confidence from warm and responsive care of others (usually mothers and other caregivers). The amount of trust derives from the quality of the maternal relationship within which mothers create a sense of trust in their children by providing sensitive care for the baby's individual needs and a firm sense of personal trustworthiness. This forms the basis in the child for a sense of identity, which will combine a sense of being "all right" with being oneself (Erikson, 1950, p. 249). Mistrust occurs when infants have to wait too long for comfort and are handled harshly.

The second stage of child development is referred to as "Autonomy vs. Shame and Doubt" (comparable to Freud's Anal stage: 1 to 3 years). Using new mental and motor skills, children want to choose and decide for themselves. Autonomy is fostered when parents permit reasonable free choice and do not force children into trying new tasks or shame them over unsuccessful attempts at new tasks or over not having the courage to try. A sense of *independence and individuation* that parents give to the child will not lead to undue doubt, fear, or shame in later life but rather fosters a sense of self-confidence and pride. On the other hand, "from a sense of loss of self-control and of foreign overcontrol comes a lasting propensity for doubt and shame" (Erikson, 1950, p. 254).

The third stage of Erikson's model of psychosocial development is termed "Initiative vs. Guilt" (comparable to Freud's Phallic stage: 3 to 6 years). Through make-believe play and intense peer interactions, children explore the kind of person they can become. Initiative, which is a sense of ambition and responsibility, develops when parents support their child's new sense of purpose. The child seems to "grow together" both in person and in body. A child appears "more himself" (Erikson, 1950, p. 255). The danger is that parents will demand too much self-control, which leads to overcontrol, creating too much guilt.

The next stage is termed "Industry vs. Inferiority" (comparable to Freud's Latency: 6 to 11). This is the stage in which school-aged children learn to win recognition by producing things. They become eager for and absorbed in productive situations. Their tools and skills teach them the pleasure of work completion by steady attention and by persevering diligence. At this stage, the child negotiates a sense of inadequacy and inferiority if her or his family life has failed to prepare her or him for school life, or when school life fails to sustain the promises of earlier stages (Erikson, 1950).

Adolescence begins with the stage "Identity vs. Identity Confusion" (comparable to Freud's Genital stage). Growing and developing youth now come to ask: "Who am I and what is my place in society?" The sense of lasting personal identity is based on the self-chosen values and vocational goals. The danger of this stage is role confusion caused by an inability to settle on an occupational identity.

In the next stage termed "Intimacy vs. Isolation," young people are eager and willing to fuse their identities with those of others. An individual is ready for intimacy, which is defined as "the capacity to commit himself to concrete affiliations and partnerships and to develop the ethical strength to abide by such commitments" (Erikson, 1950, p. 263). Because of early disappointments, some individuals cannot form close relationships and remain isolated.

Middle adulthood is the stage of "Generativity vs. Stagnation." Generativity is experienced as the concern for others in terms of child rearing, caring for other people, and otherwise productive and meaningful work. Individuals who think they have failed in these ways feel an absence of meaningful accomplishment and thus struggle with a sense of stagnation.

The last stage is "Integrity vs. Despair." In this final stage, individuals reflect on their past experiences and come to terms with self-acceptance. Integrity results from feeling that life was worth living as it happened. It is a "post-narcissistic love" of the human ego as an experience, which conveys some world order and spiritual clarity (Erikson, 1950, p. 268). Older individuals who feel dissatisfied with their lives fear death.

The irony of Erickson's theory, which has become the focus of criticism by relational theorists, is that:

> From Basic Trust we move to Autonomy, Industry, and Identity, and then, after adolescence, to intimacy driven by sexuality. Only after this sexual intimacy do we get to Generativity, the participation in the development of others. Only after self achieves a certain 'maturity,' can we learn to really relate to others. (Bergman, 1991, p. 3)

What follows in the next chapter is an elaboration of this point and of the idea that we grow in and through relationships over the life span.

Erikson's personal and professional lives involved crossing and recrossing a variety of borders or traditional lines of demarcation, and he became increasingly adept at it. These borders concerned his disciplinary and occupational allegiances, the structure of his conceptualizations, his religious and national loyalties, his "real" and "native" languages, and even the men he could call "father" (notably, Freud as his romantic image of his own father) (Friedman, 1999). In brief, the geographic, social, disciplinary, personal, and intellectual contexts of his life were constantly changing. He was always in the *process of becoming*.

Erikson considered himself a Freudian and a depth psychologist who not only continuously explored the depth of an individual's inner psyche but also took joy in mapping the crossings of social surfaces. Erikson reflected a great deal about Freud. He identified much with Freud and felt that this great "mythical figure" seemed somehow to connect with him. He recalled how Freud "was a doctor like my adoptive father" (Friedman, 1999, p. 74). Erikson's stepfather adopted him; he never knew his biological father.

Freud's powerful presence seems to have drawn Erikson into the psychoanalytic profession and was keeping him there because he substituted Freud for his stepfather and, more important, for his romanticized biological father. The "stepson" was essentially adopting the romanticized father of psychoanalysis as his own idealized, artistically talented father. This was central to the decision Erikson made to train with Freud's daughter, Anna.

Anna Freud and others of the circle around Freud took Erikson in and opened a life's work for him. He felt that he had finally secured a real "vocation in the area of child psychoanalysis and education" (Friedman, 1999, p. 70). It was these early circumstances and cultural contexts that shaped and influenced Erikson's thinking. They may also have inspired and motivated him. Despite his well-known eight-stage universal model of the human life cycle, which distributed the first two decades of life among five separate and distinct stages, he saw his own early life as a single unified stage that composed the entire flow of the events from his birth to young adulthood. He characterized this stage as both broadly constrictive and energizing. Drawing from Erikson's late-life reflections, it is easy to understand why he regarded identity development as part of a continuous and unified developmental stage characterized by a goal of *autonomy and separation*. It is also interesting to note, in an ironic sort of way, that much of his professional direction was driven by relational motivators and his sense of connectedness and belonging.

Jean Piaget's Constructivist Model

Swiss cognitive theorist Jean Piaget (1896–1980) received his academic education in zoology. Consequently, his theory has a distinct biological flavor. According to Piaget (1967, 1971), human infants do not start out as cognitive beings. Instead, out of their perceptual and motor activities, children build and refine psychological structures and organized ways of making sense of experiences that permit them to adapt more effectively to their external world. Children are intensely active in the development of these structures. They select and interpret experiences using their current structures and they modify these structures so that they take into account more subtle aspects of reality. Because Piaget viewed children as discovering or constructing virtually all knowledge about their world through *their own activity*, his theory is often referred to as a *constructivist* approach to cognitive development.

Piaget believed that children move through four stages of development: sensorimotor, preoperational, concrete operational, and formal operational. During these stages, the exploratory behaviors of infants transform into the abstract, logical intelligence of adolescence and adulthood.

At the sensorimotor stage (birth to 2 years), infants "think" by acting on the world with their physical senses. As a result they invent ways of solving sensorimotor problems. At the preoperational stage (2 to 7 years), preschool children use symbols to represent their earlier sensorimotor discoveries. Development of language and make-believe play takes place. However, thinking lacks the logic of the two remaining stages.

At the concrete operational stage (7 to 11 years), children's reasoning becomes logical. School-age children understand that a certain amount of lemonade or Play-Doh® remains the same even after its appearance changes. They also organize objects into hierarchies of classes and subclasses. However, thinking falls short of adult intelligence. It is not yet abstract. At the formal operational stage (11 years and older), the capacity for abstraction permits adolescents to reason with symbols that do not refer to objects in the real world, as in advanced mathematics. They can also think of all possible outcomes in a scientific problem, not just the most obvious ones.

Piaget's stage sequence has three characteristics. First, it's a *general* theory; it assumes that all aspects of cognition develop in an integrated fashion, undergoing a *similar* course. Second, the stages are *invariant*, meaning that they always follow a fixed order, and no stage can be skipped. Third, the stages are *universal*; they are assumed to describe the cognitive development of children everywhere.

Piaget regarded the order of development as rooted in the biology of our species and as the result of the human brain becoming increasingly adept at analyzing and interpreting experiences common to most children throughout the world. He emphasized that individual differences in genetic and environmental factors affect the speed with which children move through the stages.

Piaget's view of development was greatly influenced by his early training in biology. Central to his theory is the biological concept of *adaptation*. Just as the structures of the body are adapted to fit with the environment, so the structures of the mind develop better if they fit with the external world. In Piaget's theory, as the brain develops and children's experiences expand, their mind follows the *ladderlike stages* (often referred to as "step wise" development), from egocentric, interpersonal to more autonomous and abstract ways of thinking. Piaget's legacy of an autonomous and constructivist model has had a great impact on American psychologists since the 1950s.

Perhaps he has had the greatest influence on Lawrence Kohlberg's thinking. The conceptual foundations of Kohlberg's moral theory (discussed later) are built on Piaget's stage model of cognitive development.

Despite Piaget's overwhelming contributions, in recent years his theory has been challenged by psychologists who point out that Piaget's stages pay insufficient attention to social and cultural influence and the resulting variation in thinking that exists among same-age children (Rogoff & Chavajay, 1995). In a latter chapter the impact of trauma on development is explored. As a staggering number of infants and children are suffering varying forms of abuse it is crucial to consider the impact these experiences have on cognitive development. Just as abuse and trauma have been recognized as important factors in relation to cognitive development, so have particular child-rearing practices. Research is now showing that breast-fed babies are more intelligent (and score higher on standardized IQ tests) (Mortensen, et al., 2002) and that holding and touching infants and developing secure attachments between infants and their caregivers facilitates neuronal development (Ornish, 1998; Siegel, 1999). These critical factors related to childhood experiences certainly throw into question any theory purported to be "universal."

Vygotsky's Sociocultural Theory and Contextualistic Critique

Russian psychologist Lev Vygotsky (1896–1934) believed that children are active seekers of knowledge, but not as *solitary agents*. In his theory (1978), rich social and cultural contexts profoundly affect children's cognition. He rejects Piaget's individualistic view of the developing child in favor of a *socially formed mind* (Rogoff, 1998; Wertsch & Tulviste, 1992). The "mind" that Piaget observed within the individual exists, for Vygotsky, in the society in which that person lives, in the form of the cultural wisdoms that the child internalizes through his or her interactions with those who are already skilled in their use.

Early events in Vygotsky's life contributed to his vision of human cognition as inherently social based. As a boy in Russia, he was instructed at home by a private tutor, who conducted lessons using the Socratic dialogue—an interactive, question-and-answer approach that challenges current conceptions to promote heightened understanding. By the time Vygotsky entered the University of Moscow, his primary interest was a verbal field—literature. After graduating he was first a teacher and only later did he turn to psychology (Kozulin, 1990, cited by Berk, 2003). Vygotsky died of tuberculosis when he was only 37 years old. Although he wrote prolifically, he had little more than a decade to formulate his ideas. Consequently, his theory is not as completely specified as Piaget's. Nevertheless, the field of child development is experiencing a burst of interest and enthusiasm in Vygotsky's sociocultural perspective.

The major reason for Vygotsky's appeal lies in his rejection of Piaget's individualistic model of the developing child. He believed that all higher cognitive processes develop out of interactions with knowledgeable others in the sociocultural contexts of social interaction. Through joint activities with more mature members of society, children come to master activities and think in ways that have meaning in their culture. Special concepts such as the *zone of proximal (or potential) development, intersubjectivity,* and *scaffolding* explain how social interaction happens between children and knowledgeable others. Vygotsky's sociocultural theory focuses on how culture (the values, beliefs, customs, and skills of a particular group) is transmitted to the next generation. Accordingly, *social interaction,* or cooperative dialogues, between children and more knowledgeable members of society is necessary for children to acquire the ways of thinking and behaving that make up a community's culture (Wertsch & Tulviste, 1992).

One of the strengths of Vygotsky's theory is the notion that children in every culture develop unique attributes that are not present in other cultures. Richard Lerner (1996) points out that development must be studied in the many contexts of daily life, contexts that not only influence individual development but are themselves changed by that individual's presence. Lerner asserts that developmental influences are bidirectional, or reciprocal. Individuals are influenced by their families, friends, and teachers, and in turn exert an influence on these very same contexts. Additionally, Lerner, as well as Bronfenbrenner (1979), account for the many features of one's environment in terms of overlapping spheres of influence, simultaneously operating at the biological and physical, psychological, and sociocultural levels. In contrast to Piaget's research on children (mostly his own children, or children in lab settings), contextual psychologists suggest that we must study children and adolescents in real-life settings, in their families, neighborhoods, and classrooms. They also point out that the development of children and adolescents is associated with other *risk factors:* poverty, stress, abuse, unemployment, and crime, etc. These contextual factors, along with gender differences, have a great influence on an individual's thoughts, feelings, and behaviors.

Feminist psychologists observed that Piaget is not the only theorist to measure females against a yardstick developed by males (marbles was a boy's game) only to find them lacking. Carol Gilligan (1982), commenting on psychological theorists in general, writes: "Implicitly adopting the male life as the norm, they have tried to fashion women out of a masculine cloth. It all goes back, of course, to Adam and Eve—a story, which shows, among other things, that if you make a woman out of a man, you are bound to get into trouble. In the life cycle, as in the Garden of Eden, the woman has been the deviant" (p. 6).

Lawrence Kohlberg's Theory of Moral Judgment

Lawrence Kohlberg's (1927–1987) theory of moral development is deeply rooted in Piaget's work. In particular, it is based on ideas in Piaget's (1965) *The Moral Judgment of the Child.* Piaget's main assumption was that cognition and affect develop on parallel tracks and that moral judgment represents a naturally developing cognitive process. In contrast, the assumption of most psychologists at that time was that moral thinking was a function of other, more basic social and psychological processes.

Kohlberg's (1976, 1984) theory bases its assumption about moral development on the organismic model, stressing the importance of the inner forces that organize individual development. His theory traces moral reasoning over a number of discrete stages. Movement from one stage to the next is prompted by the need to resolve conflict. Cognitive maturity also contributes to moral development. In a sense, Kohlberg has helped finish Piaget's unfinished work. Like Piaget, Kohlberg used a clinical interviewing procedure to study moral development whereas Piaget asked children to judge the naughtiness of a character who had already chosen a course of action. Kohlberg presented men with a moral dilemma and asked them what the main actor should do and why.

From these results, Kohlberg (1976) organized his six stages of moral development into three general levels and made strong statements about this sequence. Laura Berk (2003) neatly summarized and described Kohlberg's three levels and six stages of moral development. At the preconventional level, morality is externally controlled. Children accept the rules of authority figures, and actions are judged by their consequences. Behaviors that are viewed as bad result in punishment, and those that lead to rewards are seen as good. At *Stage 1: the punishment and obedience orientation,* children find it difficult to consider two points of view in a moral dilemma. As a result, they ignore people's intentions and instead focus on fear of authority and avoidance of punishment as reasons for behaving morally. At *Stage 2: the instrumental purpose orientation,* children become aware that people can have different perspectives in a moral dilemma. They view right action as flowing from self-interest. Reciprocity is understood as an equal exchange of favors.

At the conventional level, individuals continue to regard conformity to social rules as important, but not for reasons of self-interest. They believe that actively maintaining the current social system ensures positive human relationships and social order. At *Stage 3: the "good boy-good girl" orientation,* or the morality of interpersonal cooperation, individuals' desire to obey rules as a means of promoting social harmony first appears in the context of close personal ties. They want to maintain the affection and approval of friends and relatives by being a "good person"—trustworthy, loyal, respectful, helpful, and nice. The capacity to view a two-person relationship from the vantage point of an impartial outside observer supports this new approach to morality. At this stage, the individual understands *ideal reciprocity,* as expressed in the Golden Rule.

At *Stage 4: the social-order-maintaining orientation*, the individual takes into account a larger perspective—that of societal laws. Moral choices no longer depend on close ties to others. Instead, rules must be enforced in the same evenhanded fashion for everyone, and each member of society has a personal duty to uphold them. The stage 4 individual believes that laws cannot be disobeyed under any circumstances because they are vital for ensuring societal order.

Individuals at the postconventional level move beyond unquestioning support for the rules and laws of their own society. They define morality in terms of abstract principles and values that apply to all situations and societies. At *Stage 5: the social-contract orientation,* individuals regard laws and rules as flexible instruments for furthering human purposes. They can imagine alternatives to their social order, and they emphasize fair procedures for interpreting and changing the law. When laws are consistent with individual rights and the interests of the majority, each person follows them because of a *social-contract orientation*—free and willing participation in the system because it brings about more good for people than if it did not exist.

At *Stage 6 (highest stage): the universal ethical principle orientation*, right action is defined by self-chosen ethical principles of conscience that are valid for all humanity, regardless of law and social agreement. These values are abstract, not concrete moral rules. Stage 6 individuals typically mention such principles as equal consideration of the claims of all human beings and respect for the worth and dignity of each person.

To evaluate Kohlberg's stage theory of moral development, we have observed that: first, he regarded the stage as invariant and universal—a sequence of steps that people everywhere move through in a fixed order. Second, he viewed each new stage as building on reasoning of the preceding stage, resulting in a more logically consistent and morally adequate concept of justice. Finally, he saw each stage as an organized whole—a qualitatively distinct structure of moral thought that a person applies across a wide range of situations. These characteristics are the very ones Piaget used to describe his cognitive stages.

In addition to assuming that the individual's reasoning increases with each developmental stage, Kohlberg has assumed that the highest levels of moral reasoning reflect *universal* values that are common to *all* cultures. Research on moral behavior and judgment, however, finds evidence for cultural differences in values. Cooperation, maintaining harmonious relations with others, and working for the common good are more highly valued in Asian, Hispanic, and other non–Western European cultures. Thus, social context appears to influence moral development more than Kohlberg initially had assumed (Carlo, Fabes, Liable, & Kupanoff, 1999).

In Kohlberg's early research, boys and men tended to achieve at least stage 4 moral thinking, while girls and women were more likely to stop at stage 3. In other words, women seemed to have a *less mature* and *less developed* sense of morality.

Kohlberg's early research was conducted entirely with *male* participants and their responses became the norm by which girls and women were later evaluated. The dilemmas posed by Kohlberg may have been easier for boys and men to relate to because they frequently involved men as the principal actors in the moral drama. The hypothetical dilemmas also might not reveal much about moral behavior in real-life situations. For all these reasons, as Gilligan (1982) noted, Kohlberg's theory may not be an adequate map of female development. The fact remains that, instead of adjusting his questioning to discover why women and girls did not fit into his "male" model, he simply determined that they must be "less moral" than men. In this way, Kohlberg further perpetrated the myth that women are the "weaker sex."

Carol Gilligan's Model of Ethic of Care

Carol Gilligan (1936–present) is the most well-known figure among those who have argued that Kohlberg's theory, originally formulated on the basis of interviews with males, does not adequately represent the morality of girls and women. After years of listening to women's different life experiences and struggles, she provides an extraordinarily compelling voice to the moral decision-making process of women (1982). Such a different voice is refreshing when compared to the dominant voice of traditional psychology in which disconnection between thoughts and feelings is still the norm (Gilligan & Farnsworth, 1995). As a gifted writer and psychologist with the ear of a musician, she is attuned to the nuances and texture of language in traditional psychological theories of *all* human development. She was a daughter raised "as a Jewish child during the Holocaust," and now a mother of three, taught to "always stand up for what you believe in" (Gilligan & Farnsworth, 1995). She also heard in her voice a social activist; aware of the power of voice to transmit, politicize, and to construct a culture (Gilligan & Farnsworth).

Gilligan was deeply affected by feminist voices in early history, the women's movement in the 1960s and the political movement in the early '70s, and always had questions about morality and values. When she was teaching with Erikson and Kohlberg (who had both created the developmental theories based on the experiences of men, namely themselves) at Harvard she felt their work didn't seem particularly creative. "I saw how Erik's work on identity came right out of his life. He was a man who had not known who his father was; he had named himself. And moral dilemmas ran all through Kohlberg's life, beginning in his childhood. So I saw the connection with both these men, between their experiences and their theories" (Gilligan & Farnsworth, p. 216). Then Gilligan began to connect her life and her work, and this connection turned out to be very radical and challenging to Western mainstream psychology.

She started her study of real-life moral reasoning by studying women who were pregnant and considering abortion. She chose the abortion situation because it provided examples of ethical decision making by women and also because it brings up a central conflict for women:

> While society may affirm publicly the woman's right to choose for herself, the exercise of such choice brings her privately into conflict with the conventions of femininity, particularly the moral equation of goodness with self-sacrifice.

Although independent assertion in judgment and action is considered to be the hallmark of adulthood, it is rather in their care and concern for others that women have both judged themselves and been judged. . . . (Gilligan, 1982, pp. 70–71)

One day, while sitting at her kitchen table reading through the interviews with women from the pregnancy/abortion decision study, Gilligan finally understood the counterpoint she was hearing between women's voices and the voice of psychological theory that she had not initially named as having anything to do with feminism or with women. She wrote: "The arc of developmental theory leads from infantile dependence to adult autonomy, tracing a path characterized by an increasing differentiation of self from relationships with others and a progressive freeing of thought from contextual constraints" (Gilligan & Farnsworth, 1995, p. 218).

From that point forward, Gilligan began to hear women speaking in different terms about their sense of self and their experience of moral conflict and choice, which clarified a problem that was embedded in psychological theory. She observed a set of assumptions that were belied by women's experiences. All of the preparation from her own life, from the historical time, from the women's movement and the fact that she was interviewing women and was at the time a mother of three young children, made the issues very real to her. It was a moment where she faced a choice: either listen to the women or listen to the traditional theories. She chose to go with the women.

It was the first time that she had written something just for her self based on her observations which were clearly "out of the box" in terms of Western psychological developmental theory to date. So, in the winter of 1975, Carol Gilligan sat down and wrote "In a Different Voice: Women's Conceptions of Self and of Morality" (Gilligan & Farnsworth, 1995). Like Kohlberg, Gilligan found preconventional, conventional, and postconventional levels of moral reasoning in the women she interviewed. However, the basis for each level is different. Each of the three levels reflects a different resolution to the conflict between *responsibility to self* and *responsibility to others*. Movement from one level to the next occurs in two transitional periods.

At Level 1, women are concerned with survival, and their immature responses, derived from their feelings of being alone and powerless, may seem selfish. Transition to the next level occurs when one sees caring only for oneself as selfish and at odds with responsibility to others.

At Level 2, women think primarily in terms of others' needs. The conventional morality of womanhood tells them that they should be prepared to sacrifice all for a lover or a potential baby. When their own and others' needs conflict, they face seemingly impossible dilemmas. Transition to the third level occurs when they experience problems in their relationships that result from excluding themselves from their own care.

At Level 3, women resolve the conflicts between their own and others' needs not by conventional feminine self-sacrifice but by balancing care for others with healthy self-care. Because the woman is acting on internalized ethical principles that value relationships, caring is extended both to self and others.

Gilligan's experience in listening to what women said about moral issues convinced her that the type of morality studied by Kohlberg is *an ethic of rights and more common in men,* while the type she discovered is *an ethic of care more common in women.* Rather than judge women as morally deficient by a male norm, Gilligan believes researchers should recognize that women and men have different but equally valid approaches to moral issues. For both men and women, the highest levels of development should integrate the moralities of rights and responsibilities. She believes that feminine morality emphasizes an ethic of care that is devalued and misunderstood in Kohlberg's system. According to Gilligan, a concern for others is a different, not less valid, basis for moral judgment than a focus on personal rights.

Abraham Maslow's Theory of Self-Actualization

Like Erikson, Abraham Maslow's (1908–1970) thinking was also greatly influenced by Sigmund Freud. At the same time, Maslow was highly critical of Freud's concentration on the study of neurotic and psychotic individuals. It was Maslow's belief that one cannot understand mental illness until one understands mental health(1954, 1962). Freud reached his conclusions about human nature by observing the worst rather than the best of man. In Maslow's opinion, studying only "crippled, stunted, immature, and unhealthy specimens can yield only a crippled psychology and crippled philosophy" (Goble, 1970, p. 14). The study of self-actualizing people must be the basis for a more *universal* science of psychology. It is this concept that makes Maslow's theory different from others.

Maslow introduced this study into psychology and psychiatry. He had stumbled on the idea that one could learn a great deal about humanity's potential from the study of exceptionally healthy, mature people, a segment of the human species Maslow has termed the "growing tip."

Maslow's (1962) study of outstanding examples of mental health started not as a scientific research project but as an effort to satisfy his personal curiosity. The investigation of self-actualized people resulted from Maslow's curiosity when he was a student. He was trying to understand two professors whom he respected and admired very much. His curiosity compelled him to analyze what it was about these two male educators that made them so different, so outstanding. As he made notes about them it suddenly struck him that their two personalities had certain characteristics in common. Excited by this discovery, young Maslow sought to discover whether this type of individual could be found in other places. From this search came his more extensive studies of fully mature people (the exceptional people he studied were mostly famous male figures).

The definition of the "self-actualized" person was described as the full use and exploitation of talent, capacities, and potentialities. Such people seem to be fulfilling themselves and doing the best that they are capable of doing (Maslow, 1962). The self-actualized person was the best possible specimen of the human species, a representative of the growing tip.

Maslow first called the exceptional people he studied "self-actualized people." The actualization process refers to the development or discovering of the true self and the development of existing or latent potential. A more descriptive word is "full human." Probably the most universal and common aspect of these superior people is their ability to see life clearly, to see it as it *is*, rather than as they wish it to be. They are less emotional and more objective about their observations. They are more decisive and have a clearer notion of what is right and wrong. They are more accurate in their prediction of future events. He found self-actualized people to be task oriented, creative, spontaneous, self-confident, and to have self-respect. It is only the evolved and mature human being, the self-actualized, fully functioning person who is so highly correlated that for all practical purposes he may be said to fuse into a unity.

Self-actualized people have feelings of *power* in the sense that they have feelings of self-control. The psychologically healthy individuals are highly *independent* and the most individualistic members of society. They are governed far more by *inner directives, their own nature, and natural needs* than by the society or the environment. Since they depend less on other people, they are less ambivalent about them. Self-actualizers have what Maslow calls "psychological freedom."

Maslow did the study on mostly men, especially those "exceptional" male figures in American history. The study of these individuals, their habits, characteristics, personalities, and abilities led Maslow to his definition of mental health and his theory of human motivation. Their attributes such as being rational, less emotional, and not dependent on others seemed to mirror stereotypical male traits. The concept of "self-actualization" based on "psychological freedom" (versus "dependency") has been prominent in Maslow's theory, but perhaps this is so because it seems to underlie the *healthy* development of men. As Miller (1984) pointed out, few men ever attain such self-sufficiency because they are usually supported by loved ones at home (such as wives, mothers, and daughters), and others (such as secretaries, nurses, coworkers) in the interpersonal settings. Thus, there is reason to question whether this model accurately reflects men's lives. We think Stephen Bergman (1991) best summed up the need to reevaluate these theories with respect to men's lives when he stated: "Much of what is promoted in these old theories seems inaccurate, irrelevant, worn and weird. It's time to work towards a new psychology of men" (p. 3).

As Stiver (1984) critically observed, men's sense of *masculinity* seems jeopardized by what is required to establish close interpersonal relationships. Men in Western societies usually experience the demands to be competitive, to suppress emotions, and to maintain an impersonal attitude. To acknowledge a need for intimacy with others is perceived as regressive "dependency" defined in traditional psychoanalytic and developmental theories. "In reality men also need to be attached to others and want to have their needs met, but they have considerable difficulties in acknowledging their needs openly" (Stiver, p. 4). Thus the invisibility of men's "dependency needs" distorts reality.

Relational-Cultural theorists (Miller, 1976; Jordan, Kaplan, Miller, Stiver, & Surrey, 1991) have argued that there exist different developmental patterns for women and men. The "self-in-relation" theory of female self-development considers that a woman's fundamental human motive or self-actualizing character is a relational one and her need to feel related to others is a crucial aspect of her identity. This helps us to understand why women feel so threatened when there is the danger of alienation from others (Stiver, 1984). They experience the goal of independence and autonomous "self-actualization" as lonely and isolating. In contrast to women's relational pathways for psychological growth, men are encouraged toward self-sufficiency through separation from others, even though they need to care and be cared for by others. "Thus, it is extremely important for men to hide any sign of their neediness and dependency from others" (Stiver, 1984, p. 9).

As Stiver (1984) further argued, both women and men want to be needed by others. A man's sense of self-sufficiency and self-esteem could be enhanced if the concept of "dependency" could be reframed. Stiver (1984) offered a new definition of dependency as "a process of counting on other people to provide help in coping physically and emotionally with the experiences and tasks encountered in the world, when one has not sufficient skills, confidence, energy and/or time" (p. 10). This notion of dependency would allow for experiencing one's sense of self as being enhanced and empowered through the process of counting on others for help and connection. In this definition, "dependency" would be seen as healthy, normal, and growth promoting in the process of self-development for both women and men.

CONCLUSION

In this chapter, we have given a philosophical overview of traditional developmental models and their relationship to Western culture. We have provided the cultural context out of which the respective theories emerged, the relational aspects of the respective theorists' lives, as well as insight into what their relationships were like with each other. Through a critical review of respective traditional developmental theories including those by Freud, Erikson, Piaget, Kohlberg, and Maslow, we have observed that these mainstream developmental models tend to see the goals of development as a process of separating oneself from the complex system of contextual relationships (i.e., the relationships with matrices of others, sociocultural and historical contexts, and the convergence of biosocial, cognitive, and psychosocial factors).

The traditional developmental models were framed and grounded in the Western cultural ideology of *individualism*. The central theme of traditional developmental models emphasizes a stage-wise process of ever-increasing levels of separation and personal independence. In most cases, "normal" development was abstracted from the impact and influence of relationships. In reality, human development is socially and culturally contextualized. Even the traditional developmental theorists (Freud, Erikson, Piaget, Kohlberg, and Maslow) were culturally and theoretically related to each other on a micro level. We have observed a contextual link among the theorists, noting how they were mutually influenced and impacted, both personally and professionally, by each other's work. We would even suggest that in these relational contexts these theorists grew personally and professionally. We have also noted the connections between their thought and the complex sociocultural, political, and historical contexts in which they lived on a macro level.

Challenges to traditional developmental theories have come from early feminist theorists such as Karen Horney, Carol Gilligan, Relational-Cultural theorists, and sociocultural theorists who emphasize how our "healthy" or "normal" development occurs and is facilitated through meaningful *connections with others* (instead of *isolation from others*). They argue that self and context are co-constituting and co-constituted in a variety of ways. Their own life experiences and senses of self, which are embedded in different cultural and historical moments, deeply affected and shaped their own theories in similar yet different ways. For Karen Horney, her early connection with her caring mother, disconnection with her remote father, and her experience of the early biased child rearing at home contributed to her later critical reading of Freud's *Phallic Worship,* which devalued women's own anatomy and socialization.

For Carol Gilligan, her early experiences of political struggles as a Jewish child, her mothering experience in a patriarchal society, the women's movement in the 1960s, and her observations of Erik Erikson and Lawrence Kohlberg definitely shaped her work, which came to fruition with the publication of *In a Different Voice*. Although Horney and Gilligan lived in different sociocultural and historical contexts, their capacities to de-center, or to be open to seeing the cultural and historical embeddedness of their perceptions, assumptions, and judgments, make their life and work inextricably relational.

We hope that, by telling you "the rest of the story," you are better able to engage in a critical analysis of these traditional theories and that you are better prepared to understand the critical contexts from which we *all* develop. The chapters that follow address many of these contexts. As our lives can be unimaginably complex and unique we are not able to address every context within which we may find ourselves. And certainly our development and the path it takes can always turn on a dime so, if anything, we ask you simply to join us on a journey of understanding.

REFERENCES

Bergman, S. (1991). Men's psychological development: A relational perspective. *Work in Progress, No. 48*. Wellesley, MA: Stone Center Working Paper Series.

Berk, L. (2003). *Child development (6th edition)*. Boston: Allyn & Bacon Publishers.

Bronfenbrenner, U. (1979). *The ecology of human development*. Cambridge, MA: Harvard University Press.

Carlo, G., Fabes, R.A., Liable, D., & Kupanoff, K. (1999). Early adolescence and prosocial moral behavior: The role of social and contextual influences. *Journal of Early Adolescence, 19,* 133–147.

Erikson, E. (1950). *Childhood and society*. New York: W.W. Norton.

Freud, S. (1923/1974). *The ego and the id*. London: Hogarth (Original work published 1923).

Friedman, L. (1999). *Identity's architect: A biography of Erik H. Erikson*. New York: Scribner.

Geertz, C. (1975). On the nature of anthropological understanding. *American Scientist, 63,* 47–53.

Gilligan, C. (1982). *In a different voice: psychological theory and women's development*.

Cambridge, MA: Harvard University Press.

Gilligan, C. & Farnsworth, L. (1995). A new voice for psychology. In Chester P., Rothblum, E. D. & Cole, E. (Eds.), *Feminist foremothers in women's studies, psychology, mental health*. Binghamton, NY: Harrington Park Press.

Goble, F. G. (1970). *The third force: The psychology of Abraham Maslow*. New York: Grossman Publisher.

Heidegger, M. (1962). *Being and time*. Translated by John Macquarrie & Edward Robinson. New York: Harper.

Horney, K. (1939). *The neurotic personality of our time*. New York: Norton.

Horney, K. (1922). On the genesis of the castration complex in women. In *Feminine Psychology*, Edited by H. Kelman. New York: W.W. Norton (reprint 1967).

Horney, K. (1973). The flight from womanhood. In J.B. Miller (Ed.), *Psychoanalysis and women* (pp.5–20). Baltimore: Penguin Books. (Original work published 1926).

Jenkins, Y. M. (1999). Embracing diversity in college settings: The challenges for helping professionals. In Y. M. Jenkins (Ed.) *Diversity in college settings: Directives for helping professionals*. (pp. 5–20). New York, London: Routledge.

Johnson, F. (1985). The western concept of self. In Marsella, G. Devos, & F. L.K. Hsu (Eds.) *Culture and Self*. London: Tavistock.

Jordan, J. V., Kaplan, A. G., Miller J. B., Stiver I. P., and Surrey, J. L. (Eds.) (1991). *Women's growth in connection: Writings from the Stone Center*. New York: The Guilford Press.

Kegley, J. A. K. (1984). Individual and community: An American view. *Journal of Chinese Philosophy, 11,* 203–216.

Kohlberg, L. (1976). Moral stages and moralization: The cognitive developmental approach. In T. Lickona (Ed.), *Moral development and behavior: Theory, research and social issues* (pp. 31–53). New York: Holt.

Kozulin, A. (1990). *Vygotsky's psychology: A biography of ideas*. Cambridge, MA: Harvard University Press.

Lerner R. M. (1996). *America's youth in crisis*. Thousand Oaks, CA: Sage.

Lips, H. M. (2003). *A new psychology of women: Gender, culture, and ethnicity*. (2nd edition). New York: McGraw-Hill Higher Education.

Markus, H. R. & Kitayama, S. (1991). Culture and the self: implications for cognition, emotion, and motivation. *Psychological Review, 96 (2)*, 224–235.

Maslow, A. (1962). *Toward a psychology of being*. New York: Van Nostrand.

Maslow, A. (1954). *Motivation and personality*. New York: Harper & Row.

Miller, J. B. (1991). The development of women's sense of self. In J. V. Jordan, A. G. Kaplan, J. B. Miller, I. P. Stiver & J. L. Surrey (Eds.), *Women's growth in connection*. New York: The Guilford Press.

Miller, J. B. (1984). The development of women's sense of self. *Work in Progress, No. 12*. Wellesley, MA: Stone Center Working Paper Series.

Miller, J. B. (1976). *Towards a new psychology of women*. Boston: Beacon Press.

Mortense, E. L., Michaelson, K. M., Sanders, S. A., & Reinisch, J. M. (2002). The association between duration of breastfeeding and adult intelligence. *The Journal of the American Medical Association, 287*, pp. 2365–2371.

Ornish, D. (1998). *Love and survival: The scientific basis of the healing power of intimacy*. New York: HarperCollins.

Piaget, J. (1971). *Biology and knowledge*. Chicago: University of Chicago Press.

Piaget, J. (1967). *Six psychological studies*. New York: Vintage.

Piaget, J. (1965). *The moral judgement of the child*. New York: Free Press.

Porter, N. (2002). Contextual and developmental frameworks in diagnosing children and adolescents. In M. Ballou & L. Brown, *Rethinking mental health and disorder: Feminist perspectives*. New York: The Guilford Press.

Rogoff, B., & Chavajay, P. (1995). What's become of research on the cultural basis of cognitive development? *American Psychologist, 50,* 859–877.

Rogoff, B. (1998). Cognition as a collaborative process. In D. Kuhn & R. S. Siegler (Eds.), *Handbook of child psychology: Vol. 2.*

Cognition, perception, and language (5th ed., pp. 679–744). New York: Wiley.

Sampson, E. E. (1988). The debates on individualism: indigenous psychology of the self and their role in personal and societal function. *American Psychologist, 43,* 15–22.

Sampson, E. E. (1993a). *Celebrating the other: A dialogic account of human nature,* Boulder, CO: Westview Press.

Sampson, E. E. (1993b). Identity politics: Challenges to psychology's understanding. *American Psychologist, 48 (12),* 1219–1230.

Siegel, D. J. (1999). *The developing mind: How relationships and the brain interact to shape who we are.* New York: The Guilford Press.

Slip, S. (1993). *The Freudian mystique: Freud, women, and feminism.* NY: New York University Press.

Stiver, I. (1983). The meaning of "dependency" in female-male relationships. *Work in Progress, No. 11.* Wellesley, MA: Stone Center Working Paper Series.

Vygotsky, L. S. (1978). Mind in society: The development of higher psychological processes. M. Cole, V. John-Steiner, S. Scribner, & E. Souberman (Eds.) Cambridge, MA: Harvard University Press.

Waterman, A. S. (1981). Individualism and interdependence. *American Psychologists, 36,* 762–733.

Wertsch, J. V., & Tulviste, P. (1992). L. S. Vygotsky and contemporary developmental psychology. *Developmental Psychology, 28,* 548–557.

Westkoff, M. (1986). *The feminist legacy of Karen Horney.* New Haven: Yale University Press.

2

Relational-Cultural Theory: A Framework for Relational Development Across the Life Span

By Dana L. Comstock, Ph.D., and Dongxiao Qin, Ph.D.

REFLECTION QUESTIONS

1. *What relational strengths do you feel you bring to relationships?*
2. *What are your strategies for disconnection and how do you exercise them in relationships? In other words, what provokes the feeling in you that you want to "check out" of a relationship? Does any particular issue come to mind?*
3. *During times of disconnections in your relationships how do you find yourself coping, or not?*
4. *What aspects of yourself do you find the most challenging to bring into relationships?*
5. *What difficulties and challenges have you faced while engaging in cross-racial relationships and how have you handled this aspect of difference?*

INTRODUCTION

Relational-Cultural theory was conceived after the publication of *Towards a New Psychology of Women* (Miller, 1976). After this publication, a group of scholars, namely Jean Baker Miller, Irene Stiver, Alexander Kaplan, Judy Jordan, and Janet

Surrey, began a process of reconceptualizing traditional models of human development and psychotherapy. What followed was the unfolding of what is sometimes referred to as "Self-in-Relation" theory, or the "Relational-Cultural" model of psychotherapy. The original scholarship of this model is a collection of Works in Progress published through the Stone Center at Wellesley College, Wellesley, MA. To date, the model is referred to as "Relational-Cultural theory," or RCT. For purposes of brevity within this chapter and in most parts of this text, it will be referred to as RCT.

RCT provides an alternate perspective to traditional ways of viewing the notion of human development. As discussed and demonstrated in Chapter 1, many traditional theories of development value the ideals of individuation, separation, autonomy, and generally honor the concept of the "self" (Fedele, 1994). As an alternative:

> RCT emphasizes health, growth and courage, and points to a new understanding of human and individual strength: strength in relationship, not strength in isolation. Isolation is seen as the source of most suffering, while the process of creating mutual empathy and mutual empowerment is seen as the route out of isolation. (Jordan & Hartling, 2002, p. 51)

In contrast to many traditional Western theories of development, Relational-Cultural theory espouses that we become increasingly relationally complex, rather than more individuated and autonomous over the life span. The uniqueness of Relational-Cultural theory is its focus on achieving growth by enhancing each individual's capacity to "create, build, sustain and deepen connection" as a lifelong goal (Surrey, 1987, p. 8). In addition, "this model is built on an understanding of people that emphasizes a primary movement toward and yearning for connection" (Jordan, 2001, p. 95).

Because RCT looks at all interpersonal dynamics through a relational lens, it "represents a departure from the separate-self view of development and posits that we grown in, through, and toward relationship" (Jordan, 2002, p. 234). Jordan (2001) makes the point that:

> This does not mean that we are in actual physical relationship with people at all times, but that there is an attitude of relatedness, of mutuality, of openness, of participating in experience. This can occur in solitude, in nature, when we feel connected and in relationship with our surroundings. In isolation, we are not in relationship, we are cut off, we are not in mutual responsiveness. Often we are immobilized and self-blaming. (p. 97)

One of the most important key features of RCT is the notion of mutual empathy. As will be discussed throughout this chapter, mutual empathy has to do with the experience of mutual responsiveness and influence. This can only be experienced *in* relationship and is basic and essential to one's sense of aliveness and mental health. In other words, RCT asserts that the experience of feeling alive and valued comes to us when we feel we have been heard and have impacted another. In contrast, we feel isolated and cut off, and are vulnerable to depression and self-destructive behaviors when we feel invisible, unheard, and as if we don't matter. The bottom line is that being "seen and heard" comes more easily for some than for others and has to do with the degree of privilege and marginalization one experiences. This is a complicated issue, which is explored in depth in the next two chapters. Clearly there are no developmental models to date that appreciate the courage it takes to resist the forces of oppression and marginalization across one's life span.

Along those same lines is the mantra: "Children should be seen and not heard." We feel this mantra speaks volumes to the lack of power children have in most families and in our culture in general. Indeed, facilitating healthy psychological growth in children has to do with the responsiveness of their caregivers, which is discussed in depth by Linda Hartling in her chapter on fostering resilience at the end of this text. Much of what is at the heart of this text is the need for individuals to feel seen, heard, understood, and valued. "Context" in this sense thus has to do with an understanding of how this does not always happen in an individual's life or in her or his relationships.

WESTERN SCIENCE AND THE CONTEXT OF TRADITIONAL THEORIES

In the spirit of context, it is important to look at the scientific models from which many theories of development and psychotherapy have emerged. Jordan (2000) makes the point that as psychology worked to make itself a legitimate science it modeled itself after the more rigorous of the hard sciences, Newtonian physics. Jordan (2001) states, that "Newtonian physics, rooted in Baconian models of science, emphasize the primary separateness of objects" (p. 92) and that "when Newtonian physics reigned, atoms and molecules were seen as the basic units of reality; separation was primary and relatedness was secondary" (p. 93). Psychology modeled this paradigm, which is most apparent in theories of development and psychotherapy. Interestingly:

> As psychology was attempting to model itself after Newtonian physics, the most creative minds in physics and ultimately the whole field of physics were moving in the direction of an appreciation of quantum physics, of indeterminacy, of the primacy of relationships rather than separate objects. But the bias of separation was taken up by the field of psychology, and it became preoccupied with the separate self. And clinical psychology has largely looked at pathology within the separate self. This is epitomized in the diagnostic manuals that use a medical/disease model to locate pathology in the individual. (Jordan, 2000, p. 1006)

Women, in particular, were the first to be noted as vulnerable to being labeled "deficient" when it came to separate-self models of development. Miller (1976) and Gilligan (1982) are considered pioneers with regard to the study of the psychology of women as they made particular strides in understanding how patriarchy influenced "psychology's understanding of women, and women's understanding of themselves" (Spencer, 2000, p. 5). Originally, RCT emerged out of an effort to better understand the psychology of women, but to date it also includes the experiences of girls and women, men and boys, and addresses a wide range of sociocultural influences that impact our capacity to establish and to participate in growth-fostering relationships.

Some of the core tenets of RCT with regard to psychological growth and development are summarized by Jordan (2000) and include the ideas that:

1. People grow through and toward relationship throughout the life span.
2. Movement toward mutuality rather than movement toward separation characterizes mature functioning.
3. Relationship-differentiation and elaboration characterize growth.
4. Mutual empathy and mutual empowerment are at the core of growth-fostering relationships.
5. Authenticity is necessary for real engagement and full participation in growth-fostering relationships.
6. In growth-fostering relationships, all people contribute and grow or benefit; development is not a one-way street.
7. One of the goals of development from a relational perspective includes the development of increased relational competence and capacities over the life span. (p. 1007)

Miller and Stiver (1997) suggest that while individuals yearn for connection with others, they develop a repertoire of strategies that keep them out of connection. Such strategies, for example, include withholding love and affection, withdrawing from others, criticizing loved ones, and hiding authentic feelings (Hartling, Rosen, Walker & Jordan, 2000). At worst, these strategies are destructive and could involve addictions, compulsions, abusive behaviors, eating disorders, and workaholism. Each of these strategies has the potential to keep individuals out of relationship and to subsequently evoke a deep sense of shame and to ultimately harm the relationship (Hartling et al., 2000; Jordan & Dooley, 2000).

Through the years the founding scholars began a more in-depth analysis of relationships to answer such questions as: What differentiates relationships that foster growth versus those that impede growth? What kinds of relational dynamics lead to connections and disconnections in relationships? How do our experiences of connections and disconnections in relationships contribute to our sense of agency in relationships or to experiences of chronic disconnection or condemned isolation? What does growth in connection really feel like? And lastly, how do social, cultural, and political contexts play into all of this?

OBSTACLES TO MUTUALITY
AND AUTHENTICITY

RCT posits that one of our more challenging developmental processes involves that of becoming increasingly able to fully represent ourselves both honestly and authentically in relationships (Miller & Stiver, 1997). There are many obstacles to authentic expression. Such obstacles may include different types of shame-based oppression and marginalization, whereby a person's experience is perceived as defective, or worse, invisible (Hartling et al., 2000; Jordan, 1997; Walker, 2001). Jordan (2002) makes the point that:

> A range in marginalization exists in this world from traumatic oppression to dismissal or trivialization. Some places at the margin are places of oppression. Some are also places of disconnection, fear, and pain. And all marginalization is an assault on our humanity and our dignity. Some people develop amazing capacities to resist and transform the dehumanizing, objectifying forces that marginalize. Some cannot. (p. 1)

It is important to remember that the degree of safety one feels to express authentic feelings is directly related to how much power or mutuality (which is the opposite of feelings of marginalization) one experiences in the relationship and that "opportunities for growth occur not only on the individual or familial level, but also at the sociocultural level" (Jordan & Hartling, 2002). As such, individuals feel varying degrees of freedom to express themselves and have varying expectations they will be heard, both of which are directly related to the degree of privilege or marginalization one experiences. Feelings of privilege and marginalization are the result of the stratification that occurs around "difference" in our culture. The degree to which one might experience ageism, classism, heterosexism, racism, or sexism for example, impacts one's expectations of mutuality in relationships. Jordan (2002) explains how individuals at the margin (subordinates) are treated by individuals at the center (members of the dominant culture):

> People at the margin are defined as "objects:" they are seen as being at the margin because of some essential failure of character or effort. The myth of meritocracy and the myth of the level playing field support this distorted understanding of privilege. That is, people who have not "made it" deserve the place they occupy . . . if you are from a marginalized group and are successful in terms of the center's definition, you are the exception to the rule; if you are not successful, it proves you *are* the problem and are inferior in some core way. In fact people at the margin are actively socialized to believe that they *have* failed. . . . The group at the center makes the rules and names the situations and conditions of privilege and disadvantage. The prevailing attitude toward those who do not enjoy the privilege and power in a given system is one of denigration. In mental health parlance we pathologize the experiences of people at the margin. This is obvious in blatant sexism, racism, or heterosexism, where broad strokes of negative stereotypes are aimed at individuals

with various characteristics who are deemed inferior by the naming group. (p. 2)

The use of the phrase "those people" (or "you people" if used directly) is a perfect example of this dynamic. We typically think of the objects (and we chose "object" because of the objectifying context in which this phrase is used) or victims of this phrase as stupid, less than, predictable (based on stereotypes), dangerous, at fault, and sadly, yet most importantly, different from the namer.

Along these same lines, research shows that White therapists generally give more negative diagnoses (indicating more psychological impairment) and that they "misconstrue uncooperative behavior among Hispanics as psychosis" (Sparks, 2002, p. 282). The research on the discrepancies between Whites and ethnic minorities in general is overwhelming in spite of "societal changes with regards to race/ethnicity that have occurred over the last 20 years" (Sparks, 2002, p. 282). Sadly, this same racial/ethnic bias applies to diagnostic patterns of ethnic minority children, the effects of which can be devastating (Porter, 2002).

RCT has been influenced by the works of Peggy McIntosh (1988), bell hooks (1984), and Patricia Hill Collins (2000). In her book *Black Feminist Thought*, Collins (2000) wrote about the notion of "controlling images," particularly how they impact African-American women, as being "stereotypical images" used "to make racism, sexism, poverty and other forms of social injustices appear to be natural, normal and inevitable parts of everyday life" (p. 69). In a sense, controlling images work to maintain and normalize the oppressive nature of the stratification around difference in our culture which leaves individuals at the margin seen as "less than" individuals at the center.

From a relational perspective, Walker (2000) posits that in this process "the dominant culture distorts images of self, images of other, and images of relational possibilities" (p. 3). In other words, internalized controlling images limit (a) an individual's perception of who, how, and what he or she can be in the world and (b) how and with whom one can mutually relate. As such, a RCT approach to relational development includes efforts to expand and resist the constricting nature of both relational and controlling images for individuals of all walks of life.

Power differentials, gender-role socialization, race, culture, health status, sexual orientation, and all the various "isms" have the potential to silence an individual's experience (Dooley & Fedele, 1999; Hartling et al., 2000; Miller et al., 1999; Walker, 2001). As such, these experiences are both disconnecting and shame-based and have the potential to move individuals into a place of what Miller and Stiver (1997) refer to as "condemned isolation" (p. 72).

Condemned isolation is the experience of "being locked out of the possibility of human connection" (Miller & Stiver, 1997, p. 72). In this experience, individuals carry a deep sense of shame and a belief that they are somehow defective as human beings. They are often unable to see the problem in a sociocultural or relational context and believe the problem is "in them" (Hartling et al., 2000; Jordan, 1999; Miller & Stiver, 1997). Hiding large parts of their experience and engaging inauthentically in order to reconnect in non-mutual relationships often becomes a strategy for survival (Miller & Stiver, 1997).

SHAME AND ISOLATION

Often these strategies for survival are constructed as a means to survive the devastating effects of isolation and shame, which Jordan (2000) describes below:

> Isolation is not the same as being alone or in solitude. . . . Shame often accompanies a sense of isolation; one feels unworthy of connection. In shame, one feels disconnected, that one's being is at fault, that one is unworthy of empathic response, or that one is unloveable. Often in shame people move out of connection, lose their sense of efficacy, and lose their ability to authentically represent their experience. (p. 1008)

Hartling et al. (2000) emphasize that "experiences of shame or humiliation—including experiences of being scorned, ridiculed, belittled, ostracized, or demeaned—can disrupt our ability to initiate and participate in the relationships that help us grow" (p. 1). Transforming shame involves understanding one's relational capacities in a sociocultural context and allows one to move out of shame and isolation and into the possibility for more mutually empathic and authentic connections (Hartling et al., 2000; Walker, 2001).

We cannot say enough about the impact shame has on one's developmental experiences. Certainly we all have issues in our lives we feel ashamed of, yet some individuals experience shame and/or shaming as a pervasive theme in their lives. Ongoing experiences of racism, experiences of invisibility related to physical challenges, or even a shame-filled haunting history of sexual abuse are all examples of experiences that take up a lot of space in one's psyche. Dismantling shame in one's life takes a great deal of courage and is best accomplished in communities of support. In Chapter 4, Mary Howard-Hamilton and Kimberly Frazier talk about "safe spaces" as a place for individuals to go and share their experiences of oppression and racism. Along these same lines we all know how powerful and healing support groups can be. Healing in these contexts has to do with making the invisible visible by naming it and taking the blame out of one's "self" and placing it in the appropriate context.

I (Dana) am reminded of a conversation I had with a colleague a few years ago about a consciousness-raising group on racism for White people that was going to be held at the Esperanza Center for Peace and Justice in San Antonio. There had been an ongoing debate about whether or not people of color should be a part of these discussions. I do not recall the specific outcome of these discussions but I distinctly remember how afraid the group members were about beginning these discussions and how much shame played into their fears. I think this speaks to a different kind of courage it takes to acknowledge that you have hurt someone and to bring this reality into the relationship so that it might heal. Indeed the impact of shame plays a crucial role in relationships. At best we can name our shame and move toward relational transformation or at worse, shame is immobilizing and keeps us locked out of relational possibilities. The complexities and the language of relational movement are discussed in detail in the next section.

THE LANGUAGE OF RELATIONAL MOVEMENT AND GROWTH

Given that RCT "necessitates assumptions different from those that underlie previous theories" (Miller & Stiver, 1993, p. 424), it requires a language that reflects its unique values and philosophy. RCT structures its language to promote its philosophies, values, and assumptions.

Mutual Empathy

Basic to RCT is the principle of *mutual empathy*. Miller and Stiver (1997) describe mutual empathy as the essence of relational healing and psychological growth. In this respect, mutual empathy reflects the healing that occurs when individuals believe that others have been genuinely moved or affected by their experiences. To truly be moved by another's experience, we have to be accessible, and to be accessible means we are vulnerable (Miller & Stiver, 1997).

The notion of mutual empathy is in contrast to the more common expression of "one-way" empathy. In relational terms, two people create mutual empathy when the listener is affected by the experience of the other and the other is moved by the impact they have had on the listener. The differences in these terms can appear discreet, even though the differences in the experience, particularly for that of the speaker, are quite powerful. Mutual empathy speaks to both the intra- and interpersonal healing that takes place when we express being moved by others' experiences (Fedele, 1994).

In a mutually empathic encounter, everyone's experience is broadened and deepened as we are "empathically attuned, emotionally responsive, authentically present, and open to change" (Miller, Jordan, Kaplan, Stiver & Surrey, 1991, p. 11). In other words, everyone must be vulnerable to be authentically present. Each participant must genuinely respect and be moved by the experience of the other if he or she is to be open to change (Jordan, 1999).

Relational Movement: Connections, Disconnections, and Transformation

Basic to RCT is the process of moving through connections, disconnections, and back into new, transformative, and enhanced connections (see Figure 1). In RCT there are five experiential components of connection (see Figure 1), which are often referred to as the "five good things" (Miller, 1986, p. 3). Connection is experienced when:

> each person feels a greater sense of "zest" (vitality, energy); each person feels more able to act and does; each person has a more accurate picture of her/himself and of the other person(s); each person feels a greater sense of worth; and each person feels more connected to the other person(s) and feels a greater motivation for connections with other people beyond those in the specific relationship. (Miller, 1986, p. 3)

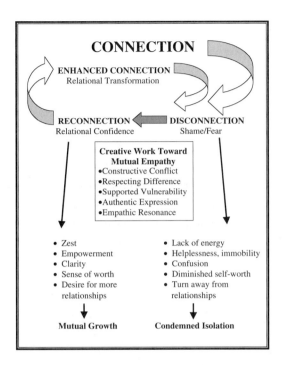

Figure 1. Relational Movement as the Process of Creatively Working through Disconnections Toward Transformative, Enhanced Connections

Based on the following sources: C. Dooley and N. Fedele, "Mothers and Sons: Raising Relational Boys," in *Work in Progress, No. 84*. © 1999 Stone Center Working Paper Services; and L. M. Hartling, et al., "Shame and Humiliation: From Isolation to Relational Transformation," in *Work in Progress, No. 88.* © 2000 Stone Center Working Papers. Adapted with permission from the authors.

The presence of these qualities is indicative of mutuality present in growth-fostering relationships (Jordan & Dooley, 2000). Disconnection is experienced as the opposite of the five good things (see Figure 1). Jordan & Dooley (2000) describe the experience of disconnection as "decreased energy; inability to act; a lack of clarity or confusion regarding self and other; decreased self worth; and we turn away from relationship" (p. 13). This "turning away" is often accompanied by a deep sense of shame. The process of moving from connection to disconnection and into reconnection is transformative (see Figure 1). Jordan (1992) articulates this by stating that:

> In cases of disconnection, transformation involves awareness of the forces creating the disconnection, discovery of a means for reconnecting, and building a more differentiated and solid connection. The movement into and out of connection becomes a journey of discovery about self, other and relationship—about "being in relation." (p. 8)

We all have different relational capacities and different tolerance levels for disconnections. Some of us are more resilient during these times, while others are prone to coping with disconnections in unhealthy ways (drug or alcohol abuse, for example). Our vulnerabilities during these times speak to how painful disconnections are and to how much courage it takes to move through them. Interestingly, we have many social rituals that serve as invitations for connections. The invitation to "have a glass of wine" or to "go get a cup of coffee" are, in fact, invitations to "connect." At a Jean Baker Miller Training Institute I (Dana) attended a few years ago, Janet Surrey talked about how she and her young daughter were working to come up with ways to "reconnect" after their mother-daughter arguments. Certainly we would all benefit from creating rituals for reconnecting following conflict!

The Central Relational Paradox

Simply stated, all individuals have a yearning for connection. The paradox is experienced when, in the face of one's yearning for connection, individuals employ strategies for disconnection. These strategies are employed to avoid the perceived (or real) risk of hurt, rejection, or violation (Miller & Stiver, 1997). This is a complicated and frustrating cycle. Individuals exercise various strategies for disconnection, and they do so to varying degrees (Miller & Stiver, 1997). Although individuals feel a yearning to connect, feelings such as shame make movement into connection difficult.

According to Jordan (1997), "shame is most importantly a felt sense of unworthiness to be in connection, a deep sense of unlovability, with the ongoing awareness of how very much one wants to connect with others" (p. 147). Dealing with relational dynamics with regard to the paradox and dealing openly with a shared sense of shame are both important components of moving past shame-based impasses and into relational transformation and empathic possibilities (Jordan, 1997; Hartling et al., 2000).

It is critical that we understand why and how individuals are "protected" by their strategies for disconnection, which, paradoxically, are the source of so much pain. When we express our vulnerabilities we need others to resonate with our feelings, rather than disconnecting or becoming emotionally disengaged (Miller & Stiver, 1997).

It needs to be noted that disconnections are an inevitable part of all relationships (Jordan & Dooley, 2000). At times, all individuals are challenged to stay in connection when they feel drawn to exercise strategies for disconnection. The process of resisting such strategies, by moving into a creative place for enlarging mutual empathy, is essential for developing relational resilience. Enlarging mutual empathy and fostering relational resilience is an ongoing task within this model. The more we are able to be creative in this way, the more relational confidence is established. This is another way of describing the oscillation of trust in relationships particularly when there is not much relational history or when the relationship is relatively new.

In spite of how hard it is to hang in there to do your part to establish trust in the relational process, we inevitably adhere to distorted expectations of how others will treat and respond to us. According to RCT, these expectations are not thought of as irrational, unfounded, or unreasonable. Rather, they are based on years of experiences, sometimes in the context of abusive relationships wrought with repeated and chronic disconnections. The challenge is for us to begin to recognize relationships in which mutual engagement is possible, in spite of our expectations. These expectations of the outcomes of relationships are what Miller and Stiver (1995) refer to as *relational images*.

Relational Images

According to Miller and Stiver (1995), relational images are expressions of our expectations and fears of how others will respond to us. In essence, they relate to our expectations of the outcomes of relationships when we make personal strides toward authenticity. If we go through our lives and continue to engage in relationships in which we are denied empathic possibilities, our relational images become confirmed. Our patterns of relating then become outdated, "fixed and difficult to alter" (Miller & Stiver, 1995, p. 3).

Outdated relational images are frustrating and binding, because they negate the emergence for new relational possibilities. In reality, individuals may have more relational possibilities than they are able to construe. Sometimes we negate the possibilities of the here and now because of our struggles with the disappointments, abuses, and violations of the past. When relational images function at their worst, they do not allow us to see other people for who they are because our ability to fully participate with them is hindered due to our expectations of how they will treat us and of how the relationship will unfold and ultimately end. I include the possibility of the relationship ending because oftentimes relational images are "relationship-defeating" (versus self-defeating).

As an indication of our growth we learn to relate more authentically and empathically, and begin to understand others and ourselves more clearly. This clarity results in the development of increased relational capacities and relational confidence. This process is something we engage in over our entire life span as we are constantly changed and challenged in a variety of relationships and relational contexts.

RELATIONSHIP-DIFFERENTIATION

Janet Surrey, one of the founding scholars of RCT, proposed a new model of development that she termed "relationship-differentiation" (Surrey, 1991, p. 36). She makes it clear that:

> By differentiation, I do not mean to suggest as a developmental goal the assertion of difference and separateness; rather I mean a dynamic process that encompasses increasing levels of complexity, structure, and articulation within the context of human bonds and attachments. Such a process needs to be traced from the origins in early childhood relationships through its extension into all later growth and development. (p. 36)

This particular theory of development does not address step-wise development, "'fixed' states, developmental crises, or one-dimensional, unidirectional goals of development" (Surrey, 2001, p. 39). Surrey offers this alternative:

1. Critical relationships would be seen as evolving throughout the life cycle in a real, rather than intrapsychic, form.
2. We would have to account for the capacity to maintain relationships with tolerance, consideration, and mutual adaptation to the growth and development of each person.
3. We would account for the ability to move closer to and further away from other people at different moments, depending on the needs of the particular individuals and the situational context.
4. We would explore the capacity for developing additional relationships based on broader, more diversified new identifications and corresponding patterns of expanding relational networks.
5. We would examine the potential problems and vicissitudes inherent in the development of relational capacities. (Surrey, 1991, 39)

Critical Relationships as Evolving Throughout the Life Cycle

The first aspect of relationship-differentiation involves *looking at how relationships change over the life span versus simply looking at how the "individual" changes over the life span*. Surrey (1991) uses the example of how we grow up "in relational" to our parents and how this experience is not static, but rather it is fluid. In other words, we don't have the same kind of relationship with our parents our entire lives. One challenge to the parent-child relationship comes during adolescence.

One thing we have all heard over and over is that during adolescence the goal of the child is to become independent, or to "break away" from the family. It is all too common that when parents interfere with this process they are seen as intrusive, enmeshed, overinvolved, and the like. The family therapy literature is full of descriptors of how difficult it is for the parents, namely mothers, to get it right. Mirkin (1994) states that jargon such as "'breaking away,' 'breaking free,' 'cutting the umbilical cord' implies a jolting disconnection" (p. 78). She goes on to state that under the old paradigm, family therapists gave the message that "we valued separation and shunned connectedness" (p. 78). An alternative to the notion of individuation during adolescence is to consider the *increased need for connection* during this time and that, in fact, much of the conflict may have to do with the pain of separation. One man, as cited in Weingarten (1998), had this to say about his experience of adolescence:

> I wanted to be close to my mother and that wish made me feel wrong and bad and stupid. At other times, I just felt confused about how I felt. I was never able to talk to my parents about what I was feeling—the fact that I felt lonely for them, but particularly my mother—because I thought I would be exposing something shameful. (p. 16)

Indeed, gender expectations do guide much of our thinking about how and when we should expect to be able to feel connected to our parents. Yet, what you will find in the chapters to follow are more painful stories of disconnections.

Weingarten (1998) describes an alternative to separation during this time and she urges us to consider how:

> Relationship, not a "clean break," is crucial for teens if they are to become able citizens of our complex world, for it is their mothers who are in the best position to assist teens in developing the independence skills they need to function competently. It is relationship, not separation, that will bring mothers of adolescents back from the periphery of their children's lives into the center, a center in which they coexist harmoniously with peers and other adults. (p. 24)

Maintaining Relationships While Adapting to the Growth and Development of Each Person

The second aspect of relationship-differentiation has to do with our *capacity to tolerate, consider, and to mutually adapt to the growth and development of each person in the relationship*. This particular aspect of relationship-differentiation reminds me of many conversations I have with families and couples who describe a process of "growing apart" in which relationships end, versus adapting to the changes one or both parties experience in their lives. Oftentimes individuals approach relationships in a very self-centered manner. When this happens, more attention is paid to what one member of the relationship needs instead of there being a mutual appreciation for what the other person or the relationship as a whole needs.

The fact of the matter is that the people in our lives are not always going to be exactly what we feel they should be or want them to be. As one party begins to bring more of who he or she is into the relationship, there may be little tolerance of this expression of new interests, for example. What much of this refers to is how well we handle conflict in our relationships. In fact, Jordan (2002) stresses that "if people bring themselves authentically into relationships, difference and conflict will arise with others" (p. 240). She goes on to state that "in unequal power situations, these potential conflicts often resolve in the direction of the more powerful person's dictating the resolution" (p. 240).

The idea of how we express ourselves, what we express, and to whom we express it can truly interfere with resolving conflict. What you will come to learn is that the expression of anger, hurt, disappointment, and fear presents different challenges for men and women based on culturally mandated sex-role stereotypes and to power dynamics in particular relationships. These cultural mandates and power dynamics clearly impact our capacity to fully represent ourselves in growth-fostering relationships. Often subordinates in relationships lose "voice," meaning we are more likely to share ourselves and feelings with someone we consider our equal versus someone we feel subordinate to. Gilligan (as cited in Mirkin, 1994) makes the point that with regard to adolescent girls, they are "more likely to speak in relationships where nobody will leave and someone will listen" (p. 81).

In fact, people do leave relationships when there is a lack of tolerance for their growth, and people in positions of power do not always hear the voices of their subordinates. Full participation during these times of disconnection (which are normal aspects of the healthiest of relationships) involves "holding the tension" between differing perceptions and experiences (Jordan, 1992, p. 6) versus opting for a complete rupture in the relationship. This is a very difficult process, which will be discussed in more detail later in this chapter.

Relational Movement and Developmental Needs

The third aspect of relationship-differentiation has to do with *our ability to move and adapt to the needs of others at different times in their lives, which often involves a particular situational context.* We all have varying degrees of comfort when it comes to the type of daily interactions we have with the people closest to us. We get preoccupied and distracted with the demands of our family and work lives. Nevertheless, from time to time, we are faced with adversities in our lives that force us to turn to others for help. "Vulnerability" is almost a taboo word in our culture and nobody likes to be thought of as needy. This is reflected in the saying "Never let them see you sweat" (Comstock & Duffey, 2003). Greenspan (1998) makes the point that our disdain for vulnerability or neediness in others or ourselves comes from "the myth of independent individualism" (p. 51). She goes on to state that:

> Through this myth we elevate a standard of autonomy that is, on close inspection, a denial of our fundamental interdependence as human beings. . . . Those we credit with rock-like independence tend to be those whose dependence is generally unacknowledged or hidden, including the dependence of the entrepreneur on the worker, the working father on the stay-at-home mother. The hidden dependencies that we revile and castigate, disavow or shun in ourselves, we find hard to take in those whose needs are visible every minute. . . . They remind us that we too are vulnerable; that we are dependent on others; that we have needs, which we consider, in the dominant masculine ideology of our society to be shameful weaknesses without merit. (Greenspan, 1998, p. 51)

Yet as difficult as it is for us to recognize and accept our vulnerabilities, they are, at times, simply unavoidable and no doubt have a tremendous impact on our relationships.

Our ability to adapt to the needs of others has to do with our responsiveness, which is dependent, in part, on how much others are willing to share with regard to their adversity. Life stressors such as divorce and health crises including chronic illness all take a toll on our relationships. In fact, a number one concern for women diagnosed with a chronic illness is the impact their illness will have on their relationships (Comstock, 2003). In order to minimize the degree of relational stressors that result from a diagnosis of a chronic or debilitating illness, many women will hide the illness from their loved ones. Men do this as well, often to an even greater degree than women. In the end, however, hiding a chronic illness only serves to increase the stress level, which, in turn, exacerbates the illness. In order for us to have some confidence our relationships can hold "contextual stressors" we need to find the courage to trust that others will be responsive, patient, and compassionate with our vulnerabilities.

Jordan (1992) also emphasizes that we have issues with vulnerability in our culture and she states that:

> Awareness of vulnerability, in fact, suggests to me good reality testing. It is the disowned vulnerability that becomes problematic. An openness to being affected is essential to intimacy and a growth-enhancing relationship; without it, people relate inauthentically, adopting roles and coming from distanced and protected places. Open sharing of our need for support or acceptance may be an essential factor in developing a sense of close connection. Therefore, part of what we are trying to transform is the illusory sense of self-sufficiency and the tendency to deny vulnerability. We need a model that encourages supported vulnerability. (p. 4)

The Development of Expanding Relational Networks

The fourth aspect of relationship-differentiation has to do with our *exploring the capacity for developing additional relationships based on broader, more diversified new iden-tifications and corresponding patterns of expanding relational networks.* I think we can all remember a time when, while discussing the particulars of certain relationships, someone commented: "We don't have that kind of relationship" or "This relationship is so different from the others I have had." In fact, this particular aspect of relationship-differentiation takes into account the impact of the relationships we have in any number of contexts that Surrey (1991) refers to as "relational networks" (p. 38). Such relational networks include relationships with "father, triangular relationships, preadolescent and adolescent friendship patterns, sexual relationships, marriage, mothering and family networks, teaching relationships, role modeling, women in work groups, and still broader reference groups" (Surrey, 1991, p. 38).

In each of these relational contexts we grow in essential, yet very different ways. What differentiates the various relational contexts, or relational constructions, are the boundaries we exercise that guide the type and degree of contact we have. Jordan and Hartling (2002), make the point that our traditional concept of boundary "arises within, supports, and reinforces the model of the separate self, which suggests that one must protect oneself from relationships," and "that safety and well-being ensue from constantly armoring oneself with invisible barriers" (p. 52). They suggest that a more helpful and relational way of considering boundaries is to think of them "as places of meeting, exchange and maximal growth" (Jordan & Hartling, 2002, p. 53).

They go on to state that in regard to boundaries:

> The emphasis in RCT is on (1) clarity in relationship (e.g., this is your experience; this is mine), (2) the right to say "no" and to exercise choice in deciding what one will share or do, (3) the importance of stating limits (e.g., "I can't do that in our relationship because it makes me uncomfortable"), and (4) redefining boundaries as places of meeting and exchange, rather than as walls of protection against others. By observing these four conditions of engaging in relationship, we essentially honor growth and safety through connections, not

through separation or imposing power over others. (Jordan & Hartling, 2002, p. 53)

Their last statement reminds us that when we have the power in a relationship (such as being the mentor of a student) boundaries serve to protect the integrity of the relationship. I also want to stress that support groups or therapy groups are integral relational networks that lend themselves to very important aspects of our relational growth and as such have their own unique set of boundaries.

Potential Problems in the Development of
Relational Capacities

The fifth aspect of relationship-differentiation involves *the examination of potential problems someone might face in working to develop relational capacities over the life span.* There are any number of issues and life experiences people can deal with that impact their capacity to participate in growth-fostering relationships. Many of these issues are dealt with in the remainder of this text. These issues and the impact they have on relational capacities are varied and complex. Yet on the other hand, a simple analysis of context moves us out of the notion of intrapyschic pathology and helps us see the bigger picture of what has shaped a person's life. As an alternative we can focus on the complexities of relational, political, and socio-cultural trauma and their direct impact on relational competence. It is important to consider that we all experience relational disconnections at times; it is when they become chronic that they have the most devastating impact.

RELATIONAL COMPETENCE AND
RELATIONAL RESILIENCE

The better able one becomes at participating in reworking disconnections and empathic failures, the better one can manage relational disconnections. According to RCT, the ability to manage disengagement is an indicator of one's relational competence or relational resilience. On an experiential level, relational competence is "the capacity to move another person, to effect a change in a relationship, or effect the well-being of all participants in the relationship" (Jordan, 1999, p. 3). Jordan goes on to state that relational competence involves:

1. Movement toward mutuality and mutual empathy (caring and learning flows both ways), where empathy expands for both self and other;
2. the development of anticipatory empathy, noticing and caring about our impact on others;
3. being open to being influenced;
4. enjoying relational curiosity;
5. experiencing vulnerability as inevitable and a place of potential growth rather than danger; and

6. creating good connections rather than exercising power over others as the path of growth. (1999, p. 3)

Notice how the characteristics of relational competence are in stark contrast to what Jordan (1999) refers to as "the dominant myths of instrumental competence which largely coincide with the myths of masculinity":

1. The myth that competition enhances performance;
2. the myth of invulnerability;
3. the myth of certainty;
4. the myth of self-sufficiency ("I did it alone, so can you");
5. the myth of mastery ("I mastered it, I *am* the master");
6. the myth of objectivity;
7. the myth of expert;
8. the myth of unilateral change (In an interaction, the less powerful person is changed);
9. the myth that hierarchy and ranking produces incentives and that people assume their places in the hierarchy by virtue of merit;
10. the myth that power over others creates safety; and
11. the myth that rational engagement is superior to and at odds with emotional responsiveness. (Jordan, 1999, pp. 2–3)

Relational competence is manifested in relationships by "enhanced empathic possibilities, capacities to stay present with a wide range of complex and difficult feelings in herself and others, and greater freedom to stay in the process and bring more and more of herself into the relationship" (Miller et al., 1991, p. 1). As a result of mutual and authentic engagement (which cannot take place in the context of the myths stated above), empathic bridges are established and all parties to the relationship have the potential to develop increased relational competence. The following questions were adapted from Comstock, Duffey, and St. George (2002) and are useful in assessing where one stands developmentally with regard to relational competence:

1. What are your strategies for disconnection and how do you exercise them in relationships?
2. Given your strategies for disconnection, how have you experienced the central relational paradox?
3. What are some of your relational images and what experiences shaped these images?
4. How have these relational images kept you out of connection and how do they impact your participation in relationships? When, and in what relationships, do you struggle more with your relational images?
5. What part(s) of yourself have you left out of relationships?
6. How has inauthentic relating affected your sense of self-worth and relational confidence?

7. In terms of authentic relating, what parts of yourself will you be challenged to bring into your relationships in the future?
8. What are some sociocultural influences that have affected your capacity/ability to develop/maintain mutuality in your relationships?
9. In response to such sociocultural influences, what types of strategies have you used for survival? For resistance? For transformation? For managing shame?
10. How does the sociocultural makeup of your community impact your sense of safety regarding authentic relating and mutual engagement?
11. What relational strengths do you feel you bring to relationships?

CONCLUSION

Developmental theorists have been writing about the influence of relationships for years. Yet the ultimate focus tends to be on "self-development," and there has been a lack of emphasis, appreciation, and value placed on the development of relational competence in lieu of autonomy and independence indicative of Western individualism.

Comstock, Duffey, and St. George (2002) used relational language to reframe and expand individualist developmental goals in order to address the essentials of relational competence. Considering these new definitions we can expand and rethink their respective meanings in our lives and come to an awareness of our own relational movement:

1. Autonomy as *increasing one's ability to represent oneself more fully and authentically in relationships. It is the growing into more complex relationships and relational networks, and not away from them, that is identified in this model as one of our most challenging developmental tasks.*
2. Self-awareness as *developing self-empathy for the purpose of increased authentic relating and the development of an understanding of one's relational movement, which is done in, not out, of engagement.*
3. Self-worth as *increasing relational confidence experienced as feeling good about one's relational capacities and one's ability to impact others.*
4. Conflict resolution as *acquiring the ability to move into healthy conflict and to resist the sources of disconnection while understanding that conflict is worked through (not resolved) only to the degree that the parties are able to move through their differences toward mutuality, and ultimately into a renewed connection.*
5. Self-knowledge as *obtaining clarity about self, other, and the relationship during the experience of connections in growth-fostering relationships. Our unique identity is defined in, not out of relationship.*
6. Trust in others as *becoming increasingly able to move into a place of supported vulnerability in mutual relationships in which we are able to build new relational images. Learning to trust oneself and others' capacity for and interest in relational movement. It is also a trust in the relational process.*

7. Self-respect as *becoming able to name and dismantle our respective sources of shame-based oppression and marginalization and to build supportive relationships that serve as a source of strength and resilience.*
8. Personal growth as *experiencing relational transformation inclusive of the growth of others, the relationship, and ourselves.*
9. Self-confidence as *developing relational confidence, which comes from the experience of going full circle in relationships and into relational transformation.*
10. Self-preservation or resistance as *understanding the playing out of the central relational paradox by respectfully honoring one's fear and vulnerability while simultaneously yearning for connection.* (p. 266)

Developing increased relational capacities as a lifelong goal requires an openness to ourselves, others, and our mutual vulnerabilities. Such an openness requires a commitment to supporting each other in our vulnerabilities and to participating in ways that ultimately foster our mutual growth. Authentic relating takes a great deal of courage, and being able to use relational language helps us to more clearly express where we are in the process of transformation. The bliss and energy of our connections can serve as a source of resilience during our times of despairing disconnections and can move us back into a place of hope over the relational possibilities that lie ahead.

REFERENCES

Comstock, D. L. (2003). The impact of chronic and debilitating illness on women's mental health. In L. Slater, J. H. Daniels & A. E. Banks (Eds.) *The complete guide to mental health for women*, pp. 177–184. Boston: Beacon Press.

Comstock, D. L., Duffey, T. H. & St. George, H. (2002). The relational-cultural model: A framework for group process. *Journal for Specialists in Group Work*, vol. 23(3), pp. 254–272. Portions from this publication are used in this chapter with permission from Sage Publications, Inc.

Comstock, D. L. & Duffey, T. H. (2003). Confronting adversity. In J. A. Kottler & W. P. Jones (Eds.) *Doing better: Improving clinical skills and professional competence*, pp. 67–83. New York: Brunner-Routledge.

Dooley, C., & Fedele, N. (1999). Mothers and sons: Raising relational boys. *Work in Progress, No. 84*. Wellesley, MA: Stone Center Working Paper Series.

Fedele, N. (1994). Relationships in groups: Connection, resonance and paradox. *Work in Progress, No. 69*. Wellesley, MA: Stone Center Working Paper Series.

Greenspan, M. (1998). "Exceptional" mothering in a "normal" world. In C. G. Coll, J. L. Surrey & K. Weingarten (Eds.), *Mothering against the odds: Diverse voices of contemporary mothers*, pp. 37–60. New York: The Guilford Press.

Hill Collins, P. (2000). *Black feminist thought: Knowledge, consciousness, and the politics of empowerment*. New York: Routledge.

hooks, b. (1984). *Feminist theory: From margin to center*. Boston: South End Press.

Hartling, L. M., Rosen, W., Walker, M., & Jordan, J. V. (2000). Shame and humiliation: From isolation to relational transformation. *Work in Progress, No. 88*. Wellesley, MA: Stone Center Working Paper Series.

Jordan, J. V. (1992). Relational resilience. *Work in Progress, No. 57.* Wellesley, MA: Stone Center Working Paper Series.

Jordan, J. V. (1997). Relational development: Therapeutic implications of empathy and shame. In J. V. Jordan (Ed.). *Women's growth in diversity.* (pp.138–161). New York: The Guilford Press.

Jordan, J. V. (1999). Toward connection and competence. *Work in Progress, No. 83.* Wellesley, MA: Stone Center Working Paper Series.

Jordan, J. V. (2000). The role of mutual empathy in relational-cultural therapy. In session: *Psychotherapy in practice,* vol. 55(8), pp. 1005–1016.

Jordan, J. V. (2001). A relational-cultural model: Healing through mutual empathy. *Bulletin of the Menninger Clinic, vol. 65(1),* pp. 92–103.

Jordan, J. V. (2002). Learning at the margin: New models of strength. *Work in Progress, No. 98.* Wellesley, MA: Stone Center Working Paper Series.

Jordan, J. V. (2002). A relational-cultural perspective in therapy. In F. Kazlow (Ed.) *Comprehensive handbook of psychotherapy,* vol. 3, pp. 233–254. New York: Wiley & Sons.

Jordan, J. V., & Dooley, C. (2000). *Relational practice in action: A group manual.* Wellesley, MA: Stone Center Publications.

Jordan, J. V., & Hartling, L. M. (2002). New developments in relational-cultural theory. In M. Ballou & L. S. Brown (Eds.), *Rethinking mental health and disorders: Feminist perspectives* (pp. 48–70). New York: The Guilford Press.

Jordan, J. V., Kaplan, A. G., Miller, J. B., Stiver, I. P., & Surrey, J. L. (1991). *Women's growth in connection: Writings from the stone center.* New York: The Guilford Press.

McIntosh, P. (1988). White privilege and male privilege: A personal account of coming to see correspondence through work in women's studies. *Working Paper, No. 189.* Wellesley, MA: Wellesley College Center for Research on Women.

Miller, J. B. (1976). *Toward a new psychology of women.* Boston: Beacon Press.

Miller, J. B. (1986). What do we mean by relationships? *Work in Progress, No. 22.* Wellesley, MA: Stone Center Working Paper Series.

Miller, J. B., & Jordan, J. V., Kaplan, A., Stiver, I. P., & Surrey, J. I. (1991). Some misconceptions and reconceptions of a relational approach. *Work in Progress, No. 49.* Wellesley, MA: Stone Center Working Paper Series.

Miller, J. B., Jordan, J. V., Stiver, I. P., Walker, M., Surrey, J., & Eldridge, N. S. (1999). Therapists' authenticity. *Work in Progress, No. 82.* Wellesley, MA: Stone Center Working Paper Series.

Miller, J. B., & Stiver, I. P. (1993). A relational approach to understanding women's lives and problems. *Psychiatric Annals, 23*(8), 424–431.

Miller, J. B., & Stiver, I. P. (1994). Movement in therapy: Honoring the strategies of disconnection. *Work In Progress, No. 65.* Wellesley, MA: Stone Center Working Paper Series.

Miller, J. B., & Stiver, I. P. (1995). Relational images and their meanings in psychotherapy. *Work in Progress, No. 74.* Wellesley, MA: Stone Center Working Paper Series.

Miller, J. B. & Stiver, I. P. (1997). *The healing connection: How women form relationships in therapy and in life.* Boston: Beacon Press.

Mirkin, M. P., (Ed.). (1994). *Women in context: Toward a feminist reconstruction of psychotherapy.* New York: The Guilford Press.

Porter, N. (2002). Contextual and developmental frameworks in diagnosing children and adolescents. In M. Ballou & L. Brown (Eds.), *Rethinking mental health and disorder: Feminist perspectives,* pp. 2262–278. New York: Guilford Press.

Sparks, E. (2002). Depression and schizophrenia in women: The intersection of gender, race/ethnicity, and class. In M. Ballou & L. Brown (Eds.), *Rethinking mental health and disorder: Feminist perspectives,* pp. 279–305. New York: The Guilford Press.

Surrey, J. L. (1987). Relationship and empowerment. *Work in Progress, No. 30.* Wellesley, MA: Stone Center Working Paper Series.

Stiver, I. P., Rosen, R. B., Surrey, J., & Miller, J. B. (2001). Creative moments in relational-cultural therapy. *Work in Progress, No. 92.* Wellesley, MA: Stone Center Working Paper Series.

Walker, M. (2001). When racism gets personal: Toward relational healing. *Work in Progress, No. 93.* Wellesley, MA: Stone Center Working Paper Series.

Walker, M. (1999). Race, self and society: Relational challenges in a culture of disconnection. *Work in Progress, No. 85.*

Wellesley, MA: Stone Center Working Paper Series.

Weingarten, K. (1998). Sidelined no more: Promoting mothers of adolescents as a resource for their growth and development. In C. G. Coll, J. L. Surrey & K. Weingarten (Eds.*), Mothering against the odds: Diverse voices of contemporary mothers.* New York: The Guilford Press.

3

Critical Thinking: Challenging Developmental Myths, Stigmas, and Stereotypes

By Maureen Walker, Ph.D.

REFLECTION QUESTIONS

1. *What social identity markers (such as race-ethnicity, gender, sexual orientation, class, religion) do you use to describe yourself?*
2. *Which ones are most salient most of the time?*
3. *How have the degrees of salience shifted over the years?*

INTRODUCTION

Memories of my early childhood are mostly filled with glimpses of very ordinary things: hummingbirds darting among hollyhocks in my grandmother's garden, a slice of pound cake in my godmother's kitchen, summer afternoons playing with paper dolls on my great-grandmother's front porch. My great-grandmother and I were frequent companions, as all of the able-bodied adults left home early each day to work in other neighborhoods. The two of us, the very old and the very young, were usually left to contend with each other. Most times our days were filled with ordinary excursions—a walk to the corner curb market or a bus ride to downtown Augusta, where she would buy seeds for her garden or perhaps a pair of stockings.

One such day, I remember standing in a five-and-dime store surrounded by adults. I don't remember asking for water, but I must have. It must have been that the "Colored" water fountain, one that I vaguely remember being within my reach, wasn't working. I can't say that I recall all that happened, but one moment stands out. A brown-skinned woman, younger and stronger than my great-grand-mother, pulled me up in her arms and held me over the taller, silver-colored "White" fountain, and told me to drink. Her words were clear and rather ordinary: "If the child is thirsty, she should have a drink of water." I also remember the nervous glances and the muffled laughter from the adult onlookers. My great-grandmother apparently told my parents, because I remember my father and mother, my father especially, enjoying a big belly laugh about my drink of water. I honestly didn't know what any of it meant, but I knew that a most extraordinary thing had happened.

Over the past several years, proponents of the Relational-Cultural model have come to increasingly emphasize the role of culture in human development (Miller, J. B., 1976; Jordan, J. V., 1992; Walker, M. & Miller, J. B., 2000; Walker, M., 2002). These writers and others have depicted culture as more than the scenic backdrop for the unfolding of development; rather, culture is viewed as an active agent in relational processes that shape human possibility. In one of her classic works, Miller (1988) has noted that healthy growth is undermined when a person must function under chronic conditions of disconnection and violation. She not only fails to get her needs met, she is likely to consider herself at fault for any violation she experiences at the hands of more powerful others. Such disconnection is most often interpreted as an interpersonal breach, either a failure of empathic attunement to or a direct assault against the legitimate needs and yearnings of another person. However a close reading of foundational premises of Relational-Cultural theory reveals a preoccupation with questions of power, specifically with structures of domination and subordination that constrict the range and distort the nature of relational options (Miller, 1976; Kaplan, 1997). Miller (1976) laid the groundwork for examining the link between culture and relational development in her explication of "power-over" systems.

> In these [power-over] relationships, some people or groups of people are defined as unequal by means of ascription: that is, your birth defines you. Criteria may be race, sex, class, nationality, religion, or other characteristics defined at birth. There is no assumption that the goal of the unequal relationship is to end the inequality; in fact, quite the reverse. . . . Inevitably, the dominant group is the model for "normal" relationships. It then becomes "normal" to treat others destructively and to derogate them, to obscure the truth of what you're doing, and to oppose actions toward equality. (p.8)

The racial apartheid culture of the American South in the 1950s represented an exaggerated version of power-over relations in a nation that categorized its citizens into presumptively permanent stations of superiority and inferiority. The "success" of this culture depended in large measure upon the effectiveness of its executive processes (institutions, laws, social norms) in communicating the normalcy of inequality. In the culture of racial apartheid, all systems were specifically and intentionally arranged to coerce compliance with terms of social stratification. In other words, *the task of human development was to internalize the precepts of inherent inequality along with any other tasks that might be deemed developmentally appropriate.*

To behave in ways contrary to the norm of inequality was to risk being characterized as deviant in some way. For example, the woman who lifted me to the fountain was likely called "crazy" (meaning reckless), albeit it with some degree of admiration by the African American women who watched her defiant act. Because she was actually breaking the law, she also risked being treated as a criminal had she been caught by any White person with authority to enforce the law. In a rigidly stratified culture, one of the developmental tasks of childhood is to learn one's social ranking—to learn not only whether one is superior or inferior but also the basis for that valuation. Within the context of racially stratified culture, my task was to "learn" that as a Black person, I was not good enough to drink cool water from the bright chrome fountain.

The history of the United States is marked by social stratification, principally but not solely along dimensions of race-ethnicity, class, sexuality, and gender. Although the resulting categories of dominance and subordination are rationalized and reinforced by a variety of social stereotypes, stereotypic thinking in and of itself does not account for the diffuse and deleterious effects of culturally derived disconnection. When one is systematically assigned to dominant or subordinate status based on a social identity marker, psychological development is shaped by a particular relationship to power, a process most elegantly defined as the capacity to produce change. The social stratification becomes a means by which one's fitness for connection is determined. The stratification also signals the type of resources to which one is entitled. In other words, the ranking itself signifies the nature and type of relationship to which one is presumably entitled. For example, Miller (1976) discusses the psychological consequences of social norms that prescribe caregiving as the primary means by which women may participate in relationship.

In a similar vein, Carter (1992) suggests that psychological, political, and economic diminishment ensues from restricted access to enabling relationships. Furthermore, Carter suggests that political and economic systems combine to produce diminished outcomes for socially marginalized people. The underlying presupposition is that such relationships determine access to education, health, and career opportunity as well as to the social validation that facilitates growth.

Consider, for example, the disparate prospects for two students: one attending school in an economically marginalized district and the other in a neighborhood where tax dollars provide premium resources to support the students' development. Let me share a more concrete example from my family life. I once visited my son's middle school science class, and at the front of the room was a 52-inch color monitor attached to an electron microscope. When the teacher asked a student what was on the screen, she promptly responded that they were red blood cells. The teacher then told the students that she had pricked her finger earlier that morning so that they could see fresh cells. Upon hearing that exchange, I was reminded of a report about a middle school in an economically depressed neighborhood less than 20 miles away. In this school, sixth grade students did not have science on a daily basis because (a) there were no certified science teachers and (b) there were not enough books for each child to have his or her own copy.

The two schools provide a striking example of the extent to which racial stratification is implicated in the construction of class (Brodkin, 1995). In my son's school, the majority population was White. In contrast, children of color, primarily African American, Afro Caribbean, and Latino, attended the school with inadequate resources. The effects of the disparities were evident seven years later. At my son's high school graduation, school officials were able to boast that a full 98 percent of the student body was college bound, many to elite universities. In the other school, less than 50 percent of the students expected to be college bound.

Jenkins (1993) explains the link between performance and cultural disconnection as an issue of social esteem: how one is regarded by the culture influences one's ability to negotiate the developmental tasks. In other words, being a member of a culturally devalued group may have negative consequences that are not completely explained by the Rosenthal effect *(for example, students living down to the expectations of their teachers).*

In a similar vein, research by Steele (1997) indicates that the mere *threat* of social devaluation has an inhibitive effect on performance. Specifically, when students believe that powerful others view them as less intelligent or less deserving of opportunities, resources, and enabling relationships, their best efforts are undermined by the resulting anxiety. In spite of the fact that these students are trying *not* to conform to expectations, they often experience less than optimal outcomes.

When power is defined as the ability to produce change, the ideas put forth by Jenkins and Steele seem to suggest a recursive link between relational development and power. Simply put, relationship creates and sustains power; power generates access to relationship. Brock (1993) maintains that persons in our culture feel present, alive, and sustained in the world through power—power to influence and participate in shaping the relational outcomes. In *Journeys by Heart,* Brock contends that power is a basic human reality precisely *because* we are related to each other.

Given the foundational premise that healthy development occurs through action-in-relationship, it follows that developmental potential is enhanced when an individual can function free of the inhibiting objectifications that limit the range of growth and possibility. When stratification functions to restrict the nature

REFLECTION ON POWER LINE EXERCISE

Using the line below, plot a point above or below to describe your own experience of power or privilege associated with your salient social identity markers. Which identities are usually more advantaged? Which less?

| race/ethnicity | gender | sexual orientation | class | other |

and range of allowable relationship, the result is a culture of disconnection and relational disempowerment.

WHO CAN BE THE CHRISTMAS FAIRY?

A stratified culture, like any other, requires a certain amount of "care and feeding" in order to maintain itself. One of the means by which the stratification is maintained is through the proliferation of controlling images (Collins, 2000). These images provide the broad parameters for ascribing the developmental expectancies and interpreting the experience of both the dominant and subordinate groups. In a racially stratified culture, controlling images typically operate according to four principles;

1. The content of the objectification is determined by the dominant group;
2. They operate without the consent of the subjugated group;
3. They protect the interests of the dominant group and exploit the vulnerability of the subjugated group; and
4. They justify or rationalize existing power relations.

Controlling images become a primary means by which the disconnections of the dominant culture are enacted. Once inferior or superior status has been conferred, the images ensure that the resulting social locations will remain fixed. Let me offer a case in point. Several years ago while attending a conference on Black women's health issues, I listened as a young woman explained what it meant to her to be Black and female while she was growing up. She said this: *"On my first day in elementary school, I looked around and I knew that with my dark skin and my nappy hair, I would never be the Christmas fairy—no matter how well I could fly."*

This story of the Christmas fairy is revelatory for many reasons. First, it illustrates how a child might experience and interpret cultural messages about worth and possibility. In a culture that tends to value and validate women based on physical appearance, this young woman quickly learned that her body was irredeemably deficient—that her appearance was so repugnant as to effectively nullify any other measure of merit. Specifically, she "learned" that whatever her other talents or qualifications might be, her appearance alone could disqualify her from the role.

The story is also suggestive of the means by which the images gain their potency. According to Collins (2000), controlling images function to make social disparity and exclusion look normal. Interestingly, this young child's prediction (which ultimately proved true) was not the result of any conversation: no mean-spirited adult announced to her on her first day that the role of Christmas fairy would belong to one of her fair-skinned, silky-haired playmates. It is in fact the absence of a conversation that contributes to the apparent normalcy of the situation: her exclusion would just look like "the way things are." Without a clarifying conversation, "the way things are" might easily be interpreted as "the ways things must be or should be." Notions of beauty, virtue, and value are systematically conveyed through the manipulation of images. Through often wordless indoctrination, the child learns who is beautiful, who is important, and who can be the Christmas fairy.

There is some evidence that suggests that persons who reside on the subjugated side of the power arrangement learn those lessons quite early (Dorcas Bowles, personal communication, 1985; McAdoo, 1970), often before they are able to adequately process the resulting shame. Janie Ward (2000) has noted the importance of conversations to discuss the shaming experiences of racism if African American children are to develop skills for relational resilience in a stratified society. Otherwise, children may become adept at relating through strategies of disconnection and withholding vital parts of themselves in order to avoid harm and humiliation. When the child views herself primarily through the eyes of a devaluing culture, her "withholding" may include holding back of her talents and capabilities, as well as her expectations and hopes for successful outcomes.

While the strategies of disconnection may provide temporary respite, they inevitably contribute to diminished capacity for clarity (*being able to distinguish what she feels and thinks from what the culture suggests she should feel and think*) and authenticity (*the ability to express her views, experience, and feeling-thoughts in relationship*). The child learn habits of disconnection, not only from potentially harmful others but also from herself. If she is unable to develop and sustain clarity about her own feeling-thoughts, she is likely to be more vulnerable to cultural interpretations of her experience. She is then at increased risk of seeing herself as the culture defines her.

Miller (1988) has written that chronic exposure to disaffirming stimuli results in self-doubt; the person begins to feel at fault and doubt her own worthiness for connection. Unable to name and interpret her experience in self-affirming ways, she is unlikely to grow in her ability to represent herself in relationship with others. She has less access to relationship as the growing medium for healthy psychological development.

There is wide consensus that the presence of a validating adult relationship is critical to healthy maturation during this developmental phase (Hartling, 2003; Ward, 2000). Describing the role of the validating adult relationship, Ward has proposed a four-step model as the antidote to the controlling images that render the experience—and sometimes the existence—of children of color as insignificant at best, or worse, problematic. Her work suggests that intentional and persistent intervention is required to counteract the insidious effects of negative controlling images. In her model, developing relational resilience skills begins with resisting the images of the dominant culture by naming them, reading them, opposing them, and replacing them. To use her terms, the job of the caring adult is to help cultivate in the young person a belief stronger than anyone's disbelief (Robinson and Ward, 1991).

It has only been as an adult that I have come to fully apprehend my mother's intense and persistent efforts to protect me from the devaluing images and expectations of our racially stratified culture. For all of my childhood, she worked as a domestic, providing cooking, cleaning, and child care for White families. Never during those years did I meet any of her employers. Her concern, as she later explained it, was that I would come to see my future through the eyes of people who expected poor African American female children to become maids. Instead, she took great care to surround me with relationships that provided both modeling and validation for my academic talents.

DOWNTOWN REDUX

On yet another foray out of our neighborhood into downtown Augusta, my great-grandmother and I boarded a city bus. While she stood at the coin machine depositing our dimes, I flopped down onto the first seat, eagerly anticipating the ride that would take us to the excitement of Broad Street shopping. Almost immediately, I started to hear laughter that at first meant little to me. Seconds later I was startled out of my 4-year-old oblivion by my great-grandmother commanding me to get up and follow her to another seat. One of the women on the bus (at this time of the day they were all Black women) laughed and said: "That child doesn't know that's where White folk sit" to the amused acknowledgement of everyone on the bus. I'm not sure that at point that I knew what "White folk" meant.

I guess I discovered that I wasn't one of the White folk, and that not being White meant not being *something* enough to claim the front seat. The problem was not just that I was *different* from White; the message was very clear that I was *less than* White. I was also aware that the problem wasn't with the White folk, or my great-grandmother, or the other women on the bus. The problem was my ignorance. It was the first time I felt very stupid.

The dominant group in a power-over culture attempts to instill in subjugated groups the belief that they are a problem people. This indoctrination may be accomplished by a variety of means, from the seemingly superficial (for example, by overvaluing a European aesthetic) to the most obviously consequential (by criminalizing innocuous behavior). An often-cited example of the former is the marketing of beige-colored bandages as "flesh-toned" with the implicit message that darker tones are outside of the realm of normal expectancy. An example of the latter would be the almost cliché experience reported by African American males of being stopped for "driving while Black."

Ayvazian and Tatum (1994) have described how schoolteachers frequently interpreted the same "roughhousing" behaviors by their same age, different race sons. Ayvazian reported that her White son was most often described as spirited. Tatum on the other hand, was advised that her Black son seemed to have problems with anger. At other times (or usually concurrently), indoctrination is accomplished by proliferating depraved images of nondominant group experience and by denying access to quality goods, services, and resources (education, health care, and the like). In either case, the members of the subjugated groups must devise ways of coping with the message from the dominant culture: that they are either an inferior being or a threat to society.

REFLECTION EXERCISE

Select a social identity that you experience as less advantaged or devalued by the dominant culture. What are some of the negative images associated with this social identity? How do you feel about these images? How do you behave? How would you describe the level of congruence between how you feel and how you behave in response to those images? Are there positive images associated with this identity? How do you feel about these images? How do you behave? How would you describe the level of congruence between how you feel and how you behave in response to those images?

When the coping mechanisms are comprised primarily of strategies of disconnection, the result is what Lipsky (1984) has termed internalized oppression. Strategies of disconnection are patterned behaviors designed to minimize the risks of being vulnerable in relationship. Relational-Cultural theorists emphasize that these strategies are also attempts to maintain some semblance of connection. They signify the paradox of both longing for and being fearful of connection (Miller & Stiver, 1997).

From a Relational-Cultural perspective, strategies of disconnection give rise to internalized oppression, a complex of relational images grounded in the distortions and disinformation required to normalize the inequalities of a power-over culture. Inasmuch as relational images shape the interpretation of experience and of expectation, they have specific implications for movement in relationship. While the behaviors associated with internalized oppression are numerous, they tend to manifest as variations on five relational themes.

First among these is "invisibility through disconnection." Some individuals attempt to cope with being a member of a subjugated group by disconnecting from the group. They may do so by mimicking the behaviors and preferences of the dominant group, sometimes to the extent of denigrating members of their own group. Their goal is to make invisible those parts of themselves that are likely to be devalued by the dominant group. For example, in an effort to gain more credibility with the dominant group, a person may use a variety of means to disassociate from the devalued group. She may be extremely critical of members of her own social identity group; she may laugh heartily at jokes that target her identity group, or she may adopt and exaggerate traits that are viewed as typical of the dominant group. The disassociation may be as fundamental as avoiding or denying a preference for certain "ethnic" foods or music in the presence of dominant group members.

The second strategy of disconnection is one that has been aptly termed by William Cross (1995) as "spotlight anxiety." This strategy defines the experience of people who are overly concerned about how they appear to members of the dominant group. They may expend exorbitant energy attempting to manage the impressions other people may hold of them, sometimes to the detriment of their personal goals and desires. As one Afro Caribbean student put it, "I get so focused on playing defense against what I think White people think of me, that I forget where I'm going."

Spotlight anxiety may give rise to many behaviors: from extreme caution to minimize risk to extraordinary attempts to assuage the anxiety of dominant group members. For example, if a student is overly concerned about the attributions made about her intelligence or her qualifications for being in a college classroom, she may not participate actively in class discussions for fear that a mistake would cast doubt not only on her abilities but also on the abilities of anyone sharing her social identity. As one student put it, "I know that if I have a bad day, somebody is going to think that the only reason I'm here is that they let Hispanics in to satisfy a minority quota."

Another example involves an African American male student known for his uproarious laughter, gifted singing voice, and his all-around good nature. As a large-sized, dark-skinned male, his need to appear nonthreatening meant that he had often laughed to mask the seething anger at a world that he never expected to validate his intellect or his professional aspirations.

The third strategy, "hyper-visibility," appears to be different, but is in reality closely aligned with the previous two. This paradoxical coping mechanism allows the individual to hide from relationship by exaggerating certain aspects of herself or himself. For example, the individual may dramatically adorn herself in ethnic attire as a way of suppressing her ambivalence about her ethnic identity, or she may prescribe and hold others to very narrow or rigid expressions of group membership.

Similarly, she may be more comfortable publicly expressing her rage than grappling with the complexities of her self-doubt and private sorrow. A young woman once described how she "tore into" a White roommate who commented that her childhood nanny looked like Aunt Jemima. Upon further conversation, it became clear that she felt saddened by her conflict with her roommate-friend, but had difficulty tolerating the sadness, fear, and vulnerability triggered by her friend's insensitive comment. These "weaker" feelings seemed incompatible with ethnic pride. Because she lacked empathy with her vulnerability, she was unable to engage the conflict in a way that was growth-promoting and productive. Instead both women warily avoided each other (and probably multiplied their projections) for the remainder of the school year.

A persona associated with the fourth strategy of disconnection is the "striving superstar." In an effort to stave off the devaluation of the dominant group, this individual may strive incessantly to prove her worth and gain the recognition, if not the acceptance, of the dominant group. In so doing, she may develop behaviors that lead to judgmentalism, especially toward herself and members of her devalued group. She may also be inclined toward unhealthy competitiveness, often of the covert variety. These persons tend to exhibit little empathy, as any indication of perceived weakness may deter advancement toward her goal. Though she may actually achieve the recognition she desperately wants, she may find that her achievements do little to assuage the pain of emptiness and isolation.

In the face of sustained and systemic oppression, some individuals learn to cope by giving up. The fifth and last strategy of disconnection is akin to "learned hopelessness." These strategies of disengagement may appear in various guises, but at root they require that the individual disconnect from her own competencies, masking or otherwise withholding them in her interactions with others. The person may come to believe that her relational possibilities are limited to those prescribed by the dominant culture. For example, if the cultural images ascribe scholastic ineptitude, she may come to accept that image as a "truth" and disavow any claim to academic advancement.

Unfortunately, this behavior may be reinforced by members of her peer group, as there is considerable anecdotal information about students who are bullied for carrying books or giving any evidence of academic aspirations. In some instances, the person may render herself irrelevant by engaging in mascotlike behavior—acting confused, using humor to deflect attention away from herself, or standing steadfastly in someone else's shadow. In other instances, the individual may adopt the self-limiting pose (acting "cool" or "fly," for example) prescribed by the controlling images of the dominant culture. In these cases, the person abdicates responsibility for interacting effectively with others in his environment, in effect becoming a "character actor" in a drama not of his own making.

When the strategy of choice is disengagement, the desired outcome is usually safety from vulnerability. Such disconnection, however, provides only the illusion of protection. Although strategies of disconnection are typically deployed to protect the person from further disconnection or harm (Miller & Stiver, 1997), when enacted on a systematic and sustained basis, the relational images generated by these strategies will conform to the controlling images of the dominant culture. In some sense, a collusive pattern develops that precludes the movement into expanded possibility. Vulnerability and conflict are active ingredients in relationship. As suggested by Ward (2000), validating relationships are required to support relational courage: the skills for accepting vulnerability and for engaging conflict. It is through this process that new relational images and possibilities emerge.

Whether enacted by the dominant group or the subjugated group, at best strategies of disconnection signify profound shame. To internalize the oppressive images of the dominant group is to internalize the shame of the dominant group. In fact, like other systems of oppression and abuse (be it interpersonal or familial), the dominant group in a power-over culture is able to "off-load" its collective shame onto members of the subjugated group. With the shame safely deposited in the "lesser" group, the dominant group may perpetrate or collude with the cultural violations that have been described as nameless, blameless, and shameless (Ward, 2000). The far-ranging impact of these disconnections cannot be underestimated. In fact, sustained and systematic violations perpetrated by the dominant culture may wreak traumatizing effects similar to those resulting from interpersonal abuse (Walker, 1999). These effects have been described by Maria Root as "insidious trauma" to connote the persistent, sometimes daily encounters with experiences that represent a symbolic threat to existence (Brown, 2003).

Phillip Cushman (1993) has made a compelling case for examining the strategies of disconnection and the resulting images that derive from the standpoint of cultural superiority. In his exegesis of the history of American psychotherapy, he discusses the politics of self in a stratified culture. In his view, two cultural movements, minstrelsy and the frontier myth, account in large part for the dichotomized constructions that justified the segmentation of society into the privileged and oppressed groups.

> The stereotypic visions of the African American [immoral, empty-headed, sexual, lazy] and the Native American [savage, lazy, sexual] was embedded in the ongoing confusion in America about the proper configuration of the White self. They were psycho-political prisons from which the oppressed groups could not escape. But these depictions provided a flawed, ill-defined configuration of the White self. (p.64)

While the subjugated group is defined through negative images (shiftless, weak, unintelligent, and so on), the dominant group is defined through contrasting images of mastery and self-sufficiency—qualities that are overvalorized in American culture. Because these privileged qualities are sustained in large measure through the continued subjugation of the "inferior" group, the dominant group may avoid the give-and-take of authentic engagement in relationship, as such connections may upend the images upon which their selfhood rests.

Like members of the subjugated group, the dominant group may be constrained by images that represent real relationship as a risk. Moreover, the culturally ascribed privileges (e.g., being viewed as the standard against which other groups are measured) may "protect" them from noticing their own vulnerabilities and may render their weaknesses and insecurities invisible. As Jordan has frequently noted (1992a, 1997), the dominant group has the power to respond to difference or conflict in ways that mask its own neediness, often by suppressing the voices of the socially denigrated group. In doing so, it loses access to relationships that offer the possibility of movement toward increased clarity, zest, resilience, and genuine self-worth. In other words, the culture of disconnection and stratification leaves no one unscathed.

While the dominant group may create the images that denigrate the subordinates, it is important to remember that such image making has a profound impact on the image maker. Whether applied to the socially privileged or the socially oppressed group, controlling images become a "stand-in" for what is sometimes thought of as identity, particularly when intertwined with the more specific relational images deriving from the personal histories of an individual within her cultural context. The South African scholar Amina Mama (2002) says it well:

"All identities have histories, and they all involve questions of power, integrity, and security, questions that have emotional as well as political currency. . . . It is about resistance . . . and citizenship, action and reaction."

It is important, therefore, in any power-over culture to examine the impact of controlling images on members of the dominant group as well. This impact of a privileged identity typically goes unregarded, since one of the "perks" of privilege is the right not to think about it. In her classic work on White privilege, MacIntosh enumerated many of the social benefits of having White skin—including a sense of automatically fitting in and not having to think about her social ranking. However, failure to examine the impact of privileged identity on relational development is to collude with the notion that the dominant group

REFLECTION EXERCISE

Select a social identity that you experience as privileged by the dominant culture.

What are some of the positive images associated with that identity? What do you enjoy about being associated with this identity group? What are some of the negative images associated with this group? How do you feel about those images? How do you behave? How would you describe the level of congruence between your feeling-thoughts and your behaviors in response to the negative images?

experience represents an ahistorical norm, that the voices of the dominant group in a power-over culture represent a decontextualized truth unmarked by the political realities of the everyday world.

WHAT "DAVID" LEARNED ABOUT CHARITY

David, a 56-year-old White male, sits in his therapy group sharing fond memories of Sister Margaret Kathleen, his first-grade teacher who sought to instill practices of kindness and generosity along with the lessons in reading, proper posture, and arithmetic. It was in her classroom that he was introduced to the weekly charity ritual at his parish elementary school. Every Friday, the young students were encouraged to bring in pennies that would be used to feed Black African babies. As David remembered the ritual, there were sometimes competitions among the classrooms, as the group who brought in the most pennies could actually *buy* a Black African baby.

It is noteworthy that none of these students had any real contact with Black Africans; their education was limited to the images promulgated by the charity ritual. David mentioned that by the time he reached fifth grade, an important innovation had been incorporated into the weekly practice. A Black painted statuette stood with a smiling face and basket in hand to receive the pennies. Once a sufficient number of pennies had been received, the statuette would bob its head in grateful acknowledgement.

As David's experience suggests, cultural images and their associated attitudes are inculcated in a variety of ways, many of which may start with benign intent. As David shared his memory with the group, he became aware that images grafted onto his consciousness at a very early age continued to influence his personal and career relationships as a middle-aged banker. Although David prided himself on his liberal politics, he came to see that the childhood ritual along with other cultural products (such as *Tarzan* movies and *Amos n' Andy* radio shows) had taught him to see Black people as needy, incapable, and dependent—a "problem" people who are at times fit objects of humor, or in the extreme, commodities for exchange and purchase. Less obvious were the valorized images of the dominant group formed in opposition to the degrading stereotypes of the subordinate group. The counterpart to images of the needy, presumably orphaned babies were the competent, generous White donors—a people capable of acquiring and controlling resources, worthy of respect and gratitude.

A term that aptly describes the images associated with David's racial socialization is internalized dominance. In his discussion of unconscious racism, Lawrence (1987) asserts that internalized dominance is rendered more potent because it exists outside of everyday consciousness, and is thereby exempt from scrutiny. Therefore, the images comprising the belief system come to represent the "rational ordering of things." Like internalized oppression, internalized dominance is a complex of relational images grounded in systemic and systematic miseducation and in the politics of social inequality.

This belief system is the result of an advantaged relationship to privilege, power, and cultural affirmation (Walker & Larkin, 1995). As is the case with internalized oppression, expressions of dominance take many guises in relationship. Some of the images associated with dominance may be enacted through relational patterns of *presumed superiority, exoticized expectation, missionary entitlement*, and *summary judgment*.

Presumed Superiority

This point of view posits the centrality of the dominant group experience. From this perspective, the dominant group is viewed as the standard against which all others are defined as diverse or minority. Moreover, the opinions, thoughts, and ideas of the dominant group are taken more seriously than those issuing from the subordinate group. This mentality is operating in settings when an idea is not "heard" or taken seriously until it is spoken by a member of the dominant group.

In addition, the standards of beauty, artistic expression, and cultural norms associated with the dominant group are considered superior to all others. An example of this cultural hegemony is the ubiquity of academic curricula that privilege the products and artifacts of Western civilizations while excluding the experiences and accomplishments of other cultures. "Cynthia," a Latina woman in her second year law program, recalled with sadness and indignation a conversation with one of her White male peers. Upon learning that she had received an A in a course, he asked, "Don't you think the professor made it a little bit easier for people like you to get good grades?" Cynthia commented that the remark would have hurt less had it been delivered by a mean-spirited person. The sad fact, according to Cynthia, was that her peer—who also received an A—could think of no other plausible explanation for her accomplishment.

Exoticized Expectation

Another strategy of disconnection associated with internalized dominance is exoticized expectation. In this case, the dominant group member may ascribe mythical qualities to a particular person who is identified with the subjugated group. Most often, these qualities function to (a) set the person apart as special—

not sharing the devalued qualities usually attributed to his or her identity group, and to (b) co-opt the special quality to serve the needs of a dominant group member. The canons of Western literature and popular media are replete with images of the special or exotic minority member who brings care and feeding (of the body and increasingly the soul) to a dominant group member. *(Some recent examples include such movies as* Bruce Almighty, Bagger Vance, *and* Bringing Down the House.)

Interestingly, these modern-day images are consistent with observations made by Miller (1976), who observed that critical nurturing functions are often assigned to the devalued members of a power-over culture. On a more ordinary note, an intake counselor reported that a woman from an upper middle class background requested a "Strong Black Woman" who could help her get over her WASP-ish neurosis. She went on to say that someone with wisdom like Oprah is what she imagined that she needed.

Missionary Entitlement

When relational images reflect this belief system, the dominant group member may feel entitled to appreciation and respect from members of the subordinate group. Dolores's dilemma provides a case in point. A White senior administrator at a northeastern college, Dolores worked relentlessly to forge relationships with Black and Hispanic undergraduates in the engineering program. In some respects, Dolores viewed herself as an ambassador, offering outreach and welcome to students of color in the mostly White institution. She was fond of lamenting with other administrators about the students' lack of social or political savvy, and would recount in detail how she helped them navigate the cultural terrain.

Over time, however, Dolores started expressing disappointment and resentment about the students' ingratitude and their lack of receptivity to her unsolicited counsel. In her own counseling she became aware of her need for the students to need her. When their real behaviors in relationship failed to conform to her image of them as hapless and dependent, her image of herself as the benevolent bestower of goods and resources was upended. At base, she was most comfortable relating to the students from a position of power-over, a position that confirmed her presumed superiority. Her relational challenge was to become more open to mutual influence. As she began to relinquish her hold on the images that locked the power differentials in place, she started to experience more genuine connection between herself and the students.

Summary Judgment

Paul prided himself on having an assortment of friends and associates from a variety of ethnic backgrounds. A popular minister in an urban church, he was known as a pastor who actively supported efforts to bridge cultural divisions and heal conflicts when they erupted in his community. He credited his relational hardiness to having grown up in what he called the "rough and tumble" of a large Italian family in a mostly Irish neighborhood. When Paul decided to enroll in university courses during a summer sabbatical, he felt a vague sense of discomfort when the tall Ugandan professor walked into the classroom. Would he receive good value for his tuition dollars? Would the professor be able to deliver an organized lecture? Would he able to learn anything from this professor? The course was the History of Colonialism in East Africa.

Paul's initial discomfort gradually turned into dismay as he slowly recognized how images of racial supremacy had shaped his reaction to his professor. His immediate and visceral reaction to his black-skinned instructor was to doubt his competence. Paul's initial skepticism illustrates that internalized dominance need not be a consciously held belief system. In fact, it may be one that is consciously and publicly repudiated. However, without active awareness, images once internalized may be suppressed, particularly if they are at odds with one's more public commitments and sensibilities. In Paul's case, the encounter with the Ugandan professor compelled him to confront feeling-thoughts and values that attenuated his capacity for connection, both within and across ethnic groups. When controlling images result in a posture of Summary Judgment, the dominant group member expects inferior performance from nondominant group members.

The expectations of inferiority may be expressed as benign surprise when a person of color is recognized as expert (discovering, for example, that a world-renowned neuroscientist is a Laotian woman) or in outright predictions of failure ("Female students don't do well in technical courses"). More overtly prejudicial expectations also may be acted out (for instance, changing doctors, refusal to hire or promote because of nondominant group status). When race is the salient social marker, a common tendency of people exhibiting the Summary Judgment pattern is to require more proof of competence from people of color. They may also attribute any shortcomings or problem behaviors to their social group membership. (The reverse is also true; more individualized attributions are made when the problem behaviors are committed by a dominant group member.)

Dominant group members exhibiting this relational pattern may use one or a few negative experiences to justify their excluding or violating behaviors. For example, had Paul persisted in his relational pattern, he may have attributed any lack of understanding or communication breakdowns to the professors' ethnicity. Furthermore, he may have refused to enroll in other courses that were taught by other faculty of color. In this case, however, Paul used the encounter as an opportunity to become curious about his own racial socialization. Engaging in relationship with the professor enabled him to confront, empathize with (not excuse), and take responsibility for his own strategies of disconnection. The relationship also facilitated movement out of constricting images that inhibited his authenticity in relationship. From a place of openness and vulnerability, he began to enlarge his capacity for mutually enhancing friendships both within and across racial groups.

The relational images comprising internalized oppression and internalized dominance give rise to complementary strategies of disconnection. While the enactment and expressions of those strategies may look different on the surface, at base each strategy is supported by profound shame. While it is fairly commonplace to associate the strategies of internalized oppression with shame, the emotional undertow of internalized dominance tends to be less obvious.

At one level, this obscured shame attests to the power of the dominant group to exempt its own practices from scrutiny. On another, the collective experience of the dominant group is explained in Miller's (1976) foundational premises about the psychological mechanisms of a power-over culture. In this work, she noted that the dominant group uses its power to ascribe disowned characteristics to members of the weaker group. Unwilling to claim the attributes regarded as shameful (weakness, dependency, and the like), the dominant group is forced to fend off any experience or encounter that threatens to unsettle the images that justify or otherwise rationalize its entitlements.

Cushman's (1995) concept of "negative identity" supports this idea: "Americans developed a concept of the self, a proper way of being human, by constructing the "other": the Negro slave, the Jew, . . . the woman—in such a way as to define and justify the White [male] self by demonstrating what it was not" (p. 346).

In other words, to enact the strategies of internalized dominance is to approach relationship from a standpoint of wariness and vigilance, rather than engagement and curiosity. Because the dominant group may use its power to avoid confronting the devalued aspects of human experience, the shame associated with a felt experience of inadequacy of any sort is "bypassed" (Hotchkiss, S., 2002), rendering it susceptible to the emotionally and politically encoded images that proliferate in a stratified society. It is as if the collective shame of the dominant group is submerged, surfacing only when a relational encounter threatens to disturb the images that hold it at bay. Rather than approaching relationship with openness to emergent experience, the person must vigilantly protect the images that provide the illusion of emotional safety.

The great myth of our culture is that separation provides safety, and correspondingly, that relational interdependence spawns weakness and vulnerability. The paradox is that the putative safety is supported by fearful reactivity, a fear that is masked/mastered only by establishing power over any aspect of experience that represents insufficiency, vulnerability, or deviation from the culturally affirmed attributes of self.

That the experience is typically accessed through an encounter with an embodied Other tends to reinforce the dichotomized frameworks that justify disconnection. Simply put, the politics of selfhood are deeply implicated in the processes of social stratification. Once configured into groups of dominants and subordinates, a culture stratified along dimensions of social identity is likely to proliferate images that support the ranked categorizations. The controlling images associated with the categories function to preserve the status quo power arrangements, thus perpetuating disconnection and attenuating relational potential.

Likewise, the images supporting internalized dominance and internalized oppression tend to fit hand in glove with each other and tend to be collusive in the processes of disconnection. Take, for example, a student who is primarily concerned with impression management or Spotlight Anxiety. She is unlikely to reveal the uncertainties that would provide the impetus for gaining new knowledge. She may not only be unlikely to authentically represent her experience, she may also be unempathic with her experience of vulnerability. If her lack of empathy causes her to disconnect from her feeling-thoughts, she loses an opportunity to gain greater clarity and self-knowledge.

If this student happens to be in relationship with a teacher whose preferred strategy of disconnection is Summary Judgment, it will be very difficult for growth-fostering movement to occur. In this case, the teacher might distance herself from the student by presuming her incompetence, a relational image that may be enacted in subtle or obvious ways. In such a relationship, the student learns that vulnerability is indeed unsafe.

If the student's revealed uncertainties are used as evidence of her intellectual *in*capacity, she is unlikely to relinquish her belief in images that seem to protect her from exposure and humiliation. Furthermore, the teacher herself fails to stretch her own capacity for learning and growing in relationship, undermining both her personal and professional potential.

Relational empowerment is the sine qua non of healthy psychological development. The ability to be present, active, and authentic in relationship is both the process and the goal of development. Such action-in-relationship suggests a certain fluidity of power, where participants in the relationship are open to mutual influence. However, power-over cultures breed reactivity and suspicion toward difference. In such cultures, "different from" is likely to be interpreted as "in opposition to."

Under conditions of stratified and oppositional power, relational practices seem like risky business indeed. However, two basic tenets of Relational-Cultural theory are that (1) humans grow through engagement with difference (Miller, 1976), and (2) that the relational practices of empathy, authenticity, and mutuality are necessary for optimizing the developmental potential that such differences offer. Empathic attunement enables connection to the complexity of the feeling-thoughts, sensations, and beliefs comprising the relational images through which experience is interpreted. Such connection enables intentionality and courage in relationship—courage to represent oneself with authenticity and awareness.

When the participants in relationship approach their encounter with attitudes of openness and discovery, they are less likely to enact the shame of the culture. They are more likely to activate mutual desire. As suggested by Miller's "five good things" (1986), mutually enhancing relationships engender desire for more relationship. In the absence of such mutuality, participants may be inclined to take refuge in images that effectively thwart possibilities for growth-enhancing movement. Relational practices promote courage—the courage to build connections that activate the shared power of human potential.

CASE STUDY

After months of seemingly unending conflict, Shelley and Veronica, a lesbian couple who had lived together eighteen months, decided to seek couples' counseling. Neither of the two women was intimidated by conflict; in fact, both prided themselves on their abilities to engage in challenging and lively conversations in their respective professions. However, in addition to becoming more frequent, they were noticing that in recent months their arguments resulted in an almost palpable emotional residue that left them feeling distant and isolated from each other. Both women were in their mid-30s. Shelley, who described herself as a "political Jew," was a clinical social worker. Her partner Veronica, who preferred to be described as Afro Caribbean or West Indian, was a college professor.

Shelley often attributed the problems in the relationship to cross-cultural issues, and suggested that Veronica was supersensitive and racially angry. Veronica described Shelley as a "pampered princess" who felt entitled to being taken care of. Veronica also felt mistrustful of Shelley's "clinical" interventions in the midst of their arguments. According to Veronica, it was not uncommon for Shelley to divert attention away from the presenting issue in the argument by crying and commenting on her (Veronica's) "abrasive" communication style. Veronica noted that this scenario was especially likely to unfold if she presented the subject of contention. If Veronica failed to soften her tone, Shelley would leave their home, once staying away for a few days. During these times, Veronica would feel bereft—certain that Shelley had abandoned her for good. Her typical response to Shelley's eventual return, however, was to make herself less available to the relationship.

Shelley for her part would attempt to engage Veronica sexually. Veronica in response would typically ignore Shelley's bids for reconnection—sometimes immersing herself in her work, or sometimes distancing herself by simply fantasizing about the possibility of dating someone new.

Shelley and Veronica began therapy with Monica, a White, heterosexual woman who was highly regarded for her work with gay and straight couples. Monica was fascinated by Shelley's interpretation of their couples' issues as being "cross-cultural." Veronica, though somewhat dismissive of the interpretation, did note that neither she nor Shelley had ever dated a woman of color. Shelley later acknowledged having had a brief fling with a Latina woman, but otherwise found the social distances between ethnic groups too wide to traverse. Veronica, who expressed great pride in her own ethnicity, found it difficult to relate to other Black women, particularly those who identified themselves as African American. For most of her childhood she had been taught to avoid relationships with African Americans because of what her parents described as their "coarse and common behavior."

During one of their therapy sessions, the following scene unfolded:

Monica: Veronica, Shelley here thinks that you are going to judge her for being too emotional. And that makes it unsafe for her to continue sharing. Are you able to offer her any assurances at this time?

Veronica: Well, I don't know about this judging thing. What's the big deal if I do judge her?

Monica: Well for sure, it's a big deal to Shelley who is just trying to share her feelings—as you did so articulately last week.

Veronica: Well, I don't mean to stifle her, but why can't I judge something as good or bad? What's this need to live without ever having someone judge what you do?

Monica: Listen, we are just trying to create a safe environment for both of you, and you're becoming very angry. What are you so angry about right now?

Veronica: It's not that I'm *so* angry.

Monica: Your body language says something different.

Veronica: I'm just trying to explain. It's not that the answer is "yes" or "no." Maybe I just want to be clear about what you mean when you say judging. I don't want to be insensitive to Shelley—but I do have a right to understand a question before I try to answer it.

Monica: What is your anger about right now?

Veronica: Because I can't explain myself!

Monica: But it really seems to push a button with you! I feel it; Shelley feels it. You are getting really worked up about this.

Veronica: Well, frankly I'm disappointed with you. I just wanted to make sure I'm answering the question accurately. I don't want to say "yes" or "no" about whether I'm judging Shelley. It's more complicated.

Monica: Well, I can take your anger at me. But something broader here needs to be looked at. You're yelling; your body is tight.

Veronica: Because you are really making me sick! Anyway, I don't think I'm yelling. This is definitely not what I call yelling. I don't think this is working . . . and I need to go.

Monica: Well, there's the door. This is not a forced situation. I am the kind of therapist that won't be bullied by you.

Veronica: I'm not trying to bully you.

Monica: Sometimes clients will do that, but I can yell right back. Is this what Shelley means when she says she's scared to talk to you?

DISCUSSION GUIDE

Discuss the above scenario noting the following issues:

- Evidence of the impact of controlling images on the relationship between Shelley and Veronica, as well as on their relationship with Monica

- Evidence of relational images shaped by internalized dominance and internalized oppression

- Interruptions in the processes of empathy, authenticity, and mutuality

- Possibilities for healing and reconnections through empathy, authenticity, and mutuality

REFERENCES

Ayvazian, A. & Tatum, B. D. (1994). Women, race, and racism. *Work in Progress, # 68.* Wellesley, MA: Stone Center Working Paper Series.

Bowles, Dorcas. Personal communication, 1985.

Brock, R. N. (1993). *Journeys by heart: A christology of erotic power.* New York: The Crossroads Press.

Brodkin, K. (1996). Race and gender in the construction of class. *Science and society, 60*(4), 471–477.

Brown, L. (2003). Women and trauma. In L. Slater, J. H. Daniels & A. Banks (Eds.) *The complete guide to mental health for women.* Boston: Beacon Press.

Carter, R. T. & Cook, D. (1992). "A culturally relevant perspective for understanding the career paths of visible racial/ethnic group people. In Lea, D. H. & Leibowitz, Z. B. (Eds.) *Adult career development: Concepts, issues, and practices* (2nd Ed.). Alexandria, VA: National Career Development Association.

Collins, P. H. (2000). *Black feminist thought: Knowledge, consciousness, and the politics of empowerment.* (2nd Ed). New York: Routledge Press.

Cross, W. E., Jr. (1995). In search of Blackness and Afro-centricity: The psychology of Black identity change. In H. Harris, H. C. Blue, & E. H. Griffith (Eds.) *Racial and ethnic identity: Psychological development and creative expression.* New York: Routledge Press.

Cushman, P. (1995). *Constructing the Self, Constructing America: A cultural history of psychotherapy.* Cambridge, MA: Perseus Publishing.

Jenkins, Y. (1993). *Diversity and social esteem.* In V. DeLa Cancela, J. Chin, & Y. Jenkins (Eds.), *Diversity in psychotherapy: The politics of race, ethnicity, and gender.* Westport, CT: Praeger.

Hartling, L. (2003). *Prevention through connection. Work in Progress, # 103.* Wellesley, MA: Stone Center Working Paper Series.

Hotchkiss, S. (2002). *Why is it always about you: Saving yourself from the narcissists in your life.* New York: The Free Press.

Jordan, J. V. (1992a). Relational resilience. *Work in Progress # 57.* Wellesley, MA: Stone Center Working Paper Series.

Jordan, J. V. (1997). *Women's growth in diversity.* New York: The Guilford Press.

Kaplan, A. G. (1997). How can a group of White, heterosexual, privileged women claim to speak of women's experience? In chapter, "Some misconceptions and re-conceptions of a relational approach" by Miller, J. B., Jordan, J. V., Kaplan, A. G. & Stiver, I. P. in *Women's growth in diversity.* J. V. Jordan (Ed). New York: The Guilford Press.

Lawrence, C. (1987). The id, the ego, and equal protection: reckoning with unconscious racism. *Stanford law review.* Vol. 39:317.

Mama, A. (2002). Gender, power, and identity in African contexts. In *Research & Action Reports* (Vol. 23, No. 2.).

Lipsky, S. (1984). Unpublished manuscript.

McAdoo, H. P. (1970). Racial attitudes and self-concepts of Black preschool children. Doctoral dissertation, University of Michigan. *Dissertation Abstracts International* 22 (11):4114.

Miller, J. B. (1976). *Toward a new psychology of women.* Boston: Beacon Press

Miller, J. B. (1986). What do we mean by relationships? *Work in Progress, # 22.* Wellesley, MA: Stone Center Working Paper Series.

Miller, J. B. (1988). Connection, disconnection, and violations. *Work in Progress, #33.* Wellesley, MA: Stone Center Working Paper Series.

Miller, J. B. & I. P. Stiver (1997). *The healing connection.* Boston: Beacon Press.

Robinson, T. & Ward, J. V. (1991). A belief in self far greater than anyone's disbelief: Cultivating resistance among African American female adolescents. In C. Gilligan, A. G. Rogers, and D. L. Tolman (Eds.) *Women, girls and psychotherapy: reframing resistance.* New York: The Haworth Press.

Steele, C. (1997) A threat in the air: How stereotypes shape intellectual ability and performance. *American psychologist, 52*(6), 613–629.

Walker, M. (1999). Dual traumatization: A sociocultural perspective. In Y. Jenkins (Ed.), *Diversity in college: Directives for helping professionals.* New York: Routledge Press.

Walker, M. & Larkin, W. (1995). Workshop series entitled *Beyond the Barrier.*

Walker, M. & Miller, J. B. (2000). Racial images and relational possibilities. *Talking Paper 2,* Wellesley College: Stone Center Publication.

Ward, J. V. (2000). *The skin we're in: Teaching our children to be emotionally strong, socially smart, and spiritually connected.* New York: The Free Press.

4

Identity Development and the Convergence of Race, Ethnicity, and Gender

by Mary F. Howard-Hamilton, Ed.D., and Kimberly Frazier, Ph.D.

REFLECTION QUESTIONS

1. *Have you ever experienced a situation that was challenging because the person was a from a different racial/cultural/ethnic background from yourself?*
2. *When did you first discover your racial identity (for example, Black, White, Latino/a)?*
3. *How do you define race, ethnicity, culture, and gender?*
4. *How might your client's race, ethnicity, or gender impact the counseling process if different from yourself? If the same as yourself?*
5. *Have you ever encountered a situation in which you were discriminated against but were unable to discern whether it was discrimination based on your race, ethnicity, and/or gender?*
6. *Discuss your first racist and/or sexist experience.*

INTRODUCTION

This chapter will review the various models of racial identity development. Additionally, the complexities of the convergence of race, ethnicity, and gender will be delineated, so as not to treat these constructs as monoliths. The many layers of identity development (including culture, class, age, sexual orientation) will be presented as a means to provide readers with a clear sense of the complex context in which we all develop. The ways in which all these factors are sources of unearned privilege or marginalization will be described and translated for counseling purposes.

Kimberly Frazier is originally from a southern state and has lived there most of her life. She lived in a neighborhood and went to a school that was predominantly African American, so most of the people she met looked like her. I (Kimberly) remember vividly the first time that I realized people were different colors. At the age of 4, I was with my mother and she introduced me to her coworker, a woman she addressed as "Ms. Judy." It was at that moment that I realized Ms. Judy was not brown like my mom and I. I remembered that I had just learned my colors and was in awe that Ms. Judy looked just like the color pink in my crayon box. At that point I exclaimed loudly, "She is pink!" From that point on, I referred to her as "the pink lady."

To understand racial identity theories, the reader needs to have a set of working definitions that will provide a context for how people describe themselves within a social system as well as how they are viewed within the social system. Definitions for race, culture, ethnicity, and gender are provided so that readers can work with a specific theme and use it as a basis for challenging or supporting previous terminology that was learned in other courses and from personal experiences that formulated biased or incomplete information.

RACE

The first definition of race was created by biologists who placed individuals in three categories: (1) Mongoloid, (2) Caucasoid, and (3) Negroid (Atkinson, 2004). The distinguishing characteristics of each race were determined by the skin color, hair, and facial features. The definition was quickly adopted by those in most academic circles, even without research validating the categorizations. It wasn't until the 1960s that researchers noted the within-group differences of varying races in addition to genetic variability and intermingling. These criticisms of the biological definition of race launched the concept of the social definition of race.

Race can be defined as a social construct intended to maintain certain societal norms. This social construct determines access to societal and within-group resources and also decides the rules by which these resources will be dispensed. In relation to culture, race can be defined as between-group differences based on cultural values, norms, attitudes, and customs (Helms & Cook, 1999).

Atkinson (2004) gives a social definition of race as how outsiders view members of a group and how individuals within a group view themselves, members of their group, and members of other racial groups. Atkinson states that race, in the social definition, provides a convenient way of categorizing people and a way for self and group identity and empowerment. For example, we, the authors of this chapter, are African American and this is the way people may view us and how we make connections with persons who are like us. Characteristics that describe other attributes of racial groups are defined as ethnicity and culture and can be easily misinterpreted or confused. The descriptions of these terms are provided in the following sections.

CULTURE

Culture refers to the values, beliefs, language, rituals, and traditions of various groups that are transmitted from generation to generation or individually learned (Atkinson, 2004). Therefore, culture can be collective/communal or individualistic/personal in nature. An individualistic culture is rooted in a person's needs or desires. Behaviors in individualistic cultures are driven by a person's need for personal fulfillment and self-interests, not based on the entire group's survival. Collective cultures place value on the survival and maintenance of the entire group, rather than an individual within the group's needs or self-fulfillment. Behaviors of individuals within the group are motivated by what is best for the entire group (Helms & Cook, 1999). Values as well as behaviors are key components to all cultures (Atkinson, 2004). It is the confluence of all these characteristics that make up one's culture.

ETHNICITY

Ethnicity is broadly defined as shared physical and cultural characteristics (Atkinson, 2004). It is also "a social identity based upon a person's historical nationality or social group" (Torres, Howard-Hamilton, & Cooper, 2003, p. 6). A more elaborate definition states that ethnicity can include a group of individuals who interact, have a form of social structure, and a system of governing norms and values, are biological and cultural descendants of a cultural group, and identify as members of the group. It has been found that "because this broad definition includes physical characteristics, ethnicity is often used interchangeably with race" (Atkinson, 2004, p. 9). The preferred terminology is ethnicity rather than race "in references to groups of people who are distinguished by their regional ancestry or unique culture" (Atkinson, 2004, p. 9).

The realization that there were different ethnic groups became part of Mary Howard-Hamilton's reality when she was in the second grade.

> I attended a segregated African American elementary school in a small midwestern city until the second grade. That is when I was transferred to an integrated school and was placed in a classroom with all White children and one African American boy from my former school. This transfer process was devastating because I left the comfortable confines of my school that had teachers and staff members who looked like me. Also, my African American classmates were all shifted to another class in the integrated school and they were classified as 'slow learners.' As I struggled to adjust and make friends with my White

classmates, I missed my reference group, the African American children and my teachers.

GENDER

The definition of gender has become clearly defined and is "viewed as a category holding social and political meaning about individual behavior and interrelationships" (Ridley, Hill, Thompson, & Ormerod, 2001, p. 197). Furthermore, the roles, behaviors, and attitudes that are expected of persons on the basis of their biological sex are connected with gender socialization, or society's perception of what is appropriate for men and women (Robinson & Howard-Hamilton, 2001). Biological sex is defined by the chromosome pair found in men, XY, and women, XX, along with physiological, hormonal, and anatomical body parts (Robinson & Howard-Hamilton, 2001).

In terms of gender and the roles prescribed to men and women, the best examples are those that pertain to career paths. Men are mechanics, construction workers, and pilots. Women are secretaries, housekeepers/maids, and flight attendants. Society covertly prescribes and plants the seeds of gender role socialization via the media, advertisements, and textbooks. Additionally, the voices of our teachers, parents, and clergy often guide girls and boys into traditional roles and behaviors. A behavioral example is the independent nature of boys and the communal spirit of girls.

It is important to understand how the intersection of race, ethnicity, culture, and gender leads to complex issues for the individual as well as perceptions that are held by members of society that may or may not be supportive of one's confluence of characteristics. Specifically, one author is an African American female who is in a commuter marriage, has no children, and is in her late 40s. She is also an upper-level academic administrator and professor and has worked in higher education for over 25 years. These are not typical gender and cultural roles for this racial/ethnic author. If she were your client, it would be important to recognize how these characteristics impact her personal and professional development.

PRIVILEGE

The process of granting the dominant group opportunities that covertly and overtly honors them systemically, politically, economically, and culturally is called privilege (McIntosh, 1988). People from dominant groups are automatically given access to power, resources, and benefits "that are denied others and usually gained at their expense" (Goodman, 2001, p. 20). The unequal distribution of these privileges makes it difficult for the oppressed groups to gain any economic, personal, or cultural momentum in society. "Privileges do not need to be desired—we get them whether we want them or not and whether we are aware of them or not.

Privileges can be both material and psychological. They can include concrete benefits as well as psychological freedoms; often these are interrelated" (Goodman, 2001, p. 21).

The unearned privileges granted to men are freedom to move about their neighborhood, city, region, or the entire country anytime of the day or night with relatively little fear for their safety. Women are not granted this privilege and, as you will read in a later chapter on women's development, fear of sexual assault is a central theme of women's development. In addition, men are often given more time to speak at meetings and feel free to interrupt others (particularly women) if they wish to make a point. Perhaps some dynamics would differ if the man were Black, Latino, or Asian. However the privileges of gender would still be covertly granted and he would benefit from this treatment socially, psychologically, and economically.

Those who are privileged based upon socioeconomic status or class have access to material goods that not only elevate their standard of living but allow them physical longevity. If you are a white-collar worker earning nearly six figures per year, you have certain health advantages that you would not have if you worked construction for an hourly wage your entire life. You could afford to go on vacation to reduce stress, be cared for by the best medical doctors and facilities, and retire comfortably without having to worry about a fixed income that places you at or near the poverty level.

People who are heterosexual are privileged because they can establish relationships that are viewed as "healthy" and can be comfortable with public displays of affection. This is exhibited by placing a photo of a loved one on your desk at work or sharing your vacation photos with colleagues, friends, and family members. You can identify your loved one by name rather than hiding the use of personal pronouns that indicate this person is the same gender as yourself. Another privilege of heterosexuals is that you do not fear someone may want to harm you or kill you because of your sexual orientation. This is something that is also explored in a later chapter on bisexual, gay, lesbian, and transgendered individuals.

Privilege is granted to all persons who are able-bodied and do not have to find a sign language interpreter or the accessible door, elevator, or ramp in every building. I (Mary) understood this privilege through personal experience when I had a student who was hearing impaired. She was a graduate student in a higher education program and I was her professor for two years as well as advisor. I learned to give better lectures because there were interpreters in class and my presentation had to be clear and concise. When she traveled to conferences, we contacted the host so that interpreters were available at all sessions and she had facilities that were accessible such as telephone devices, computers, and interpreters that could assist with check-in/out of hotels or conference registration.

Racial privilege is a dynamic that is embedded everywhere in our society. When Whites do not have to think about their race, culture, or ethnicity on a daily basis they are experiencing this form of privilege. When the dominant group is validated on college campuses by seeing themselves in the advisors, professors, and administrators they come in contact with most of the time, this is racial privilege. Most people from the dominant racial group can walk around a campus, corporate structure, or prominent landmark and be validated because the sculptures, paintings, and other artifacts will present their history or contribution to the creation or development of the structures and environment.

The voices and artifacts from oppressed groups are rarely seen as major contributors to the establishment of major landmarks. I (Mary) struggle each time there is a special occasion that creates a need for me to search for a greeting card because there are so few that have people who look like me on them. For my female readers, have you wondered what it would be like to enter every store and find that the predominate number of cards have male characters on them while only a few feature women?

It is difficult to dismantle the invisible system of privileges unless people are willing to recognize and acknowledge these benefits and understand that it leads to an unequal system that deepens the gap between the haves and the have-nots. A bridge should be built between the privileged and oppressed because when there is an identification with superiority and privilege, people from an advantaged group retreat from allying with people from one of the disadvantaged groups (Goodman, 2001). People who have a keen sense of their privileges and racial identity are the persons who can affect change in the world. Without a heightened sense of awareness in both areas, there will be very little movement to let go of the benefits that come so naturally with unearned privilege. Have you given some thought to the daily covert privileges you have been granted? Take a moment to reflect on what they may be and how you could become more sensitive and aware of the impact on others who do not have "access" to these unearned gifts. If you find it difficult to identify sources of privilege in your life you will find the exercise in chapter 3 to be helpful.

When I (Kimberly) went to graduate school at an urban southern university I was the only African American in my cohort. Being the only African American in my cohort was a learning experience because I spent my undergraduate academic career at a historically Black institution. Throughout my graduate program, I looked to be the voice of the African American experience. In one class setting, my professor, the only African American male on the faculty in the college of education, lectured on the topic of White privilege and sparked a discussion about the history behind this issue. I noted that all of my classmates seemed to share similar points of view on numerous subjects, so when asked to discuss their feelings on the subject of White privilege, one student echoed what most of the students were thinking and that was the issue of 'if privilege happened so long ago, why should Whites be held accountable for this issue now?' This comment and the negative feelings the students had regarding privilege spoke volumes about how much freedom and flexibility they had to control a class and the topic. Furthermore, the professor who taught the class was often thought of as being too sympathetic to his own race and not focusing on the important tools needed to be an effective counselor. These perceptions are those of Whites who hold a privilege status and as students can judge a faculty member as well as voice what they feel are the important (or safe) topics that should be discussed. It is interesting to consider how the students might have responded to the topic and the faculty member if he had been White. It is crucial to consider how oppressive teaching and counseling experiences can be if the issue of White privilege is not addressed and processed to understand that some of the difficulties discussing White privilege have to do with the fact that "naming" it is the first step to dismantling it.

RACIAL IDENTITY MODELS

Significance of Culturally Centered Models

Culturally centered models have developed out of the necessity for counselors to understand the culture of their diverse population of clients. Such models enable counselors to examine the concept of culture as related to the beliefs, forms, and traits of a racial or ethnic group within the context of mental health issues. An understanding of culture will assist in overcoming any possible barriers that may exist between the client and the counselor and/or the client and society (Lee, 1997).

Racial identity models began to emerge in the counseling and psychology literature in the 1970s (Hardiman, 1994). Models that described the emotional and psychosocial development of Black Americans were written by Cross (1971, 1991) and Jackson (2001). Models and theories that described the racial identity development of White people were written by Hardiman (1979, 1994, 2001).

Carter (1995) noted that Whites have not had any models that offered steps to develop their racial identity, "nor are they presented with opportunities to understand the meaning of their race if they choose, to abandon their racist perspectives" (p. 100). The struggle to get Whites to identify their racial identity is exacerbated by the fact that very few view themselves as racial beings. For example, if a White person were asked, "What race are you?" that person would most likely say "American." "Although knowing and appreciating one's ethnic heritage is as important as other social identities (e.g., gender, socioeconomic class, religion), race as an aspect of identity, is minimized by Whites" (Carter, 1995, p. 100). Additionally, White identity development closely parallels the history of racism in the United States (Helms, 1990).

The methods used to describe people different from Whites was generally borne out of reinforced attitudes or beliefs that have been learned, taught, or reinforced by society. There are three types of racism prevalent in our society: (a) institutional, (b) cultural, and (c) individual (Carter, 1995). Institutional racism is maintained by the establishment of rules, regulations, laws, and policies that provide advantageous status (economically, politically, and socially) for one group over another. Cultural racism is the celebration of customs, beliefs, values, and symbols that clearly communicates the superiority of White cultural mores over other visible racial ethnic groups. Individual racism is when a person espouses personal ideas, attitudes, and behaviors that promote the superiority of Whites over racial/ethnic groups.

In order for a White person to acquire a healthy racial identity, the eradication of all three forms of racism must occur. "This means Whites must accept their

Whiteness, understand the cultural implications and meaning of being White, and develop a self-concept devoid of any element associated with racial superiority" (Carter, 1995, p. 101).

White Racial Identity Models

Helms originally created the White Racial Identity Model to aid in the description of the psychotherapy process between a White individual and a person of a different race. The racial identity models in general are intended to describe the various processes of development by which individual members of various groups shift their versions of internalized racism toward self-affirming and realistic racial group or collective identity. The descriptors in Helms's model are called statuses rather than stages because "Statuses are cognitive—affective-cognitive intrapsychic principles for responding to racial stimuli in one's internal and external environments" (Helms & Cook, 1999, p. 84). Movement or evolution into more advanced racial identity statuses means the individual is able to perceive and respond to racial information in both their external and internal environment in increasingly complex ways (Helms & Cook, 1999).

There are seven statuses and two phases of the White Racial Identity Model (Carter, 1995; Helms & Cook, 1999). The first three statuses, contact, disintegration, and reintegration fall into the first phase, the abandonment of a racist identity. In the second phase, the establishment of a nonracist White identity, are the final four statuses of pseudoindependence, immersion, emersion, and autonomy.

The first status, "contact," finds the individual in complete denial of having a racial identity. Moreover, there is a tremendous amount of satisfaction and comfort with the racial status quo (Carter, 1995; Helms & Cook, 1999). Individuals see people of color as "different" and treat them with trepidation, wonderment, or fear. In this status, a White person experiences a Black person or Black people either directly or vicariously, and the information process that takes place by White people at this stage is "obliviousness." At contact status, Whites are more comfortable with their racism when compared to those at other identity statuses because they have not discovered the reality of racial inequality.

In the next status, "disintegration," the individual wrestles with emotions of angst, anxiety, and dissonance affiliated with the realization that racial differences do exist in our society. White individuals move into this stage when there is increased exposure to discriminatory practices perpetrated on Blacks and other people of color, by taking a course on racism, or recognizing the discourse between Whites and other people of color. This status is highlighted by the recognition of other Whites and White institutions that ignore and suppress the reality of racial inequality in order to maintain their privilege in society.

In this status individuals may attempt to overcompensate for their conflicted feelings by becoming preoccupied with people of color. These individuals are also looked down upon by their White peers for questioning the status quo and being disloyal. They may attempt to build exclusive alliances with Blacks and other persons of color and exclude other Whites. When Blacks and other people of color question their genuineness regarding the desire to build alliances "within group" and not with Whites, individuals in the disintegration status distance themselves from people of color because they notice the discriminatory behavior.

In the third status, "reintegration," the White person realizes and acknowledges the internalized belief that people from racial ethnic groups are inferior and Whites are superior. Individuals move to this status once they realize they will benefit from association with Whites only. They consciously acknowledge that cultural and institutional racism are what is due White people because they have earned such privileges and preferences. Whites at reintegration stage begin to selectively reinterpret information to conform to stereotypes of Blacks and other people of color.

The highlight of this status includes selective perception and negative distortion of Blacks and other persons of color. Whites at this particular status rationalize that they are not racist because they have a Black or person of color as a friend or that some in the minority groups have similar beliefs as them. Overall, the first three statuses (first phase) must be resolved in order to abandon a racist attitude and move into the second phase, the establishment of a nonracist identity.

The last four statuses are very difficult because Whites must challenge themselves to open their minds and process the information they have been conditioned to believe. The "pseudoindependence" status begins the process for defining a positive White identity by reviewing "ideas and knowledge about Blacks and other people of color" (Carter, 1995, p. 106). The White person is awakened to the fact that Blacks may not be inferior to Whites and that the responsibility for changing societal attitudes rests upon the dominant group.

During this status, Helms and Cook (1999) state that gaining a more advanced status for Whites characteristic of the "reintegration" status is difficult because it requires them to abandon their racist identity. This status starts the process of building a positive White identity. Dominant group members in this status worked through major dilemmas and have abandoned racist views and acknowledge that minority group members are not inferior to the dominant group. Persons at pseudoindependent status have not internalized the distortions that perpetuate racism and continue to unknowingly perpetuate racist views. Once they have reflected, these Whites realize they still harbor racist views and become uncomfortable with their White identity.

The fifth status, "immersion," finds Whites revising "myths about Blacks and Whites by incorporating accurate information about the present with the historical significance and meaning of racial group memberships" (Carter, 1995, p. 107). The self-exploration process contained in immersion has the person asking the question "Who am I racially?" The grappling with this and other tough questions about race, racism, privilege, and Whiteness involves the search for a new and nonracist definition of Whiteness. Also at this status, the individual begins to search for an understanding of what racism is. Whites begin to recover from distorted racial socialization and seek accurate information about race and racism as pertinent to themselves.

The penultimate status, "emersion," focuses on affective behavior because emotionally the person is moved to tears of joy and gratitude that a breakthrough has occurred. This status is rejuvenating and pushes the person to reestablish goals that will promote growth toward self-knowledge about diverse groups. The emersion status is begun by seeking to be involved and immersed into a community of other reeducated Whites. At this juncture, Whites solidify their goals of seeking new self-knowledge.

The final status, "autonomy," finds the person establishing personal definitions about race and ethnicity without the influence of societal brainwashing. This is the most advanced status and "to evolve permits complex humanistic reactions to internal and environmental racial information based on realistic, nonracist self-affirming conception of one's racial and collective identity" (Helms & Cook, 1999, p. 93). Individuals at the autonomy level are able to recognize the actions of themselves and other Whites in perpetuating racism and no longer feel the need to oppress, idealize, or denigrate people based on race.

Hardiman (2001) designed the five-stage White Identity Model (WID) that describes the way Whites come to terms with Whiteness and racism, and race privilege. The first stage is "No Social Consciousness of Race," or Naivete, about race and unawareness of racial differences. During early childhood the person moves into the "Acceptance" stage, in which the societal programming efforts to instill beliefs of superiority are beginning to take effect and are stored unconsciously.

The information about racial superiority is difficult to release and there is "Resistance" to do so. In this stage the person questions the internalized messages of elitism and racial superiority but enjoys the privileges and thus does not wish to let go of the programmed information. Some individuals can see how their behaviors harm others so they begin the "Redefinition" process to work against racism and take responsibility for their Whiteness rather than deny it. Once they have increased their levels of racial consciousness and identity, there is an "Internalization" process to have this new frame of reference embedded in all aspects of their lives.

People of Color (POC) and Minority Identity Models

The underlying premise behind Helms's POC model is that "in the United States the symptoms or consequences of racism directed toward one's racial group are a negative conception of one's racial group and oneself as a member of that group" (Helms & Cook, 1999, p. 85). POC refers to the experiences American racial/ethnic groups face regardless of the origins of their ancestry. Specifically, Asian, African, Latino/Latina, and Native Americans are POC because they have been, in their collective experiences, subjected to appalling economic and political circumstances for not being similar to White people. POC have historically faced similar racial group discrimination that has forced them to internalize negative self-denigrating ego identity concepts. The desired outcome for POC is to abandon the societal prescribed behaviors and attributes and create their own self-affirming identity.

There are five statuses in the POC Model. The beginning process is very basic and primitive in that the person interprets and responds "to racial information in a manner that suggests negative own group identification, endorsement of societal prejudices towards one's group, and uncritical esteem for the White group"

(Helms & Cook, 1999, p. 89). The POC Model's final status shows an evolutionary process that empowers the person to resist the negative collective group identity and societal oppression to forming one's own identity.

The summative impact of the movement through each status is to find the person equipped with the personal and cognitive abilities to resist delimiting information and cope with the constant bombardment of racist assaults on the psyche. The process is to create an emotionally mature individual who can manage the complexities of living in a world that places limitations on one's ability to be successful by systemically creating barriers that oftentimes seem impenetrable.

The first status is called "conformity" and it has two modes according to Helms, active and passive. Active mode occurs when the person of color is overt in his/her display of culture denigration and White idealization. In the passive mode the person deflects his/her internalized racism by ignoring the sociopolitical impact of race on daily interactions with Whites and other people of color in forming opinions on social issues. In the passive stage the person actively and unconsciously filters out information that refutes his/her distorted reality, while maintaining the belief of racial equality.

Once these people realize they cannot attain full access or advantage in White society they enter the "dissonance" stage. This stage is characterized by several personal or social events that are inconsistent with the belief of racial equality. These events leave people feeling confused and they come to the realization that no matter how they conform to White standards, most Whites will still perceive them as inferior. Gradually the person experiences an epiphany that leads them to seek out a Black, Asian, or Hispanic identity. The primary strategy of this stage is to repress anxiety-provoking racial information.

Persons of color enter the "immersion/emersion" stage by immersing themselves into their culture in an effort to construct a new definition of what it is to be Black, Asian, or Hispanic. The first phase of this racial identity stage involves idealization of their own culture and devaluation or rejection of the White culture and people. When people enter this stage they may distance themselves from their White peers due to feelings of anger toward Whites and peers of the same race because they perceive them as being unenlightened or having conformity worldviews.

In immersion, persons of this status define themselves in a rigid manner, relying on stereotypes and misperceptions regarding race. A person at this stage often idealizes her or his own culture while expressing anger toward the dominant race. This individual is eager to differentiate her or his culture and race from that of the dominant race and culture. In the emersion phase, the person begins to learn about her or his culture by understanding strengths and weaknesses. A new cultural perspective comes from an informed and realistic understanding of her or his culture and the sociopolitical implications surrounding it. People operating from this racial status are hypervigilant regarding racial stimuli and operate from dichotomous thinking.

The next stage in the model, "internalization," is reached when the person is able to integrate his or her fully evolved identity into his or her everyday being. The individual is able to reject racism and other forms of oppression because she or he can cognitively recognize how these serve to exploit all humanity. At this stage the person is able to establish relationships with Whites and is able to distinguish the strengths and weaknesses of Whites and White culture.

The final stage in the model is integrative awareness and is considered the most sophisticated status. At this stage one is able to express a positive racial self and recognize and combat the multiplicity of practices in his or her environment. The person is also able to accept, redefine, and integrate aspects of himself or herself that may be characteristic of other cultures and groups. This stage involves complex processing of people and environmental events that show healthy intrapsychic and interpersonal functioning.

Sue and Sue (1990) first introduced the Minority Identity Development Model (MID) in the early 1980s. This model arose out of a need for a framework in which mental health professionals could better understand and assess non-European clients. In creating this model, Sue and Sue hoped to accomplish three things: (a) to diminish the stereotypes and misconceptions about minority groups fashioned by researchers, (b) to correlate the strength of the counselor-client relationship to the degree that the client identifies with the therapist, and (c) to develop a model that could be applied cross-culturally (Sue & Sue, 1990).

Racial/Cultural Identity Development Model

The Racial/Cultural Identity Development Model (R/CID) takes into account the client's self-esteem and identity issues. It attempts to help the client explore and come to terms with such issues. It asks the following questions: (a) who do you identify with and why; (b) what minority cultural attitudes and beliefs do you accept or reject and why; (c) what dominant cultural attitudes and beliefs do you accept or reject and why; and (d) how do your current attitudes and beliefs affect your interaction with other minorities and people of the dominant culture?

Most importantly, this theory seeks to confront clients based on their level of acculturation and conformity into the dominant culture and their rejection of their own culture. According to the Sues' (1990) model, an individual's experiences and/or level of acculturation determine how well or how poorly he or she accepts his or her own identity (Ponterotto, Casas, Suzuki, & Alexander, 1995).

Although the R/CID model is effective in working with nonconventional clients, there are some limitations to this theory. First, most of the work depends on the model and the stage of the client's development. Changes in stages depend on specific circumstances and whether they are dynamic or static. Second, the counselor choosing this theory must have reached a level of autonomy prior to its use. Otherwise, the counselor's stage of development could possibly be negatively projected onto the client. Lastly, in order to effectively conduct this theory of cultural identity development, the counselor must be culturally aware and accepting of various cultures. Specifically, the counselor must be aware of the client's cultural background.

Conformity Stage—Individuals in this stage of the model have an extreme preference for dominant culture values over their own. Values, norms, and beliefs that most mirror those of the White society are highly valued, while those most like their own minority group are looked down upon or repressed. Attitude and beliefs toward self are self-depreciating. Physical and cultural characteristics that are common to the individual's racial or cultural group are perceived negatively and as something to be avoided, denied, or changed. In this stage, the person may attempt to mimic "White" speech patterns, dress, and goals. A person at this stage has low internal self-esteem. Attitude and beliefs toward members of the same minority are group depreciating. The person holds the beliefs and attitudes that the dominant culture has about that particular minority group. The individual may have even internalized the "White" stereotypes about their particular group and in turn have the concept of herself or himself as "not being like the rest," "I'm the exception." Attitudes and beliefs toward members of different minorities are discriminatory. In an attempt to identify with the dominant White culture, the individual also holds the same beliefs as the dominant culture toward other minorities. Minority groups that are most similar to Whites are viewed more favorably than those dissimilar. Attitudes toward members of the dominant culture are group appreciating. The individual believes that White cultural, societal, and institutional standards are superior. Members of the dominant culture are admired, respected, and emulated.

Dissonance Stage—The movement into this stage is a slow process. Here individuals are experiencing a conflict with information or experiences that challenge his/her current concept of himself/herself. Movement into this stage is generally characterized by a traumatic event. Attitudes and beliefs toward self are conflict driven and occur between self-depreciating versus self-appreciating attitudes and beliefs. The individual becomes aware that racism does exist, and that not all aspects of minority or majority culture are good or bad. For the first time the individual begins to entertain thoughts of possible positive attributes in their culture and encounter a sense of pride in self. Shame and pride begin to tug at the person, hence a conflict emerges. Attitudes and beliefs toward members of the same minority trigger a conflict between group-appreciating and group-depreciating attitudes and beliefs. Dominant held views on the individual's minority group strengths and weakness are questioned. Certain cultural aspects in the minority culture begin to have appeal, since contradictory information about the culture is received. Attitudes and beliefs toward members of a different minority begin to change as a conflict of dominant-held views arises. Stereotypes associated with other minority groups are questioned and an understanding of common circumstance is discovered. All energy is focused on resolving conflict between members of the dominant group, same minority group, and self. The attitudes and beliefs directed toward members of the dominant group result in a conflict. The conflict exists between group appreciating versus group depreciating. The person begins to discover that not all dominant cultural values are beneficial to him/her. A growing suspicion and distrust begins to develop for the dominant culture group members.

Resistance and Immersion Stage—Movement into this stage is characterized by the resolution of the conflicts and confusions that occurred in the previous stage, hence allowing a deeper understanding of social forces and how they play a role in the individual's existence in society. The person also begins to wonder why people should feel ashamed of themselves or their culture. In this stage the individual begins to reject views held by the dominant culture and embrace minority-held views. Here the person completely rejects White societal, cultural, and institutional beliefs because they hold no validity for him/her. The person has a deep desire to eliminate the oppression of his/her particular minority group and that becomes a very important motivation in her or his behavior. Dominant feelings in this stage include: *shame,* because the person feels he/she has "sold out" his/her own racial or cultural minority group; *guilt,* because the person feels she or he has been a contributor and participant in the oppression of her/his people; and *anger,* because of the realization of being oppressed and brainwashed by the dominant White society. Attitudes and beliefs toward self begin to take a self-appreciating turn. Individuals encountering this stage are focused on self-discovery of his/her particular history and culture. He/she begins to actively seek out information that gives him/her a sense of identity and worth. Cultural symbols and characteristics now evoke pride and honor. Attitudes and beliefs of members of the same minority are also self-appreciating. The person begins to have a strong identification with members of his/her own minority as more information about the group is acquired. For the first time, feelings of connection with other members of the racial or cultural group emerge and a new identity begins to form. Members of the minority group are admired and respected. The values, norms, and beliefs are now accepted without question. Attitudes and beliefs toward members of different minority groups evolve into a conflict between feelings of empathy for other minority-group experiences versus feeling of culturocentrism. Alliances may be formed based on convenience factors of a large group coming together to confront a larger perceived enemy. Attitudes and beliefs toward members of the dominant culture are group depreciating. The individual begins to perceive the dominant group as the oppressor, which is responsible for the current plight of minorities in the United States. A large amount of anger and hostility is also directed toward White society. There in turn is a feeling of dislike and distrust for all members of the dominant group.

Introspection Stage—The factors that move a person from the resistance/immersion stage into the introspection stage include: the person discovers that this level of intensity of feelings is psychologically draining and does not allow time to devote energy into understanding one's racial/cultural group; the individual senses the need for positive self-definition and a proactive sense of awareness. A feeling of disconnection emerges with minority group views that may be rigid. Group views may start to conflict with individual views. Attitudes and beliefs toward self are still self-appreciating; however the person feels that he/she too rigidly held onto minority-group views. Now a conflict arises between responsibility and allegiance to the individual's minority group versus personal independence. Attitudes and beliefs toward members of the same minority are still group-appreciation. The person might feel that positions their minority group are taking are too extreme. This can lead to resentment if the person feels the group is attempting to pressure them into making various decisions that are inconsistent with personal values, beliefs, or norms. Attitudes and beliefs toward members of a different minority turn from culturocentrism into an ethnocentrism basis for judging others. The individual may reach out to other minority groups to find out what types of oppression they experience and how they have handled that oppression. The importance lies in understanding potential differences in oppression that other groups have experienced. Attitudes and beliefs toward members of the dominant group begin to shift from group-depreciation. Conflicts arise between attitudes of complete trust for the dominant culture versus attitudes of selective trust and distrust. The person experiences the conflict because she or he discovers there are many aspects of American culture that are desirable and functional, yet the confusion lies in how to incorporate these elements into the minority culture.

Integrative Awareness Stage—Minority persons in this particular stage have developed an inner sense of security and can appreciate various aspects of their culture that make them unique. Conflicts and discomforts experienced in the previous stage are now resolved, hence greater control and flexibility are attained. Individuals in this stage recognize there are acceptable and unacceptable aspects in all cultures and that it is important for them to accept or reject aspects of a culture that are not considered desirable to them. Attitudes and beliefs toward self are self-appreciating. A positive self-image and a feeling of self-worth emerge. An integrated concept of racial pride in identity and culture also develop. The individual sees himself or herself as a unique person who belongs to a specific minority group, a member of a larger society, and a member of the human race. Attitudes and beliefs toward members of the same minority are group appreciating. People in this stage experience a strong sense of pride within the group, without having to accept group values without question. There is no longer a conflict over disagreement with group goals and values. Strong feelings of empathy associated with group experience and awareness that each member of the group is an individual are discovered. Attitudes and beliefs toward members of a different minority are also group appreciating. There is a reaching out to different minority groups in an attempt to understand their cultural values and norms. A belief that the more one understands other cultural values and beliefs, the greater the understanding among various ethnic groups arises. Attitudes and beliefs toward members of the dominant culture are selective appreciation. The individual begins to selectively trust and like members of the dominant group who seek to eliminate oppressive activities of the dominant group. People in this stage also begin to be open to the constructive elements of the dominant culture. Emphasis is on whether White racism is a sickness in society and on the possibility that White people are victims who are in need of help as well.

The racial identity models provide a framework to understanding how embedded and infused racism and issues related to racial/ethnic identity are woven into the societal fabric. Thus, our immutable characteristics (such as race and gender) are part of who we are and how people recognize how different we are from each other. Can we really focus on only one part of our identity and leave the others behind? Can we focus only on our race and not reflect on how gender role socialization keeps women relegated to low income positions? Thus, one's race, gender, and socioeconomic status are all part of how society marginalizes persons who have multiple identities that are not favored.

Another racist incident occurred at my (Kimberly) southern Montessori school in the second grade, this time it was the issue of color symbolism. My school was composed of mostly affluent White children and only a few African

American children sprinkled throughout the school. A well respected White teacher began disciplining a beloved classmate's handling of white boxes: "Be careful with those pretty white boxes, you don't want to get them black and ugly." I remember the other two Black girls and I in the classroom glanced at each other and then looked down at the floor after the teacher's statement, as if reading each other's minds, "We must be black and ugly." I remember being wound with anxiety the rest of the day and going home distraught and confused asking my mother "Does this mean that I am ugly since I am Black?"

Critical Race Theory

There are very few theories that can effectively provide a context for the varying situations experienced by women of color (Howard-Hamilton, in press). Many traditional theories are very general and there is an inherent assumption that most people have similar personal, professional, and developmental issues so there is a "one size fits all" frame of reference by those creating these models. One theory that takes into consideration the multifaceted and dynamic identities, roles, and experiences of women of color is Critical Race Theory (Delgado & Stefancic, 2001).

Critical race theorists understand that it is the laws and legal policies that keep oppressed people subjugated in our society (Delgado & Stefancic, 2001). The reason for this is that the dominant culture designed laws that were supposed to be race neutral but still perpetuate racial and ethnic oppression. Critical Race Theory "emphasizes the importance of viewing policies and policy making in the proper historical and cultural context to deconstruct their racialized content" (Villalpando & Bernal, 2002, pp. 244–245). Thus, Critical Race Theory:

1. Recognizes that racism is prevalent in our society
2. Expresses skepticism toward dominant claims of neutrality, objectivity, color blindness, and meritocracy
3. Challenges ahistoricism and insists on a contextual and historical analysis of institutional policies
4. Insists on recognizing the experiential knowledge of people of color and our communities of origin in analyzing society
5. Is interdisciplinary and crosses epistemological and methodological boundaries
6. Works toward eliminating racial oppression as part of the broader goal of ending all forms of oppression. (Villalpando & Bernal, 2002, p. 245)

For people from marginalized groups and in particular women of color, this is a liberating frame of reference because it speaks to the complexity of their multiple identities and the personal issues they face on a daily basis. Methods used to heighten the sensitivity and consciousness of people of color are exposure to microaggressions, creation of counterstories, and development of counterspaces (Delgado & Stefancic, 2001; Howard-Hamilton, in press). Microaggressions are the covert and overt verbal, nonverbal, and visual forms of insults directed toward people of color and they are a form of racist psychological battering. In order to minimize the impact of these microinsults, the use of counterstories is employed. Counterstorytelling is used to reframe and dispel the myths being told by those casting aspersions on people of color. A safe and comfortable counterspace should be provided for marginalized groups to share their counterstories so they can be open and honest about living in a nonsupportive environment and what that means for their daily survival. Solorzano, Ceja, and Yosso (2000) write:

> When the ideology of racism is examined and racist injuries are named, victims of racism can find their voice. Further, those injured by racism discover that they are not alone in their marginality. They become empowered participants, hearing their own stories of others, listening to how the arguments are framed, and learning to make the argument themselves. (Solorzano, Ceja, and Yosso, 2000, p. 64)

As an undergraduate at a predominantly White university in the Midwest, Mary found it comforting to be able to find refuge in the African American Cultural Center. Here she could meet with women and men who could empathize with the stories being told regarding faculty who were biased and ignored the raised hands of students of color (an example of a microaggression) and students who showed blatant disregard for people who did not look like them. The cultural center was the counterspace where there were voices of support and encouragement as well as counterstories being told so that the students of color could reframe the negative energy into positive power and perseverance.

Black Feminism

African American women's lives and careers are placed at the intersections of race, class, gender, and sexual orientation, which shape their experiences (Collins, 1990, 1998). According to Collins (1990, 1998) this perspective supports the "weaving together of personal narrative, stories, and critical social theory" (p.119). Black feminist research, or a Black women's standpoint, is primarily aimed at "maintaining dialogues among Black women that are both attentive to heterogeneity among African American women and shared concerns arising from a common social location in the U.S. market and power relations" (Collins, 1990, 1998 p.73).

On the other hand, mainstream research published by the dominant culture tends to ignore the experiences and voices of women, ethnic minorities, and those speaking from low socioeconomic backgrounds. According to Black feminists, the metaphor of voice breaks the silence for oppressed persons, develops self-reflexive speech, and confronts elite discourses. Collins (1990, 1998) writes:

> Self-reflexive speech refers to dialogues among individual women who share their individual angles of vision. In a sense, it emphasizes the process of crafting a group-based point of view. In contrast, I prefer the term self-defined standpoint, because it ties Black women's speech communities much more closely to institutionalized power. (p. 47)

While African-American women have race and gender in common, they are not part of a monolithic culture. The Afrocentric perspective and Black feminism support the notion that each group member of the race is unique but share core values and philosophical assumptions that have their origins in African history (Patitu & Hinton, 2003; Asante, 1987; Collins, 1990, 1998). In developing our own unique characteristics we also develop a sense of belonging to the cultural and ethnic group. Because of racism and sexism African American women see themselves as both racialized and sexualized individuals and perceive being treated differently because of the group status (Hinton, 2001). Patitu & Hinton (2003) asserts that African American women are situated at the bottom of society's White male-dominated hierarchy, which results in the belief that this dominance is the socially accepted style to think, act, speak, and behave. Take, for example, the situation of African American women on university campuses. Promotion and tenure, and academic standards and activities, are historical designs of White male dominance and African American women pursuing graduate degrees, tenure, and promotion through the ranks of administration have to learn and "reproduce the ways of the dominant group" (p. 115) in order to receive career stability, promotion, and financial benefits.

Speech that defines a standpoint in higher education environments is often difficult for Black women to establish because of the lack of a critical mass on most campuses. However, as Black women get opportunities to "talk" to each other offstage they share the challenges faced being the only or one of few persons of color on predominantly White campuses as students, faculty, or administrators. The offstage, private speech or otherwise underground testimonials are often empowering for African American women because it allows them to "disrupt public truths about them" (Collins, 1990, 1998, p. 48) developed by the "elite discourses" (p. 49). These offstage opportunities to share are referred to as "safe spaces" or a "homeplace" (Patitu & Hinton, 2003, pp. 79–93) and allow African American women to move from the margins to the center, from object to subject. African American women pursuing graduate degrees find the strength of other African American women and men as safe places through support, advice, encouragement, and collaborative scholarship opportunities (Patitu & Hinton, 2003, pp. 79–93; Hinton, 2001).

Womanist Identity Development

The Womanist Identity Development process is comparable to Helms's White Racial Identity Development Model (Ossana, Helms, & Leonard, 1992). There are four stages that occur in a stepwise developmental process. Stage one is "preencounter," in which there is an acceptance of the socialized roles for girls and women and denial of gender bias. Stage two, "encounter," finds the person experiencing significant confusion and bewilderment about women's prescribed gender roles in this society. This leads to questioning the outside sources who have created these restrictive career, academic, and work boundaries for girls and women. The next stage finds the woman rejecting these rigid roles and rules, thus entering the third stage of an "immersion-emersion" developmental process. The rules designed and created by men are rejected and conversely women are idealized. The fourth stage is the woman's formation of internalized womanist identity attitudes, based upon culture and personal context, without the major influence of traditions that dictate the roles for women in this society.

Feminist Therapy/Theory

Laura Brown (1994) provides the following comprehensive definition of feminist therapy/theory:

> Feminist therapy is the practice of therapy informed by feminist political philosophy and analysis, grounded in multicultural feminist scholarship on the psychology of women and gender, which leads both therapist and client toward strategies and solutions advancing feminist resistance, transformation, and societal change in daily personal life, and in relationships with the social, emotional, and political environment. (p. 22)

Along with the definition there are six key components and behaviors that are essential to understanding and practicing the art of feminist therapy theory. They are, according to Brown (1994, p. 23):

1. an understanding of the relationship of feminist political philosophies to therapeutic notions of change;
2. an analysis and critique of the patriarchal notions of gender, power, and authority in mainstream approaches to psychotherapy;
3. a feminist vision of the nature and meaning of psychotherapy as a phenomenon in the larger social context;
4. concepts of normal growth and development, distress, diagnosis, boundaries, and relationships in therapy that are grounded in feminist political analysis and feminist scholarship;
5. an ethics of practice tied to feminist politics of social change and interpersonal relatedness;
6. a multicultural and conceptually diverse base of scholarship and knowledge informing this theorizing.

There are several meta-assumptions that are the underpinnings of the theory in addition to the six components. These assumptions are the importance of having a clinical usefulness for the selection and implementation of this theory, having a deep understanding of diversity and the complexity it adds to the lives of women, always viewing women in a positive frame of reference and placing them as key central figures in every conversation, making sure that the stories told arise from the experiences of women, understanding the details of the women's experiences, knowing that what happens around women (the outside environment) impacts and intersects with their inner world, making sure that the therapist is not restricted to the use of traditional theories, and there is a personal philosophy to support feminist methods of psychotherapy.

Overall, feminist therapists should continually strive to demystify the process of therapy for the client. The therapy process is open and collaborative with the client being empowered to be an active participant in her self-healing. The therapist explains the therapeutic methods and along with the client can design the intervention strategies together so that there is no assumption on the part of the therapist that he or she is the person who dictates how the developmental process will ensue (Espin, 1994).

CONVERGENCE OF MULTIPLE IDENTITIES

At the beginning of the chapter you were presented with several reflection questions that related to how you conceptualize race, ethnicity, and gender and its impact on your growth and development. One important item is to realize that racial identity development not only intersects with every aspect of who you are and what you believe but is also a primary construct throughout your life span. There is no one period or age in your life that will be the "racial identity moment" because these "moments" will constantly occur due to the lack of awareness and sensitivity for those who have not moved beyond the basic levels or statuses of racial sensitivity. Moreover, those who continually challenge persons who believe in inclusiveness have not dealt with or relinquished their baggage full of societal privileges.

The convergence of race, ethnicity, gender, and other characteristics should always be reviewed in depth first and foremost before making assumptions about the direction of the interaction or considering intervention strategies. When working with males, consider variables such as age, birthplace, gender role socialization, parental influences, and cultural concepts or values that may be aligned with the difficulty of recognizing one's privileges. Women of color should not be viewed as either/or but by the confluence of their immutable and mutable characteristics. For example, if you are working with a biracial woman who is in her late 20s, a single parent with two children who are also mixed lineage, then the time should be taken to review the individual and family history very closely to understand how the dynamics of multiple racial groups impacts the daily lives of the mother as well as the children. Overall, the collection of contextual data is extremely important in addition to understanding the importance of cultural connectedness for people of color. If the counselor is not familiar with the history of multicultural people in America, there is a need to extend the knowledge base further so that assumptions coming from a traditional frame of reference are not applied to nontraditional people.

Even though we have moved into the 21st century, there has been very little change or movement to translate information from a traditional perspective to one that is more inclusive of others. There is still a reliance on counseling theories and interventions that have been designed and validated using a monolithic White heterosexual model because it is assumed that they represent the normative group (Constantine, 2001). The critical reminder in this chapter is to think outside the box at all times. It is best to see everyone as a racial/gendered being because the first step in empowerment is affirming their identities and being comfortable with naming yours.

CONCLUSION

The number of theories and therapeutic techniques provides a strong indication that there are complex issues involved when counseling women and women of color because of the multiple identities, societal influences on gender roles, issues of privilege, power, and status by the dominant group that impede the progress of women politically, economically, cognitively, and interpersonally. The tantamount issue presented in this chapter is that the counselor must continually be aware of her or his own personal agenda and not let that be the guiding force in the therapeutic process. Gaining understanding of the sociopolitical and relational issues as they relate to the role of women in this society will provide human services helpers with more empathy and insight in the counterspace so they can listen to the stories about the microaggressions, racism, and other atrocities then create the counterstories for empowerment and enlightenment.

CASE STUDY #1

Sandra is an African American university doctoral supervisor in her early 20s and has been assigned to her first master's level practicum student, Carla. Carla is a master's practicum student assigned to Sandra for supervision at a predominantly White university. Carla is European-American in her late 30s, married, and the mother of two children ages 11 years and 5 months. Sandra's initial contact with Carla is cordial and she arranges to meet Carla at a coffeehouse to get better acquainted. When Sandra arrives for the meeting, Carla is prompt and is carrying a rocker with her infant. When Sandra inquires about the infant Carla apologizes for the inconvenience, but explains that when Sandra gets older and grows up she will understand the woes of being an adult. As the meeting progresses and Sandra begins discussing the logistics of how supervision works, Carla is unable to commit to a specific time and place for their weekly supervision meetings, but promises she will call in the next two days to let Sandra know. Before leaving the meeting Sandra gives Carla a list of ten different time openings she has available for supervision.

The next day Carla calls back as promised, but says none of those days and times fit into her busy schedule. She comments that she has a husband and real life, so the best fit for her would be for Sandra to drive over to her home around 10:00 at night once a week. Sandra explains that supervision is a venue to model professionalism and counselor identity, hence meeting at her home would not be possible. Carla becomes agitated and says she will have to call back after she speaks with her husband.

Two days have passed and Carla has not called, so Sandra decides to call her and ask if she has arrived at a decision. Carla states that she has arrived at a decision, but she must bring her baby to this meeting because she has yet to make arrangements with a babysitter. Though hesitant, Sandra agrees to the arrangement and sets up a time for the next supervision meeting. The next supervision meeting Carla arrives late with her fussy infant. As she rocks the car seat to quiet the infant, Carla explains that she cannot afford additional daycare and must bring the infant to all future supervision meetings. Sandra again explains that it is not appropriate for these meetings to bring an infant and Carla apologizes for the inconvenience. This case illuminates how race, gender, and privilege intersect and lead to complex dynamics between people and institutions. The supervisee, Carla, in this example sees Sandra not as the supervisor but as a person of color and a woman. Thus the differences are viewed as deficits and hinders the supervisory process because there is no common connection between them. Carla has a skewed view of Sandra's personal life as well because she has not lived a real life as a wife and mother.

CASE STUDY # 2

Frances is an African American first year university doctoral supervisor in his mid-20s and has been assigned to his first master's level practicum student in counseling. In his first meeting with his supervisee, Briana, Frances finds that Briana is a biracial woman in her late 40s and has decided to start a second career in counseling. Frances explains what his expectations are for the supervision experience and expresses his desire to model professionalism and counselor identity. Frances inquires about Briana's expectations regarding the supervision process and specific expectations she might have for Frances. Briana quickly comments that she can't think of anything, but will have suggestions next week. Before the close of the first supervision session, both the supervisor and supervisee exchange numbers and e-mail addresses and agree on a time and place to meet. Briana promises to have her first tape for the second session.

At the second session, Frances waits for 20 minutes for the supervisee to show. After calling the number given by Briana and leaving a voice message he decides that apparently the supervisee will not be coming to session. The supervisee finally shows for the third session without a tape or an explanation for her prior absence. At the close of the session Briana states to Frances, "See you next week, kid!"

DISCUSSION GUIDE

Discuss the above case studies using the following issues:

1. What are the feminist therapy and theory techniques evident in the cases?
2. What are the gender role socialization issues in the cases?
3. Which multicultural theories would you choose to help guide your understanding of Briana's concerns? Carla's concerns? Identify the stages of development that are prevalent in each case.
4. Which multicultural theories would you choose to help guide your understanding of Frances's concerns? Sandra's concerns? Identify the stages of development that are prevalent.
5. How would Critical Race Theory be used in each case?
6. What would be the best intervention methods for each case?

REFERENCES

Asante, M. K. (1987). *The afrocentric idea.* Philadelphia: Temple University Press.

Atkinson, D. (2004). *Counseling American minorities* (6th ed.). New York: McGraw Hill.

Brown, L. S. (1994). *Subversive dialogues: Theory in feminist therapy.* New York: Basic Books.

Carter, R. T. (1995). *The influence of race and racial identity in psychotherapy: Toward a racially inclusive model.* New York: Wiley.

Collins, P. H. (1990). *Black feminist thought: Knowledge, consciousness, and the politics of empowerment.* Cambridge, MA: Unwin.

Collins, P. H. (1998). *Fighting words: Black women & the search for justice.* Minneapolis, MN: University of Minnesota Press.

Constantine, M. G. (2001). Addressing racial, ethnic, gender, and social class issues in counselor training and practice. In D. B. Pope-Davis & H. L. K. Coleman (Eds.), *The intersection of race, class, and gender in multicultural counseling,* (pp. 341–352). Thousand Oaks, CA: Sage.

Cross, W. E. (1991). *Shades of Black: Diversity in African American identity.* Philadelphia, PA: Temple University.

Delgado, R., & Stefancic, J. (2001). *Critical race theory: An introduction.* New York: New York University Press.

Espin, O. (1994). Feminist approaches. In L. Comas-Diaz & B. Greene (Eds.), *Women of color: Integrating ethnic and gender identities in psychotherapy* (pp. 287–318). New York: Guilford.

Goodman, D. J. (2001). *Promoting diversity and social justice: Educating people from privileged groups.* Thousand Oaks, CA: Sage.

Hardiman, R. (2001). Reflections on White identity development theory. In C. L. Wijeyesinghe & B. W. Jackson (Eds.), *New perspectives on racial identity development: A theoretical and practical anthology.* (pp. 108–128). New York: New York University Press.

Hardiman, R. (1994). White racial identity development in the United States. In E. P. Salett & D. R. Koslow (Eds.), *Race, ethnic-ity and self: Identity in multicultural perspective* (pp. 117–142). Washington, DC: National Multicultural Institute.

Helms, J. E. (1990). *Black and White racial identity: Theory and research.* Westport, CT: Greenwood.

Helms, J. E., & Cook, D. (1999). *Using race and culture in counseling and psychotherapy: Theory and process.* Boston: Allyn and Bacon.

Hinton, K. G. (2001) "The experiences of African-American women administrators at predominantly white institutions of higher education." Unpublished doctoral dissertation, Indiana University, Bloomington.

Howard-Hamilton, M. F. (in press). *African American women in higher education.* New Directions in Student Services. San Francisco: Jossey Bass.

Jackson, B. W. (2001). Black identity development: Further analysis and elaboration. In C. L. Wijeyesinghe and B. W. Jackson (Eds.), New perspectives on racial identity development: A theoretical and practical anthology (pp. 8–31). New York: University Press.

Lee, C. (1997). *Multicultural issues in counseling.* American Counseling Association. Thousand Oaks, CA: Sage.

Ossana, S. M., Helms, J. E., & Leonard, M. M. (1992). Do "womanist" identity attitudes influence college women's self esteem, and perceptions of environmental bias? *Journal of Counseling and Development, 70,* 402–408.

Patitu, C. L., & Hinton, K. G. (2003). The experiences and issues of African American women faculty and administrators in higher education: Has anything changed? In M. F. Howard-Hamilton (Ed.), *African American Women in Higher Education.* New Directions in Student Services (pp. 79–93). San Francisco: Jossey Bass.

Ponterotto, J. G., Casas, J. M., Suzuki, L. A. and Alexander, C. M. (1995). (Eds.) *Handbook of multicultural counseling* (pp. 181–198). Thousand Oaks, CA: Sage.

Ridley, C. R., Hill, C. E., Thompson, C. E., & Ormerod, A. (2001). "Clinical practice guidelines in assessment: Toward an idiographic perspective." In D. Pope-Davis and H. Coleman (Eds.), *The intersection of race, class, and gender in multicultural counseling* (pp. 191–211). Thousand Oaks, CA: Sage.

Robinson, T. L., & Howard-Hamilton, M. F. (2001). *The convergence of race, ethnicity, and gender: Multiple identities in counseling.* Upper Saddle River, NJ: Prentice Hall.

Solorzano, D., Ceja, M., & Yosso, T. (2000). "Critical race theory, racial microaggressions, and campus racial climate: The experiences of African-American college students." *Journal of Negro Education, 69*(1–2), 60–73.

Sue, D. W., & Sue, D. (1990). *Counseling the culturally different: Theory and practice.* New York: Wiley.

Thompson, C., & Carter, R. (1997). An overview and exploration of Helm's racial identity development theory. In C. Thompson & R. Carter (Eds.), *Racial Identity Theory: Applications to individual, group, and organizational interventions,* (pp.15–32) London: Lawrence Erlbaum Associates.

Torres, V., Howard-Hamilton, M. & Cooper, D. (2003). Identity development of diverse populations: Implications for teaching and administration in higher education (Vol. 29, number 6). San Francisco: Jossey-Bass.

Villalpando, O., & Bernal, D. (2002). A critical race theory analysis of barriers that impede the success of faculty of color. In K. Smith, P. Altbach, & K. Lomotey (Eds.), *The racial crisis in American higher education: Continuing challenges for the twenty-first century* (pp. 243–269). New York: Suny Press.

5

The Developing Counselor

By Marilyn Montgomery, Ph.D., and Jeffrey Kottler, Ph.D.

REFLECTION QUESTIONS

1. *What motivated you to pursue a career in counseling?*
2. *What "turning points" or "critical incidents" have you experienced in your development as a counselor?*
3. *How have normative life events (such as marriage, parenthood, or aging) and nonnormative life crises (such as divorce, serious illness, or the untimely death of a loved one) shaped your life? How do you think these experiences might impact your ability to form and sustain therapeutic relationships?*
4. *How has your cultural background (including gender, ethnicity, socioeconomic status, geographic location, sexual orientation, and other factors) influenced your development of trust, empathy, compassion, security, and self-confidence as a helping professional?*

INTRODUCTION

Picture yourself sitting within a circle on the sand, comfortably perched on a beach chair. You are within sight of the Huntington Beach Pier, watching the surfers and the sunbathers in the distance, but still in a secluded spot. The crashing surf is just steps away, but not so close that it drowns out the conversation. You've come to join with other counselors who want to spend the last day of a conference at the beach, participating with others in exploring issues related to development as a professional and a human being. The group leader invites you to ponder several questions:

1. Where are you at this point in your career—and your life?
2. Where are you headed next?
3. What meaning does being a counselor hold for you?

We were the leaders of such a group, facilitating the process of 15 counselors at various stages in their development. Some had been practicing for many decades, while others were enrolled in their first counseling class at the university. What all the participants had in common was a hunger to understand more thoroughly what sustains them in their work, why they had chosen counseling as a "calling," and how they might increase their enjoyment and satisfaction with the profession.

We will return to these same questions later in the chapter, as well as tell you how counselors at various stages in their development responded to them. First, however, we will take you on a guided tour through several models of counselor development in order to stretch your ways of thinking about these issues.

DEVELOPMENT: WHAT DEVELOPS NEXT?

In a way, the topic of "counselor development" is as hard to capture as the topic of "adult development." We like to think that we gain experience and knowledge, and *develop,* somehow as we accrue years on this planet—hence the birthday platitude, "You're not getting older, you're getting better!" Development implies movement or flow, particularly in a positive direction. It is easy to see that children grow, become more clever, and gain capacities for many things, including peer relationships, problem solving, emotion and impulse management, morality, and thinking in abstract ways. Adolescence—the last period of rapid, visible, and pervasive change—transforms the child into an "emerging adult" (Arnett, 2000). What changes, or develops, after that, besides the slow but inevitable physical decline that begins after we've peaked at age 25?

Physical changes aside, many adults talk as though they have experienced remarkable development—in its most positive sense of *growth*. They refer to ways in which they think or feel or relate to others distinctly differently than they did earlier in their adult life (usually with relief). Other adults talk about themselves as being pretty much the same person, inside and out, as they were at age 21 (usually with pride). So, do adults develop—or not? What is "development" anyway, and is it useful for talking about what happens to counselors who move along the path from entering graduate training to being seasoned practitioners?

During the last century, many developmental theories were proposed to explain human change over the life span. The most notable of these theories include Freud's (1924) psychosexual theory, Erikson's (1950) psychosocial theory, Piaget's (1926) cognitive development theory, Bridges's (1930) emotional differentiation theory, Loevinger's (1976) ego development, Kohlberg's (1969) moral development theory, Super's (1957) theory of career development, Sullivan's (1953) interpersonal theory, Sue's (Sue, Ivey, & Pedersen, 1996) cultural identity development, Helms's (1990) racial identity development, Basow's (1992) gender development theory. The concept of "development" in the context of these theories means that human change happens in predictable, invariant sequences of enlarging capacities (for thought, emotions, social perspective-taking, and the like). These qualitative changes emerge through the interaction of the growing biological organism and an environment that presents the organism with challenges, and result in movement through qualitatively different stages of development across the life span—hence, these are sometimes called "organismic" theories. Observations of the biological world inspired this metaphor of human life: tadpoles become frogs, mollusks develop differently according to the temperature and chemical composition of their surrounding waters, and human lives predictably unfold.

Recently, respect has grown for other metaphors for human change, and developmental and counseling theorists have borrowed and applied these as ways to understand counselors' and clients' growth. For example, postmodern theorists (Kelly, 1955; Foucault, 1979; Gergen, 1994) have given us the metaphor of a *narrative* as a way of thinking about how we write—and rewrite—life stories, and have created whole new schools of counseling based on these ideas (see White & Epston, 1991; Monk, Winslade, Crocket, & Epston, 1997; Neimeyer & Mahoney, 1999, as examples). Physicists (Wiener, 1948; Gleick, 1987; Cohen & Stewart, 1994) have given us the metaphor of chaos theory and the notion of everything that exists, from galaxies to atoms, as complex, *dynamic systems* of energy in which change is both patterned and random. While the literature on adult development and counselor development has historically emphasized the first model, and we agree that it is often useful to conceptualize change in terms of predictable, qualitatively different stages, we prefer to use this metaphor, though with a degree of caution. Looking at our own lives, we think that chaos theory more accurately describes much of the changes that we've experienced! Nevertheless, thinking about some common patterns of change can also be useful, and sometimes in this chapter we call these common patterns "stages," even though we don't think they are necessarily sequential and invariant.

A Personal Illustration

It may help to illustrate these initial concepts by telling you a bit about our own development as counselors, a process that has hardly proceeded in a predictable, sequential way. I (Jeffrey) initially started out as a business major, at least until the first day of class. I looked around the room during orientation and noticed that it was filled with eager young men, mostly dressed in the uniforms of success. There was not a single woman to be seen. I thought to myself: "These are definitely not my people!"

I fled the room and wandered the halls, looking for an alternative major. I couldn't help but notice that in one room across the hall there was at least an even split between the genders. I didn't know what they were doing in there but I was willing to give it a try. It turned out it was psychology.

I actually ended up as an English major eventually, at least until I experienced major trauma in college when a love relationship ended unexpectedly. I went into a tailspin, became seriously depressed and dysfunctional, and only recovered after some hard work in therapy at the university counseling center. While intrigued with the work of counselors, and certainly grateful for my own counselor's help, I still hoped to find a more practical career.

When the time came to graduate from college, jobs were scarce. I decided to go to graduate school mostly as a means to "hide out" and escape the realities of having to get a "real" job. I started work as a preschool counselor, mostly because it was the only job I could find at the time. Once I discovered that I loved working with kids, I was hooked. But it was more serendipity than intentional strategy that led me into this field to begin with. Later, when I studied the various career development theories of Donald Super (1957), John Holland (1973), Anne Roe (1957), and others I felt even more alien than usual: what they described as a relatively orderly and sequential developmental process did not fit my life experience at all. Rather, it seemed to me that random factors, various opportunities that arose, and people I met along the way all influenced where I ended up.

Now, when I look back on my own development as a counselor, I still don't understand how it all happened the way it did. I know that my instructors had in mind a plan for us that proceeded along a curricular route. Supposedly, by taking certain classes in a particular order, reading books and writing assigned papers, we would somehow *become* competent counselors. I don't mean to sound like the class work was not important, but I learned far more outside the classroom from (1) being a client in counseling, (2) informally talking to faculty about ideas, (3) interacting with peers to make sense of what we experienced, and (4) seeing clients under intensive supervision. More recently, almost all of my growth and learning has come not from reading books or attending seminar but through transformative travels to foreign lands where I am required to stretch myself in new ways (Kottler, 1997, 2001, 2003). It is one thing to *talk* about cultural diversity issues, but quite another to *live* them when forced to meet one's basic needs while living in rural villages in Nepal, Thailand, Greenland, or Bali.

My development as a counselor, like yours as well, was influenced by a number of factors and dimensions. Some of these are planned and intentional (assigned class work). Some are trials and tribulations (traumas you have suffered and challenges you have faced). Some involve serendipitous or fortuitous events (meeting the right people or being in the right place at the right time). And some were affected by own internal needs and interests as they developed over time.

WHAT DEVELOPS FOR COUNSELORS IN
A TRAINING PROGRAM?

As you reflect on the growth and circumstances that brought *you* to your enrollment in a counseling program, you can also imagine the great variety of stories that would be told if your classmates were polled. However, from the moment of the acceptance letter until graduation, a great deal of *planned* change and development also happens for the student. Indeed, one purpose of a counselor training program, which can be thought of as an "intervention," is to evoke comprehensive positive changes and development in attitudes, knowledge, and skills. These are often geared toward professional domains yet, as often as not, counselors are impacted personally in profound ways as well. In fact, one of the greatest benefits of our job is that everything we learn as counselors helps us to become more effective human beings; likewise, everything we learn in our personal lives—all the experiences we log and books we read and conversations we have—helps us to be more understanding and wise professionals (Kottler, 2004).

Once in a counselor training environment, students find that particular critical issues—personal, clinical, and supervisory—tend to emerge. This is particularly true with respect to confronting certain fears—of failure, of ineptitude, of mediocrity, of shattered illusions, of losing control, of rejection, and of one's own limitations (Kottler & Jones, 2003)—and areas of cultural mistrust (Vinson & Neimeyer, 2000). Interviews with students in programs across regions show that, upon beginning their training, students predictably become concerned about interpersonal aspects of relationships with other students, faculty and supervisors, and clients (Morrissette, 1996). These issues include self-disclosure, emotional overinvolvement, sexual attraction, countertransference, being competent enough, and conflict. When they respond to counselor development questionnaires, their answers reveal underlying clusters of themes of (a) anxiety and doubt, (b) autonomy and independence, (c) learning methods and gaining skills, (d) feeling validated in choosing counseling work, (e) feeling ambivalence about commitments, and (f) the challenge of learning respectful confrontation (Reising & Daniels, 1983; see also Gladding, 2002).

With progress through the program, and with some experience with real clients, students begin to feel less stressed and more self-efficacious and self-confident (Borders, Rainey, Crutchfield, & Martin, 1996; Melchert, Hays, Wiljanen, & Kolocek, 1996). Alas, if only that feeling would last! Often, however, when entering a new environment (for example, a post–practicum community internship or a postgraduation position in an agency), we find that the old feelings of anxiety, doubt, and ambivalence and similar interpersonal concerns (conflict, attraction, self-disclosure, etc.) tend to recur. For this reason, some think that counselor development—indeed adult development as a whole—is more like a cycle than an upward trajectory. One such cyclical model of counselor development, "Becoming Empowered," was proposed by Sawatzsky, Jevne, and Clark (1994). You fit their model if you've found yourself (a) experiencing dissonance (due to recognizing gaps in skills, knowledge, and experience, and feeling emotional turmoil); (b) responding to dissonance (by acquiring new skills and experiences); (c) seeking and responding to supervision (by changing attitudes and perspectives); and (d) feeling empowered—at least until the next new challenge!

Skills

If you are like most students, the kind of development you probably had in mind as you began a counselor training program was adding skills to your current repertoire. After all, it is not as if you began your counselor preparation as a blank slate. To a certain extent, you came to the program with a lifetime of experiences in a variety of settings and situations—some as a victim of abuse, some in a helping role, and some others that taught you a variety of useful skills. Indeed, many students feel impatient with courses that are aimed at nurturing their ability to think critically about theories or challenging their cultural empathy, and want to "get on with it" by learning the how-to tricks of the trade! Actually, with time in a program, students do show gains in basic listening skills, multicultural skills, and influencing skills (Torres-Rivera, Wilbur, Maddux, Smaby, Phan, & Roberts-Wilbur, 2002). In addition to adding microskills, counselor trainees learn to combine them into complex abilities, such as showing greater efficacy toward difficult client behaviors than less advanced trainees (Stoltenberg, 1981). With experience, students also show increased quality and clarity of conceptualizations about a case (Borders, Rainey, Crutchfield, & Martin, 1996) and multicultural competency (Vinson & Neimeyer, 2000).

You have probably already noticed the infinite ways that you have already been able to apply counseling skills to other relationships in your life. We are not talking about attempting to counsel your family and friends (a frequent but misguided effort of beginners), but rather those instances when you demonstrate greater empathy and compassion for others. You will notice ways that you become a more attentive listener, a more articulate and persuasive speaker, and a more effective communicator. If you can ask good questions with clients, then you can do the same with others in your life. If you can confront clients nondefensively, you can do the same with family members. This is the "gift" that comes with the territory of being a counselor—you are encouraged to become a more effective human being, more loving and respectful, and more relationally competent with every new skill you acquire and practice over time (Kottler & Brew, 2002).

Thinking

Some who observe students' development think that the most important changes occur not just in *what* or *how much* they know but in *how they think about what they know.* A Harvard scholar named William Perry (1981) elaborated a model of how people develop with respect to their epistemological assumptions about knowledge, and this has been applied to the changes that counselor trainees undergo (Granello, 2002). Essentially, least developed students accept the pronouncements of authority in an uncritical way. Often, as students encounter so many potentially viable versions of "truth," they become more relativistic, which means that they decide that what is right can only be an individual matter and that all viewpoints are "relative" and relevant, but there is no way to decide which is right or better. Those who move on from this stage retain their respect for a broad range of perspectives, but believe that positions *can* be judged for their relative value, and commitments can be made on the basis of chosen ethics—so it is again possible to take a stand, while recognizing others' rights to do so, even if they arrive at a different conclusion from you.

This kind of development in thinking may have happened for you in your first "Theories of Counseling" class. *At last,* you might have thought, *I'll learn from the experts how to really understand people and how to help them.* For a while, you might have thought, "Well, if Carl Rogers said so, it must be right" or "If my professor thinks so, that must be the truth." Then as the weeks progress and you read example after example of equally plausible (but often contradictory) explanations of human nature and change, you might have found yourself lost in the sea of great ideas without a compass or a paddle! Perhaps then you decided that they all must be right; picking a theory was just a matter of flipping a coin, so "anything goes"—especially after cultural variations come into play. Eventually, however, you might have moved on to a point of "committed relativism" (Belenky et al., 1986), when you decided that there are premises for making evaluations and weighing off theories, and that certain theoretical notions are more useful for the kind of professional life to which you aspire than others—and that *you* (not an external authority) have the ability to judge knowledge claims.

In a similar vein, some scholars have focused on the kind of thinking that the wisest counselors use, and characterize their cognitive habits as "dialectical think-ing," or thinking in a way that seeks clarification through the understanding of opposites, with the goal of reconciliation and deeper understanding (Basseches, 1997a, 1997b). In this view, opposing concepts are seen as built into the structure of thought, and indeed nature itself. Rather than being dismayed by contradictions and ambiguity, students who develop the ability to think dialectically appreciate that opposing polarities derive their meaning from each other and complement each other—similar to the yin-and-yang concept in Eastern philosophy. Develop-mentally mature students are those who appreciate seeming contradictions, within themselves or others (Hanna, Bemak, & Chung, 1999), and who can work with them and even enjoy them.

Cognitive science studies of expertise offer yet another perspective for under-standing the transition from novice to expert counselor (Etringer, Hillerbrand, & Claiborn, 1995; Sakai & Nasserbakht, 1997). These studies of professionals such as physicians, nurses, and counselors show that experts and novices differ in the way they encode, organize, and use information. With experience, professionals get better at pattern recognition (which helps with diagnosis), forward reasoning (from problem to goal, which helps with treatment planning), and at translating declarative knowledge—the "facts" about counseling—into procedural knowl-edge of how to move a client from A to Z. Perhaps you have already seen changes in yourself along these lines.

EMPATHY AND RELATIONAL CAPACITIES

Empathy has been nominated as a core condition for providing counseling, but how does one develop empathy? Empathy is one of those concepts that makes a lot of intuitive sense, but is difficult to define, measure, or study, though some have attempted to do so. It does seem clear that the deepest empathy involves both the "head" and the "heart"—both the cognitive capacity to project what the world must look like from someone else's shoes, and the emotional capacity to feel and express resonance with another's emotion (Wang et al., 2003). Although there has been some debate about whether empathy is essentially a learned skill or inborn trait, there is little doubt that a developmental process takes place as part of your participation in a program in which you are forced to get outside yourself and appreciate the viewpoints of multiple others who come from quite different backgrounds (Lyons and Hazler, 2002).

Is there such a thing as "advanced empathy," and is this what is needed for our really tough cases? How does one develop the "unflinching empathy" (Marotta, 2003) that allows us to share the emotional experience of abused children, suicide attempters, sex offenders, or tortured refugees, if their experiences were never ours? How can we cognitively project what the world looks like from our client's shoes, if that client has a cultural background that we, though well-intentioned, know next to nothing about? As North America becomes increasingly diverse and the world becomes more globalized, these abilities will be more and more important for counselors to develop (Chung & Bemak, 2002).

Promising guidance for facilitating the development of the capacious, unflinching empathy that will be needed in this century is offered in Relational-Cultural theory of psychotherapy and relational development. In this model, interdependence and connections with others are emphasized. In fact, individual development is thought to proceed *only* by means of connections in relationships (Miller, 1986). A particularly healing phenomenon is something called *mutual empathy*, which occurs when one person describes a problem, and the other not only listens closely but shares her own feelings, helping both to understand their reactions and also to take action. When both people let themselves be affected by the discussion, something greater happens: there is a sense of vitality, self-understanding, and insight about what to do in response. This has also been referred to as "clarity in connection."

In contrast to the connections that mutual empathy fosters are *disconnections*, which result in opposite feelings, such as disempowerment, confusion, and decreased self-worth and energy, which in turn lead to withdrawal (Jordan & Dooley, 2000). Although both connections and disconnections are normal parts of human relationships, being able to name disconnections and subsequently move toward re-connections promotes growth and development, and deepening capacities for empathy. This ability to notice, name, and transform disconnections into connections may be particularly important for counselors, who must seek and establish support for their own growth through relationships, while they simultaneously encourage others to do the same (Comstock, Duffey, & St. George, 2002, 2003; Kottler, Montgomery, & Marbley, 1998).

The metaphor of "connections" is one that I (Marilyn) find particularly help-ful when reflecting on my own growth in counseling and supervising across cultures. After my first several exposures to models of racial identity development, analyses of oppression, and accounts of White privilege, I was a big ball of shame and guilt. I was aware that these feelings got in the way of positive connections—with myself, my forebearers, and with "persons of difference"—but could not imagine a way past them. One chance conversation with a senior colleague (a Polish-American woman who spent more time consulting on nearby Indian reservations than on campus) was helpful; one day I confessed my feelings and in her no-nonsense, no-coddling way she told me that these feelings were doing no one any good and I would certainly have to lose them if I ever hoped to get along on a reservation. I realized that in trying to "own" the disconnections imposed upon this world by history and my own (witting and unwitting) participation in it, I was identifying with the problem rather than "being the solution." So I relaxed a little and hoped that life would offer me some opportunities to move on to a new stage of racial/cultural development.

It did. A literal move to a new city (Miami, Florida) brought me opportunities to balance the "disconnections" with joyful, energizing connections across cultural differences on a daily basis. My perspectives underwent radical change in a new context where my department chair, university president, and 90 percent of my graduate students were of races and ethnicities different from the ones I inherited by birth. I was caught off balance by my invisibility while shopping in my neighborhood (until I learned a new set of nonverbal skills); "privilege" took on new meaning for me as I realized that no one of my phenotype was electable to public office in my county. I unwittingly made my cross-cultural participants angry by asking about their race or ethnic identity—I learned that not only is there "no right answer," there is no right question! ("Latino/a," "Hispanic," "Spanish," and "of color" are all offensive terms to different people in this hemisphere, and for different reasons.) I learned that our salient and distinct categories of American life—or stages of racial identity development—are not the same in the Caribbean or South American communities, where, as one student taught me, "You can't say anything bad about another group, because you'd be talking about your grandmother."

The open, relaxed, and sometimes playful ways that my students and I have evolved for sharing our culture-themed stories and selves have fostered the healing described by Relational-Cultural theory: there is for us a sense of vitality, self-understanding, and insight about what to do in response. Not every counselor-in-training can make a radical geographic move to develop more cultural empathy. But even in a homogenous community there is diversity, if you look for it; and the conversations (and banter, teasing, and challenges) that occur in an open, safe environment—whether between a counselor and a client, or a supervisee and a student, or one peer to another—are the most likely to be instructive and growth-promoting (Garcia & Van Soest, 1997; Hays & Chang, 2003).

Professional Growth and Relational Autonomy

In our profession, counseling is an expertise that is assumed to best develop through relationships with teachers, mentors, and supervisors, which is why training requires so many hours of supervised practice. Initially, students are rewarded for learning their mentor's favorite theory and techniques well. The whole learning process is certainly more meaningful if your supervisor or mentor is someone whom you truly admire and from whom you believe you have much to learn. Like all relationships, the supervisory relationship grows and ideally a degree of mutual influence emerges. Good supervisors, like good mentors and parents, adapt the type of structure and support that they provide to the changing needs of the developing person in their charge (Montgomery et al., 2000). In time, however, counselors who are developing professionally come to feel more competent

Table 1 Life Span Stages of Counselor Development

Contemplation	*Looks great; could I be one?*
Entry	*What if I don't have what it takes?*
Mentorship	*If only I could be like you!*
Eclecticism	*I'll try one of these, and one of these, and . . .*
Experimentation	*Let's see if I can help clients better, faster.*
Self-Assurance	*Hey, I'm really good at this!*
Mid-Career Doubts	*But what's the use, really?*
Burnout	*Another day, another dollar; got to pay the bills.*
Revitalization	*I've found a whole new level from which to work.*
Empowerment	*Watch out, world . . . now I mean business.*
Mentorship	*I'll share some of my best with you.*
Review & Appraisal	*Looking back . . .*

Adapted from *The Imperfect Therapist: Learning From Failure in Therapeutic Practice* by Jeffrey Kottler and Diane Blau (1989) and *On Being a Therapist* by Jeffrey Kottler (3rd ed., 2004). These stages are not necessarily sequential; some individuals may never experience some of these stages, and some will experience all of them several times.

(at least to some degree) and experiment with adaptations and variations that seem more pragmatic from their perspective (see Table 1).

Establishing a therapeutic identity of one's own is an important developmental task for trainees. Granting yourself the permission to do this—to feel confident in your own creative and relational capacities—requires you to muster up enough courage to take small risks in trusting your competencies. Students are sometimes depicted as gradually replacing an external supervisor with an internal one, as experience is gained and self-confidence improves (Tryon, 1996). This doesn't happen in one semester; it may take years or decades. But gradually, you'll move from reliance on external authorities to reliance on your own internal authority. What helps you to develop your courage to work more autonomously is your engagement and interaction with multiple sources of influence over time (Skovolt & Ronnestad, 1992), finding useful tools in many different places, and in essence, putting together your own counseling "tool kit."

Once out of school, one begins to question how to promote continued growth in oneself throughout the marathon of life. According to Relational-Cultural theory, "autonomy" can be reframed and thought of as "one's ability to represent himself or herself more fully and authentically in relationships. It is the growing into more complex relationships and relational networks, not away from them, that is identified . . . as one of our most challenging developmental tasks" (Comstock, Duffey & St. George, 2002, p. 266). Fortunately for counselors, this goal is consistent with both personal and professional growth!

The Person as Professional—and Vice Versa

How *do* the personal and the professional aspects of one's life interact for a developing counselor? In some ways, the different dimensions of ours lives are inseparable. Although we are admonished to make certain there is no destructive "leak-age" that takes place in our work, such that we end up meeting personal needs in sessions, or experience countertransference reactions that affect our personal lives, in one sense such distinctions are artificial, if not impossible to separate. If you are truly paying attention, every client you see, and every session you conduct, will have a profound impact on who you are, and who you become as a person; like-wise, every experience you have in life (the books you read, movies you see, con-versations you have, events you live through) will affect how you develop as a counselor (Kottler, 2004a, 2004b).

Just yesterday I (Jeffrey) saw a client who was complaining in a very critical way about the materialistic plasticity of the lifestyle in Southern California where I live. I was aware of feeling defensive and even commented to him about his cyn-icism and judgmental attitude. The client felt scolded and immediately apolo-gized. But afterward, I realized I was speaking as much to myself as to him. It was a classic projection in which I was seeing the most negative aspects of myself (crit-ical judgment of qualities in others that I have in myself). So the client leaves the session with much to think about, and reflect on, and so do I. And often that internal processing of counselors has as much to do with our own behavior as it does with our clients.

Many think that positive personal development is a *prerequisite* to counseling skills development. Others think that personal characteristics impact a counselor's *rate* of development (or even the extent of his or her development). For example, personal maturity and the ability to think in complex ways can impact a student's rate of professional growth (Stoltenberg & Delworth, 1987). This is a conclusion you have probably already reached, based on your own experiences in a training program! But even extrinsic personal characteristics can interact with a student's experiences in a training program. For example, one study of supervision tran-scripts found that supervisors used different types of questioning behavior with their male and female supervisees (Granello, Beamish, & Davis, 1997). Questions they asked male supervisees were of the type that tends to advance critical think-ing and complex reasoning, whereas questions put to female supervisees were directed at lower levels of complexity. Even subtle (and unintentional) training differences like this can mean that some students get more growth-promoting experiences than others. Of course, other external personal characteristics of race or ability would be equally likely to be associated with subtle (and even uncon-sciously fostered) differences in the growth-producing potential of the training environment.

In many other ways, intrinsic aspects of one's self clearly interact with the growth experiences of a training program. For example, one's stage of ego develop-ment (based on the complexity of one's thoughts and judgments about one's inner, subjective, world; outer, objective, world; and social, normative, world; Loevinger, 1976) restricts the attainable level of counseling skills (Borders, 1998). On the other hand, counseling training and skills development can promote personal ego development. Similarly, counselor training has been found to lead to increased self-other awareness (Tyron, 1996). This is a valuable skill for most social contexts, and certainly an important aspect of personal development that counseling training fostered for both of us (see Marilyn's story in "Finding a Way to Take the Inter-personal Heat").

Finding a Way to Take the Interpersonal Heat

While driving to and from work, I (Marilyn) listen admiringly to radio traffic report personalities. How is it, I wonder, that they are so facile with words, communicating so much in only 15 seconds? I, on the other hand, often grope for the words that don't readily come, or take the slow and awkward route to conveying a relatively simple notion. I respect verbal simplicity and parsimony, and I continually try to be more succinct and clear, in both personal and professional contexts.

I was thrilled, then, while at a recent public event, to learn that one of the traffic report personalities I admire would be giving a brief treatise on philosophical aspects of traffic. As he took the microphone, I went through moments of surprise, "Oh, that's what he looks like!" and pleasure—what a lovely idea, to draw reflective attention to the deeper meaning of one of life's perpetual irritations. However, this was followed by soberness as I realized that as this traffic personality began speaking without the usual 15-second limitation, more was forthcoming than anyone had anticipated.

He announced that he was going to be leaving the traffic desk at the radio station, and was going to instead do public service announcements. Why? It had gotten to him. He was not emotionally well. He found that as time went by, he was less and less able to provide a neutral report of the daily "injury accidents" without becoming emotionally involved. It took a few years, but he had reached a point where encountering, on a daily basis, the deceptively simple reports that signal life-changing crises for individual mothers, fathers, partners, friends, and children had brought unbearable anxiety and sadness into his own life. "And the worst of it," he confided, "was the animals. The pig on the interstate; the chickens that fell off the truck at the cloverleaf. I barely restrained myself from saying, 'People, would you just slow down? The animals were here before us! They're innocent victims of traffic, it's not their fault!'" . . . At this point, which was well beyond the succinct humor and inspiration we listeners had been expecting, someone else got up and politely edged him away from the microphone with a few saving light-hearted quips. It was clear to everyone listening that doing the traffic reports had "got to him." The line between his own wholeness and the brokenness of accidents had blurred.

My developmental struggle in learning to be a counselor has been about learning how to be close and intimate with other people's acute needs and pains (and other kinds of life accidents) while remaining whole and healthy myself. This struggle was a part of my life even before I entered the profession of counseling. In fact, I think one reason I did return to school was because I knew I needed more tools and practice in maintaining this balance in close relationships. Even in my first career as a junior high school teacher, I struggled with feeling emotionally raw at the end of each day, just from encountering more than 100 young adolescent "Thous" (Buber, 1970), each one so brimming with poignant individuality. I also knew that previously, during times of deep grief and anguish, I had felt very alone because those close to me became more distant, presumably to avoid feeling sad or overwhelmed themselves. I wanted to learn to provide that which I had most deeply needed but had missed, by building up a tolerance for being truly open and resonant to others' poignant individuality, while savoring their quirkiness and joining them in their tender, aching places.

During my early training and first years as a therapist, I watched myself get better at being with people in these ways I had hoped for. However, something else was happening: the empathy and "hungry listening" (Hurston, 1937/1978, p. 23) that I brought forth when I was with each client was leaving *me* feeling hungry and lonely. I noticed that while I had gotten good at providing others with that which I most deeply needed, I hadn't gotten good at making sure that I was provided for myself. The hollow feeling inside me intensified, and I began to feel desperate. I resonated strongly with my clients' neediness, and there wasn't enough of a "me" there to keep my balance.

I had to let go of the security of being a "provider" (as a way to feel like a real, grown-up person) and explore some of my own core feelings of anguish with the curious attention and tenderness I had learned to offer clients. Finding a therapist helped me make a space and time in my busy life to do this. I also had to learn to speak up and put my own wishes and desires and views out there for others to hear, so that they would have more to work with when joining me in mutual empathy. For me, this was—and still is—very hard work, and where my admiration for clear, succinct radio personalities comes into play!

A metaphor that I happened upon at a continuing education seminar, somewhere through this dark time of chaos and growth also helped: that of the therapist as a crucible. A crucible is a vessel that is used to take the heat and hold the reaction of elements that are being combined in ways that are not ordinarily possible. A crucible also connotes for me the alchemy and the magic that can happen for an individual, couple, or family who sign on with me in search of change. Of course, I, too, am changed by my interactions with clients or others . . . but I feel now that my essence is—usually!—steady, strong, and whole. I am able to be fully present with great depths and reaches of emotion—both my own and others'—and still ring true. I have learned to be more appropriately intimate with myself, which allows me to be more appropriately intimate with others.

TRAINING EVOKES ONE'S DEEPER ISSUES

Counselor training programs offer students the opportunity to do an almost overwhelming amount of personal reexamination. The proverbial "life review" that people in extreme crisis experience may not be too different from what some students experience over the course of training! Perhaps for this reason, graduate students have a common reputation for shaking up their lives with radical changes in life trajectories. Regardless of the level of intensity of the personal upheaval that training may invoke, it makes sense that encountering a plethora of views of a good life, and how people come to have one (or not), leads to lots of questions and reexamination of previously worked out answers—in other words, the cognitive development discussed earlier. At the emotional level, too, the experience of being in a counselor training program offers many opportunities to encounter one's own and others' strong feelings that result from fresh encounters with ghosts from developmental past histories. Those who experienced trauma of any sort (physical, psychological, or sexual) may be particularly likely to find history repeating a new version of itself (see "The Personal Becomes Professional—A Case Study"). At the same time, these training experiences can be an opportunity to rework old emotional connections into more healthy ones, when mentoring is both supportive and challenging.

THE PERSONAL BECOMES PROFESSIONAL—A CASE STUDY

When Shandra was accepted into the counseling program, she was joyous. She had been out of school a few years, had married, and had children. She had tried several different lines of work and volunteerism and had found herself happiest when she was helping others. Throughout her first semester of classes, she fairly beamed; she told her advisor with great assurance, "I found out this is where I need to be."

Over the course of the next semesters, she began looking less contented and self-assured. She had received feedback in her introductory skills class that despite the tremendous empathy she felt for others, she looked a little "blank" at times. She found that she had trouble wrapping up sessions, structuring, and setting limits with clients. In the group counseling class, she had received feedback that while her empathy skills were extraordinary, her self-disclosures seemed tangential. In prepracticum, her supervisor confronted her for breaking the rules and taking a child client outside the clinic to buy her a snack during the session. Shandra was surprised at this incident herself; she had felt compelled to do it because her client had seemed listless and she thought she might be hungry. The supervisor's response, "Being a little hungry doesn't hurt the client," seemed cold to Shandra, but reflecting on the incident, she became aware that she had a very strong need to provide comfort, caring, and nurturance to clients in a way that might not be best for them. Nevertheless, she began packing extra food on the days that she saw clients and sometimes secretly offered it to them during sessions.

When practicum and internship rolled around, Shandra looked stressed and forlorn, and she found herself slipping into a pattern of being late to class and group supervision sessions. In theories and techniques classes, she and her classmates had discussed the concept of "resistance," and she began to wonder if it applied to her. She realized that she did find it hard to motivate herself to get to group supervision. She also realized that as she tried to remember everything about structuring a session and using appropriate techniques, she felt uncomfortable, inadequate, and "frozen"—hence the "blank" look—but she didn't know how to change what she was doing. She also realized that she felt especially despondent when working with children. She confessed to another student how much she ached to take her child clients home, comfort them, and feed them warm chicken soup.

Soon after that, Shandra found herself in tears during an individual supervision session, saying, "I want so much to please you!" to her older female supervisor whom she very much liked and respected. Then all the pieces started coming together: As Shandra and her supervisor explored Shandra's early developmental experiences, Shandra disclosed that she had lost her mother when she was 2 years old. Her father had remarried a woman with five other children. Subsequently, while Shandra grew up in a noisy and busy family, she nevertheless felt lonely and overlooked for the remainder of her childhood. Just at puberty, she was sexually abused by one of her older stepbrothers, and no one noticed. Yearning for special attention from her supervisor while competing for it with other interns was evoking the feelings of loss, inadequacy, and shame from her early years. Her compulsion to feed clients was a kind of "vicarious nurturing" that revealed her own unmet needs for nurturance and special care.

Shandra's supervisor and the others in her practicum group encouraged her to take her sense of longing for more attention seriously and to explore its interpersonal meaning. She took stock: She was a busy wife and mother who nurtured others unceasingly. Graduate school had taken away time that she had previously spent with supportive friends. Because her husband, who was normally a responsive partner, had recently developed a chronic disease, the balance of mutual support and care had been disrupted in their marriage. Shandra was running on "empty" in her personal life, and this was evoking emotions and unmet needs from her developmental history that spilled over into her current professional training. She realized how she needed to find more sources for her own nurturance, and ask for more caring attention from those she loved. After another semester, Shandra was able to see her own empathic capacity as a strength, and to interpret setting appropriate limits with herself and with clients as another form of caring. As she began to see her "developmental wounds" as the root of her professional strengths, her sense of exuberance at having found the right profession returned.

COUNSELOR DEVELOPMENT: OVER THE LONG HAUL

There are some common "life span" issues in being a counselor for many years. For example, you, like many who have gone before you, may find that you have a tough time counseling people with same life-course issues that you currently face (such as infertility, partner unfaithfulness, divorce, terminally ill parents, empty nest, grief, or retirement). You might find, as have many counselors, that you have a tough time working with people the same age as your children. For example, parents of young children often do not want to do play therapy; parents of teens have a tough time serving oppositional defiant youth; parents of young adults sometimes have a hard time working with other young adults, when therapeutic issues of materialism, identity crisis, and lifestyle choice strike close to home. For most counselors, the need to "take a break" from certain types of clients is just that—temporary, and over once their own life has moved on.

Your clients change you too; over the long haul, they are major sources of influence and become your primary teachers of counselor development once you are out of school (Freeman & Hays, 2002). Ironically, the clients whom we find most challenging, or who "stump" us, are those who have the greatest power to nudge our growth, if we are open to entering the cycle of uncertainty, searching, and resolution once more. The greatest power of challenging clients may be in the opportunities that they offer us for personal development, by calling into question our most comfortable self-conceptions and habitual ways of approaching others (Kottler & Carlson, 2003).

It is important to plan ways to get the nurturance you need in order to transform these challenges of the counseling life into continued growth and development. One of the most helpful things you can do to sustain your personal engagement is to find a professional "family." The professional community formed in graduate school, or assembled later, can act as an informal but significant source of continued learning, and as a supportive family that sustains its members as they mature and seek their own independent pursuits. Your professional family can also provide support and peer mentoring for getting through the predictable stages and not-so-predictable career crises that arise through the professional life course (Anderson, Kottler, & Montgomery, 2000).

CONCLUSION

We return to the beach from where we began. Settled around the circle in the sand, experienced and beginning counselors have been talking about their development over time. They have spoken about the challenges they have faced, as well as the fears they have struggled with—fears of failure, of being a fraud, of hurting someone, of making a mistake, of being mediocre, of not really helping anyone, of wasting their lives. They have also spoken about the joys and elation they have experienced in their work related to fostering others' growth and development. They have told stories about the privilege they have felt being part of clients' journeys.

Yet when all has been said, no consensus has been reached. There is no place to "arrive," developmentally speaking. Some have seen a neat pattern of stagelike development unfolding in their lives; others are proud to have survived a lot of personal and professional chaos.

Even if we wanted to find a comfortable niche and stay there, our families and our friends and our bodies keep changing and prodding us into new territory. Besides, while the periods of consolidation are a joy, when they are sustained they lead to "feeling myopic" (in the words of one group participant) and eventually restless, prompting us into a new phase of active searching for new experiences and new knowledge. One of the things it means to be a counselor is that we are constantly sifting and sorting through new stimuli, new insights and understandings, new challenges and mysteries, all of which (hopefully) encourage us to push onward to the next stage of development, wherever that might lead.

REFERENCES

Anderson, B., Kottler, J., & Montgomery, M. (2000). Collegial support from a Maōri perspective: A model for counsellor kinship. *Canadian Journal of Counselling, 34,* 251–259.

Arnett, J. J. (2000). Emerging adulthood: A theory of development from the late teens through the twenties. *American Psychologist, 55,* 469–480.

Basow, S. A. (1992). *Gender: Stereotypes and roles.* Pacific Grove, CA: Brooks/Cole.

Basseches, M. (1997a). A developmental perspective on psychotherapy processes, psychotherapists' expertise, and "Meaning making conflict" within therapeutic relationships: A two-part series. *Journal of Adult Development, 4,* 17–33.

Basseches, M. (1997b). A developmental perspective on psychotherapy processes, psychotherapists' expertise, and "Meaning making conflict" within therapeutic relationships: Part II. *Journal of Adult Development, 4,* 85–106.

Belenky, M., Clinchy, B., Goldberger, N., & Tarule, J. (1986). *Women's ways of knowing.* New York: Basic Books.

Borders, L. D., Rainey, L. M., Crutchfield, L. B., & Martin, D. W. (1996). Impact of a counseling supervision course on doctoral students' cognitions. *Counselor Education & Supervision, 35,* 204–217.

Borders, L. D. (1998). Ego development and counselor development. In P. Westenberg, A. Blasi, et al., (Eds.) (pp. 331–345). *Personality development: Theoretical, empirical, and clinical investigations of Loevinger's conceptions of ego development.* Mahwah, NJ: Lawrence Erlbaum Associates.

Brandtstaedter, J. & Lerner, R. M. (1999). *Action & self-development: Theory and research through the life span.* Thousand Oaks, CA: Sage Publications.

Bridges, K. M. B. (1930). A genetic theory of the emotions. *Journal of Genetic Psychology, 37,* 514–527.

Buber, M. (1970). *I and Thou.* New York, Scribner.

Chung, R. C-Y., & Bemak, F. (2002). The relationship of culture and empathy in cross-cultural counseling. *Journal of Counseling & Development, 80,* 154–159.

Chung, R. C.-Y., Bemak, F., & Kilinc, A. (2002). Culture and empathy: Case studies in cross-cultural counseling. In P. R. Breggin & G. Breggin (Eds.), *Dimensions of empathic therapy* (pp. 119–128). New York: Springer.

Cohen, J., & Stewart, I. (1994). *The collapse of chaos.* New York: Viking.

Comstock, D. L., Duffey, T., & St. George, H. (2002). The relational-cultural model: A framework for group process. *Journal for Specialists in Group Work, 27,* 254–272.

Comstock, D. L., Duffey, T., & St. George, H. (2003). Gender issues in counselor preparatory programs: A relational model of student development. *Journal of Humanistic Counseling, Education, and Development, 42,* 62–78.

Erikson, E. (1950). *Childhood and society.* New York: W. W. Norton.

Etringer, B. D., Hillerbrand, E., Claiborn, C. D. (1995). The transition from novice to expert counselor. *Counselor Education & Supervision, 35,* 4–17.

Foucault, M. (1979). *The archaeology of knowledge.* London: Penguin.

Freud, S. (1924). *A general introduction to psychoanalysis.* New York: Washington Square.

Garcia, B., & Van Soest, D. (2000). Facilitating learning on diversity: Challenges to the professor. *Journal of Ethnic & Cultural Diversity in Social Work, 9,* 21–39.

Gergen, K. J. (1994). *Realities and relationships: Soundings in social constructionism.* Cambridge, MA: Harvard University Press.

Gladding, S. T. (2002). *Becoming a counselor: The light, the bright, and the serious.* Alexandria, VA: American Counseling Association.

Gleick, J. (1987). *Chaos: Making a new science.* New York: Penguin.

Granello, D. H. (2002). Assessing the cognitive development of counseling students: Changes in epistemological assumptions. *Counselor Education & Supervision, 41,* 279–293.

Granello, D. H., Beamish, P. M., & Davis, T. E. (1997). Supervisee empowerment: Does gender make a difference? *Counselor Education and Supervision, 36,* 305–317.

Hanna, F., Bemak, F., & Chung, R. C.-Y. (1999). Toward a new paradigm for multicultural counseling. *Journal of Counseling & Development, 77,* 125–134.

Helms, J. E. (1990). *Black and White racial identity: Theory, research, and practice.* New York: Greenwood.

Holland, J. (1973). *Making vocational choices: A theory of careers.* Englewood Cliffs, NJ: Prentice-Hall.

Hurston, Z. N. (1937/1978). *Their eyes were watching God.* Urbana: University of Illinois Press.

Jordan, J. V., & Dooley, C. (2000). *Relational practice in action: A group manual* (Jean Baker Miller Training Institute Project Report, No. 6). Wellesley, MA: Stone Center Publications.

Kelly, G. (1955). *The psychology of personal constructs.* New York: Norton.

Kohlberg, L. (1969). *Stages in the development of moral thought and action.* New York: Holt, Rinehart & Winston.

Kottler, J. A. (1997). *Travel that can change your life.* San Francisco: Jossey-Bass.

Kottler, J. A. (2001). The therapeutic benefits of structured travel experiences. *Journal of Clinical Activities, Assignments, and Handouts in Psychotherapy Practice, 1*(1), 29–36.

Kottler, J. A. (2003). Transformative travel: International counselling in action. *Journal for the Advancement of Counselling, 24*, 1–4.

Kottler, J. A. (2004). *On being a therapist* (3rd ed). San Francisco: Jossey-Bass.

Kottler, J. A. & Blau, D. (1989). *The imperfect therapist: Learning from failure in therapeutic practice.* San Francisco: Jossey-Bass.

Kottler, J. A., & Brew, L. (2002). *One life at a time: Helping skills and interventions.* New York: Brunner/Routledge.

Kottler, J. A., & Carlson, J. (2003). *The mummy at the dining room table: Eminent therapists reveal their most unusual cases and what they teach us about human behavior.* San Francisco: Jossey-Bass.

Kottler, J. A., Montgomery, M. J., Marbley, A. F. (1998). Three variations on a theme: The power of pure empathy. *Journal of Humanistic Education and Development, 37*, 39–46.

Kottler, J. A., & Jones, W. P. (2003). The natural and unnatural evolution of therapist development. In J. Kottler & W. P. Jones (Eds.), *Doing better: Improving clinical skills and professional development.* New York: Brunner/Routledge.

Kottler, J. A. (2004a). *On being a therapist* (3rd ed.). San Francisco: Jossey-Bass.

Kottler, J. A. (2004b). *Introduction to therapeutic counseling* (5th ed.). Pacific Grove, CA: Brooks/Cole.

Loevinger, J. (1976). *Ego development.* San Francisco: Jossey-Bass.

Lyons, C., & Hazler, R. J. (2002). The influence of student developmental level on improving counselor student empathy. *Counselor Education and Supervision, 24*, 119–130.

Marotta, S. A. (2003). Unflinching Empathy: Counselors and tortured refugees. *Journal of Counseling & Development, 81*, 111–114.

Melchert, T. P., Hays, V. L., Wiljanen, L. M., & Kolocek, A. K. (1996). Testing models of counselor development with a measure of counseling self-efficacy. *Journal of Counseling & Development, 74*, 640–644.

Miller, J. B. (1986). *Toward a new psychology of women.* Boston: Beacon Press.

Monk, G., Winslade, J., Crocket, K., & Epston, D. (1997). *Narrative therapy in practice.* San Francisco: Jossey-Bass.

Montgomery, M., Marbley, A. F., Contreras, R., & Kurtines, W. (2000). Transforming diversity training in counselor education. In McAuliffe, G., & Eriksen, K. (Eds.), *Preparing counselors and therapists: Creating constructivist and developmental programs* (pp. 148–169). Washington, DC: Donning Publishers.

Morrissette, P. (1996). Recurring critical issues of student counselors. *Canadian Journal of Counselling, 30*, 31–41.

Neimeyer, R. A., & Mahoney, M. J. (Eds.) (1999). *Constructivism in psychotherapy.* Washington, DC: American Psychological Association.

Piaget, J. (1926). *The language and thought of the child.* New York: Harcourt Brace Jovanovich.

Perry, W. G. Jr. (1981). Cognitive and ethical growth: The making of meaning. In W. Chickering and Associates (Eds.), *The modern American college* (pp. 76–116). San Francisco: Jossey-Bass.

Reising, G. N., & Daniels, M. H. (1983). A study of Hogan's model of counselor development and supervision. *Journal of Counseling and Development, 30*, 235–244.

Roe, A. (1957). Early determinants of vocational choice. *Journal of Counseling Psychology, 4*, 212–217.

Sakai, P. S., & Nasserbakht, A. (1997). Counselor development and cognitive science models of expertise: Possible convergences and divergences. *Educational and Psychology Review, 9*, 353–359.

Sawatskzsky, D. D., Jevne, R. F., & Clark, G. T. (1994). Becoming empowered: A study of counselor development. *Canadian Journal of Counselling, 28*, 177–192.

Skovolt, T. M. & Ronnestad, S. H. (1995). *The evolving professional self: Stages and themes in therapist and counselor development.* Chichester, UK: Wiley & Sons.

Stoltenberg, C. D. (1981). Approaching supervision from a developmental perspective: The counselor complexity model. *Journal of Counseling Psychology, 28*, 59–65.

Stoltenberg, C. D. & Delworth, U. (1987). *Supervising counselors and therapists: A developmental approach.* San Francisco: Jossey-Bass.

Sue, S., Ivey, A., & Pederson, P. (1996). *A theory of multicultural counseling and therapy.* Pacific Grove, CA: Brooks/Cole.

Sullivan, H. S. (1953). *The interpersonal theory of psychiatry.* New York: W. W. Norton.

Super, D. E. (1957). *The psychology of careers.* New York: Harper and Row.

Torres-Rivera, E., Wilbur, M. P., Maddux, C. D., Smaby, M. H., Phan, L. T., & Roberts-Wilbur, J. (2002). Factor structure and construct validity of the counselor skills personal development rating form. *Counselor Education and Supervision, 41*, 268–278.

Tryon, G. S. (1996). Supervisee development during the first practicum year. *Counselor Education and Supervision, 35*, 287–294.

Vinson, T. S., & Neimeyer, G. J. (2000). The relationship between racial identity development and multicultural counseling competency. *Journal of Multicultural Counseling and Development, 23*, 177–182.

Wang, Y. W., Davidson, M. M., Yakushko, O. F., Savoy, H. B., Tan, J. A., Bleier, J. K.(2003). The Scale of Ethnocultural Empathy: Development, validation, and reliability. *Journal of Counseling Psychology, 50*, 221–234.

White, M., & Epston, D. (1991). *Narrative means to therapeutic ends.* New York: Norton.

Wiener, N. (1948). Cybernetics. *Scientific American, 179*, 14–18.

6

Women's Development

By Dana L. Comstock, Ph.D.

REFLECTION QUESTIONS

1. *When you think of the "ideal woman" what characteristics come to mind?*
2. *What kinds of activities have you done in your community to combat sexual assault? What motivated you to undertake these efforts? If you have not engaged in any activities, how much thought do you generally give to this issue and why?*
3. *When you think of women's strengths, what kinds of things come to mind?*
4. *When you think of women's weaknesses, what kinds of things come to mind?*
5. *Think of the most important women in your life and consider how they have contributed to your growth. What have they taught you? What do you know about the sources of strength in their lives?*

INTRODUCTION

Never have I been so aware of the impact of "gender" as I was at the time of my daughter's birth. The meaning of those three little words, "It's a girl!" hit home with a myriad of emotions including everything from elation and joy to fear and trepidation. In the chaos of her arrival I also felt some relief due to a sense of familiarity. My relief however, was short-lived. As a mother, overwhelming questions emerged out of my sense of fear and a need to protect her. How will I encourage her to be authentic in a world where girls are supposed to hide so many parts of themselves? How will I teach her to feel good about herself and to resist the constant bombardment of images of the feminine cultural ideal? How will I teach her to celebrate her sexuality in a world that will objectify her and discourage her from having her own desires? How will I protect her from sexual violence and harassment at the hands of boys and men, the majority of whom will not be strangers to her? And how will I deal with my own sense of failure at not being able to single-handedly change any of this for her, or for myself?

It strikes me at this moment that most of the passages in my own life, and in the lives of many women, are wrought with fear; a fear of failure, to be specific, and my passage into motherhood was no exception. My fear of failure has to do with unattainable cultural feminine ideals and all the confusing messages about what and how we are supposed to be in the world. As a White, heterosexual, educated woman of middle-class socioeconomic standing I have a head start in that I share the same skin color with most of the images of the "ideal woman" I see portrayed in the media. In some ways this is an advantage, a "skin color" privilege, and in other ways the pressure is more intense as the expectations are higher and unattainable. I suppose this is a "paradoxical privilege" if there is such a thing, yet a privilege nonetheless, as are my lifestyle, relationship construction, education level, and physical health.

In the grand scheme of things my privilege has given me freedom from some forms of oppression in my life that other women, particularly women of color, have to deal with. With regard to women's development we need to understand the complexities of not only gender and race, but sexual orientation and class and the ways in which each contribute to the experiences of oppression and the ways in which they intersect with each other to shape a woman's respective gender identity (Abrams, 2003).

With regard to the experiences of sexism in particular, which is an essential context of women's development, "many women of color may understand gender oppression differently than white women, and may assign it a different priority in their lives" (Greene, 1994, p. 337). In fact, if we fail "to take into account the interaction of multiple, overlapping, and conflicting aspects of a woman's oppression, . . . we cannot appropriately understand or contextualize the range of dilemmas confronting women as an oppressed but diverse group" (Greene, 1994, p. 337).

An essential challenge for women, particularly in relation to the various contexts of development, has to do with the process of perceiving aspects of "strength based on their own life experiences, rather than believing they should have the qualities they attribute to men" (Miller, 1986, p. 36) or be force fit into theories of development based on men's experience. In fact, males have historically been at the center of our research efforts not in just mental health but in medicine as well. Men have been studied more often than women in the sense that men's experiences were thought to represent the "human experience." As a result, women were often force fit into various models of mental health, models of development, and paradigms related to physical health and biophysiology. When the fit wasn't "neat and clean," inferences were made that the differences were due to "deficiencies" of sorts. This is a noted trend that applies to *all* kinds of differences and is characteristic of patriarchal cultures "wherein differences are viewed hierarchically" (Robinson and Howard-Hamilton, 2000, p. 62).

A FEW "LITTLE KNOWN FACTS"
ABOUT WOMEN

A few years ago I was having a discussion with a supervision group when one of my students, who was doing a practicum in a school setting, began telling a story about a female teacher in her school. Apparently this particular teacher had a child who attended this school and to everyone's shock and dismay her child had unexpectedly died. Initially, we all shared our feelings around the horror of a child's sudden death, yet as our conversation continued it shifted to the details of how this teacher, and mother, seemed to be handling herself in the face of her grief. Apparently, instead of going home to retreat into her grief (which seemed to be an expected course of action by both my students and this teacher's colleagues), she had returned to the school the day after her child's death bringing food and treats for her child's classmates out of concern for them having lost a peer and a friend.

My students, bewildered by her behavior in the face of her grief, hypothesized that she must be in "denial," that she had "repressed" her feelings and clearly couldn't face the reality of her loss, that she was being "avoidant" by not facing her family and by not spending her time at home and that clearly she was not "taking care of herself" by tending to the needs of others. And because of all of this, she was surely in for a crash landing. Knowing what I am about to share with you meant I didn't so much as raise an eyebrow with regard to what was going on with this mother, and I certainly knew there wasn't anything "pathological" or "deficient" in the way she was responding to the enormity of her acute stress and grief. Once again, there was an expectation of "self-preservation" or "self-protection" evident of the "individualistic" nature of our culture.

Perhaps the most striking example of this self-preserving and self-protective philosophy is the "flight or fight" paradigm, which has been used to describe "human beings'" response to stress. This theory, which was first described by Walter Cannon in 1932, has basically gone unchallenged for the past five decades. I think it is fair to say that we all understand the basic premise of this theory: A basic human response in the face of danger (including both physical and emotional threats to our safety and well-being) is to size up the threat and to respond in a manner that assures survival by either fighting or fleeing. But as Taylor et al. (2000) remind us, "A little known fact about the flight or fight response is that the preponderance of research has been conducted on males, especially male rats" (p. 412) and that this tendency has gone unnoticed due to the fact the researchers have basically failed to report the sex of their participants and their rats in published reports (Taylor, 2002).

In fact, prior to the mid–1990s, women represented only 17 percent of human participants in stress response studies designed specifically to look at biophysical responses to stress, in spite of the fact that a plethora of research on some aspects of *behavioral* responses indicated there were significant gender differences (Taylor, 2002). One of the rationales given for women (and female rats) not having been included in biophysical stress response research was because of cyclical variations in neuroendocrine responses due to the reproductive cycle, which inevitably impact the data making it difficult to understand or interpret (Taylor et al., 2000). It wasn't until recently that Shelley Taylor and her colleagues at UCLA asked the question: "What if the equivocal nature of the female data is not due solely to neuroendocrine variation but also to the fact that the female stress response is not exclusively, nor even predominantly fight or flight?" (Taylor, 2000, p. 412).

In their research, Taylor et al. (2000) found that women have very different behavioral and physiological responses to stress that are best described as a tendency to "tend and befriend" versus to "fight or flight." The differences in stress response have been found to be due to the impact of the hormone oxytocin, which is secreted by both males and females, but in significantly different levels. For example, in women, high levels of oxytocin are produced during labor and childbirth; oxytocin is also produced by both women and men postorgasm. The effects of oxytocin are inhibited by androgen hormones (responsible for the development of aggressive behavior) in males and enhanced and regulated by estrogen, endogenous opiod mechanisms, and prolactin in females (Taylor et al., 2000; Taylor, 2002). I will spend some time discussing the implications of this research, in part because mental health research and paradigms have yet to mainstream this knowledge into actual practice and, more importantly, understanding it gives us a solid foundation to explore the context of women's lives, their development and relationships, as well as some insight into their needs with regard to the difficulties they face.

The impact of oxytocin on female behavior during times of stress prompts nurturing or "tending" behaviors (particularly to offspring, biological or otherwise) and tendencies to affiliate or "befriend" (seeking support from others, particularly from other women). In essence, these hormones "serve both to calm the female who is physiologically aroused by a stressor and also to promote affiliative behaviors including maternal behavior toward offspring" (Taylor, 2000, p. 416). The physiological impact of these hormones is complex and in human females oxytocin produces a sense of calm and well-being, an almost sedative effect indicative of the bliss new mothers feel after giving birth when oxytocin levels are at their highest (Taylor, 2002).

A fascinating example of this "tending behavior" in times of stress was found in Rena Repetti's (1997, as cited in Taylor, 2000, 2002) research. Repetti, a developmental and clinical psychologist at UCLA, designed a study to assess how well men and women (namely fathers and mothers) managed their family lives with the demands of work. During the course of her study, men and women were given questionnaires designed to rate the degree and type (i.e., was it "conflicted" or "busy") of stress they felt on a particular day as well as what kinds of things happened that evening at home. Children were given questionnaires designed to assess their parents' behavior without knowing what kind of day their parents reportedly had.

What Repetti found was that there was a correlation between the degree of stress their mothers' experienced (regardless of type) and the quality of their tending behaviors. In other words, on the mothers' worst days their children described them as more affectionate (they gave more hugs), were more loving, patient, and spent more time with their children. According to Taylor (2002) the "mothers don't seem to realize they've grown more affectionate, but hugging and loving their children seem to work well for mothers ready to shed a bad day" (p. 23). It is important to note that in a similar study Repetti and Wood (1997) found that mothers who are *chronically* stressed or distressed (i.e., poverty, verbal abuse, domestic violence) may show an increase in withdrawal versus tending behaviors and that these conditions are factors in the lives of many women. It is important to note that the women in her latest (2000) study experience *intermittent* work related stress and are relatively free of other *chronic distressing* experiences such as poverty and verbal and physical abuse characteristic of the lives of the women in the earlier (1997) study.

Withdrawal behaviors may be due in part to the impact of shame or humiliation, which serve to "alienate and silence individuals," particularly women who are made to feel responsible for their victimization (Hartling, Rosen, Walker, & Jordan, 2000). It is worth noting that women's responses to shame on a behavioral level involve "avoiding eye contact, withdrawing, and hiding" and on an emotional level women are vulnerable to extreme self-consciousness, they blame themselves, and experience a felt sense of unlovability (Hartling et al., 2000, p. 3). In addition, these emotional responses to shame are more likely to be experienced in dominant-subordinate relationships characteristic of heterosexual relationships.

As for the fathers, those who reportedly had stressful days basically withdrew from their families. On those days they reported as being highly conflicted the fathers were reported as being picky, impatient, critical, and crabby toward both their children and their wives (Repetti, 2000, as cited in Taylor, 2002). These findings are consistent with those of other researchers that indeed indicate that in response to shame and humiliation, typical of conflicted relational interactions, men are more likely to respond aggressively and, in the worst-case scenario, violently (Gilligan, 2001).

This isn't to suggest that human females are never physically aggressive and never fight; rather, incidences of female aggression are more related to self-defense and are often noted in response to "rape, assault, homicide, and abuse of offspring" (Taylor et al., 2000, p. 418) which are more likely to come from acquaintances or from their own partners. In general, human females are more likely to display indirect forms of aggression such as gossiping, spreading rumors, isolating female peers from cliques, and other behaviors referred to as "relational aggression" versus physical aggression (Simmons, 2002; Wiseman, 2002). Another "little known fact" is that female aggression of any sort is more prevalent in patriarchal cultures than in matrilineal or matrilocal cultures or family constructions in which cooperation and affiliation dominate the relational styles of women (Glazer, 1992).

THE CENTRALITY AND DEVALUATION OF RELATIONSHIPS IN WOMEN'S LIVES

Relationship, affiliation, and "befriending" are central to women's lives, so much so that women have been socially designated as the "keepers of connection" (Jordan, Banks, & Walker, 2003). Miller and Stiver (1997) make the point that:

> If we observe women's lives carefully, without attempting to force our observations into preexisting patterns, we discover that an inner sense of connection to others is the central organizing feature of women's development. By listening to the stories women tell about their lives and examining these stories seriously, we have found that, quite contrary to what one would expect based on the governing models of development emphasizing separation, women's sense of self and of worth is most often grounded in the ability to make and maintain relationships. (p. 16)

Yet in spite of what we have come to learn and know about the centrality of relationships in women's lives, women still pay a price for their relationality. Miller and Stiver (1997) make the point that bookstore shelves are lined with pop psychology books designed to help women who "love too much," or who are "addicted to relationships," or who have a "Cinderella complex" (p. 15). Inevitably the message to women is that they just don't seem to get it right and that they must move out of either their "dependency" or "codependency," which are Catch-22 phrases used to pathologize women's central role in our culture: Tending to and participating in activities that foster the psychological growth and development of others, "a form of activity that is essential to human life" (Miller & Stiver, 1997, p. 17).

Miller, Jordan, Kaplan, Stiver, and Surrey (1997) make the point that even though we have made some strides in understanding and valuing women's experiences and relational ways of being in the world "most of us have internalized a deficiency model of women" (p. 26) and that "women, ourselves, still have trouble claiming our own strengths and values. We can still find it hard to believe that how we tend to think or feel, or what we tend to want or like is valuable and important—as compared to some of the things we supposedly should be striving for which often don't feel as congenial" (Miller et al., 1997, p. 27).

An internalized model of deficiency as mentioned above is otherwise known as "internalized misogyny," which is prevalent in patriarchal cultures or in any context where a sexist ideology exists. According to hooks (2000), "Between men and women, sexism is often expressed in the form of male domination, which leads to discrimination, exploitation, or oppression. Between women, male supremacist values are expressed through suspicious, defensive, competitive behavior. . . . Sexism teaches women woman–hating, and both consciously and unconsciously we act out this hatred in our daily contact with one another" (p. 48).

Other scholars suggest internalized misogyny is at the core of women's self-destructive behavior such as chronic dieting, anorexia, bulimia, substance abuse, and other forms of self-mutilation including plastic surgery (Saavitne & Pearlman,

1993). It is important to note that women marginalized by race and sexual preference, for example, may also struggle with internalized racism or internalized homophobia while privileged women may have internalized domination central to their gender identity, all of which make relating across difference particularly challenging whether we are conscious of it or not. In spite of the challenges related to periodic divisiveness, women's relationships with other women are enriching and in them "women receive validation, liberation, empowerment, and a greater sense of their own worth and ability to be self-determining in spite of societal pressures" (Sanchez-Hucles, 2003, p. 120).

SOLIDARITY, DIVERSITY, AND CHALLENGES TO CONNECTION

The dynamic of adolescent female relational aggression has been a popular topic of discussion due to several new books, namely *Odd Girl Out* (Simmons, 2002), *Queenbees and Wannabees* (Wiseman, 2002) and *Girlfighting: Betrayal and Rejection Among Girls* (Brown, 2003). With regard to these publications, it is most important to remember that *there is a sociopolitical context for relational aggression* between women and girls and that a discourse on this issue is missing when this topic has been the focus of public discussion, particularly on news programs and popular talk shows. The result has been a propensity for women and girls to fear each other. Bell hooks (2000) argues that this dynamic serves a political purpose by undermining the notion of "Sisterhood" necessary for political solidarity among women. She goes on to make the point that "unless we can show that barriers separating women can be eliminated, that solidarity can exist, we cannot hope to change and transform society as a whole" (p. 44).

Perhaps the women who face the biggest challenges to solidarity are Palestinian and Israeli women. Yet despite their sociopolitical challenges, these women have forged a sense of solidarity like no other in the world. Every year, a feminist group called Bat Shalom organizes an event in northern Israel called Succat Hashalom. Succat Hashalom, or "peace tent," brings together Israeli and Palestinian women dedicated to promoting a peaceful coexistence between Arabs and Jews. At the time of the annual event in October of 2000, the political tension in Israel had peaked once again, yet in spite of the tension Bat Shalom decided to hold the event anyway. Terry Greenblat, director of Bat Shalom, reported to the editor of *Ms.* magazine that: "Right-winged Israelis tried to intimidate the women participants, calling us 'whores' and 'witches' and threatening to bomb the succah [tent] because Jews were meeting with Arabs" (*Ms.*, Feb./March, 2001, p. 30).

Despite the intimidating conditions, these women, mothers and activists, joined together to face the grim task of dismantling war "waged between their two peoples" (Benevento, 2001, p. 32). It was reported that as they sat together in unity, the tension, fear, and confusion turned to hope because for them, "the simple act of getting into a car, or onto a bus, is a powerful act of defiance. We, the women, they are saying, will not let you destroy our future. We will not let racism triumph" (Benevento, 2001, p. 32). Benevento (2001) also reported that on the last evening as the women were saying their good-byes and collecting posters for the highway protest the next day, she noticed one woman holding a sign, "We have tried war already" in one hand and a red rose in the other.

I think it takes unspeakable courage for these women to come together year after year despite political setbacks and threats to their lives. Interestingly, the editors of *Ms.* magazine remind us, "The mainstream media, both in Israel and in the U.S., have virtually ignored this seemingly unimportant 'women's work'" (p. 30). I can't help but wonder what the effects on politics would be if we were to focus on what women were willing to risk for peace versus what is sacrificed in war. In this case, the "women's work" was to instill a sense of hope where hopelessness seems to fuel acts of violence. I also can't help but wonder what type of a media hype would have ensued had this been a meeting for peace between Arab and Jewish Israeli men. Irrespective of my musings, I think this is an excellent example of how, in times of stress, women tend and befriend, overcome challenges to their solidarity and, once again, how these efforts go unnoticed. I have to ask: How do we identify women's strengths when they are so often made invisible?

SILENCING THE SELF: A PARADOXICAL DILEMMA AND MILESTONE

A sense of feeling devalued is a salient developmental theme in women's lives. Feelings of devaluation and invisibility are also a source of anger and distress. The expression of anger is a complicated issue for women, but in general anger in women is seen as unnatural, inconsistent with maternal behavior and caregiving, and is generally not well received (Miller & Surrey, 1997). In the spirit of relationship, anger would ideally "function in the service of moving a relationship toward a better connection" yet in spite of this, women often fear "that anger will create a disconnection or end a relationship" (Miller & Surrey, 1997, p. 202).

In order to negotiate potential conflicts in relationships and to preserve a felt sense of intimacy, women will paradoxically hide their true feelings, particularly anger, and engage in what is referred to as "silencing the self," a style of relating that begins in early adolescence (Witte & Sherman, 2002). Brown and Gilligan (1992) wrote of this phenomenon as a rite of passage from adolescence to womanhood. What they found was that as adolescent girls began to truly struggle with feminine sex-role expectations they "give up relationships for the sake of 'relationships'" (p. 216).

Brown and Gilligan (1992) describe a process of a "loss of voice" whereby girls tend to dissociate from other women, their mothers, and ultimately themselves in order to learn to be "with men." Rather than risk losing relationships by saying what they are feeling, they observed that:

> girls enacted this disconnection through various forms of dissociation: separating themselves from their psyches or from their bodies as to not know what

they are feeling, dissociating their voice from their feeling and thoughts so that others would not know what they were experiencing, taking themselves out of relationship so that they could better approximate what others want and desire, or look more like some ideal image of what a woman . . . should be. (p. 218)

They also make the analogy that this developmental milestone is similar to what boys go through in early childhood when they are socialized out of relationship with their mothers. They also observe that for both boys and girls these experiences are incredibly painful and wounding.

A sad truth is that one in four girls shows symptoms of depression (Deak & Barker, 2002). Other writers have long felt that depression is a normative experience for women that stems from a lack of "opportunity to participate more fully in relationships with both authenticity and a sense of empowerment" (Stiver and Miller, 1997, p. 218). This is precisely what happens with the notion of silencing the self and loss of voice. What can follow is a lifelong sense of loss that comes out of the efforts women make to preserve an *illusion* of a connection (Jordan, Kaplan, Miller, Stiver, and Surrey, 1991). Wooley (1994) makes the point that we really never lose what we know, and that "we know much more than we have put into words" and she warns, "silence must be recognized as a danger" (p. 319). Miller and Stiver (1997) make the point that most women experience a degree of sadness in their lives that leads to a type of affective flattening, a "nonfeeling" state. Over time this can lead to more severe forms of depression, thus the danger in silence.

As so much of the devaluation of women and the many forms this takes in our culture are insidious, our sadness in response to any of it is treated or seen as what is referred to as "nonevents" (Miller & Stiver, 1997). If women perceive their feelings are in response to either nonevents or to things they have "brought on themselves" then the consequence is that women see something inherently wrong with their experience or worse, with themselves as human beings. My sense is that in the larger culture women are thought to be "obsessed" with their appearance, weight, image, and the like, and that this is viewed as something that they do, in fact, "bring on themselves" and that "nobody *makes* them feel bad about themselves." It seems to me that if the effects of cultural mandates are insidious to women why would anyone, including men, notice them in the first place? In fact, Wiseman (2002) reported that adolescent boys are outspoken on the issue, with one 16-year-old stating: "Girls are messed up, especially over looks and weight. It always surprises me that the most popular girls would have the most self-image problems and a lot of them looked like they were 20" (p. 185).

A recent study suggests there is a correlation between the acceptance of sex role stereotypes and depression, particularly in White middle-class adolescent girls and women (Witte & Sherman, 2002). Other theorists speculate that because women of color are more removed from images of the "white feminine ideal," they have a lower risk of developing eating disorders or engaging in other self-destructive behaviors, particularly African American women (Ward, 1991; Molloy and Herzberger, 1998, as cited in Abrams, 2003). Part of their resilience has to do with the degree to which their respective families and culture support a wider range of emotional expression and, in essence, these girls are able to relate more authentically and even express that most unfeminine emotion, namely "anger."

Families of color and low socioeconomic status (SES) White families are reported to value the expression of anger in girls because it prompts them to "act quickly and forcefully to protect themselves" in the event they "are hurt or treated badly" (Brown & Jack, 2003). A dilemma exists for these girls if they express anger in other contexts, like their schools, "where they are judged against white, middle-class norms of femininity" (Brown & Jack, 2003, p. 128). These same girls must often choose between standards of behavior that are accepted (and expected) by their own families and communities and those values deemed necessary in order to be successful in academia. The fact that they experience a forced choice is cause for more anger and distress (Brown & Jack, 2003) in that they are not only required to give up "voice" as a developmental milestone, they are forced to give up aspects of their racial identity, which is their greatest source of resilience.

With the enormous amount of developmental challenges and vulnerabilities in the lives of girls and women it is no wonder that depression, anxiety, disordered eating, addiction, poor body image, confusion, and feelings of "I'm just not right" are normative aspects of women's experiences. Much of our sense of well-being has to do with how vulnerable we feel, how supported we feel in our vulnerability, and in what context. While there are a myriad of vulnerabilities any of us can experience in our lives, our capacity to cope has to do with our sense of support and personal power. I chose to use the language of "vulnerability" to address the context of women's difficulties that arise in a culture that values strength ("You've got to be strong"), autonomy, control (especially emotional control), and where we are taught the best defense is "self-defense." This particular paradigm goes against strength in connection, relationality, and supported vulnerability as a condition for relational growth and emotional resilience, which are central to women's emotional well-being.

My goal for the remainder of the chapter is make visible the insidious and the invisible, so that we can begin a process of critical thinking about the influences that shape *all* of our lives. To get us started I'd like to share this about the context in which girls and women are challenged to forge their lives and relationships:

1. The mass media and advertising have a tremendous impact on women's body image and self-worth, and perpetuate the destructive myth of the feminine ideal.

2. Because we are raised to consider the needs of others before our own, we have difficulty naming and acknowledging our desires, including our sexual desires. This leads to a complicated dilemma as we are then sexualized and objectified for the pleasure of others, namely men, and, as a result, we are sent a message of compulsory heterosexuality.

3. As a backdrop to our development, girls and women deal with an ongoing fear of sexual assault or sexual abuse, which is a sad reality for one in four girls. We also fear this for our daughters, sisters, mothers, partners, and other loved ones in our lives.

MASS MEDIA, ADVERTISING, AND THE
MYTH OF THE FEMININE IDEAL

If you have never considered the impact of advertising, consider this: "The average American is exposed to at least three thousand ads every day and will spend three years of his or her life watching television commercials" (Kilbourne, 1999, p. 58). Jean Kilbourne, one of the most noted researchers on the impact of advertising on women's self-worth, body image, and addictive and relational patterns makes the following point: "Advertising is our ENVIRONMENT. We swim in it as fish swim in water. We cannot escape it. Unless, of course, we keep our children home from school and blindfold them whenever they are outside of the house. And never let them play with other children. Even then, advertising's messages are inside our intimate relationships, our homes, our hearts, our heads" (Kilbourne, 1999, p. 58). She goes on to state:

> Today the promise is that we can change our lives instantly, effortlessly—by winning the lottery, selecting the right mutual fund, having a fashion makeover, losing weight, having tighter abs, buying the right car or soft drink. It is this belief that drives us to buy more stuff, to read fashion magazines that give us the same information over and over again. (Kilbourne, 1999, p. 68)

In essence, we strive to buy and attain what is simply a fantasy or an image. It is important to consider what the costs to our self-image are as we repeatedly fail to purchase, if you will, something that is simply unattainable. I believe that the endless striving for "attainment" becomes a lifestyle, a way of life that is filled with frustration and disappointment. And to find some relief, we are encouraged to buy more and more stuff, to eat this particular food, to smoke this cigarette, and perhaps the most costly to us, to drink this alcohol. I believe that the trick, or the seduction of advertising, is that we are lured into the idea that we can buy a degree of control or power in a culture in which many women simply feel invisible and inadequate.

It is important to emphasize that most of the images of the feminine ideal are almost exclusively of White women of European descent. If women of diverse backgrounds are presented you can usually expect to see these women share the European characteristics with their White counterparts. Lately, I've been seeing a TV ad for a medication to treat "IBS" (irritable bowel syndrome) that basically focuses exclusively on the bellies of white women (who are all wearing crop tops with low rider pants) with the same skin shades and body type. I keep noticing how I am drawn to the shape of the women's belly buttons (centered cleverly in the middle of the screen), that I take in very little about the drug until the end of the commercial (when you are actually shown the women's faces) and am repeatedly struck with the proportion of time that was spent on images of beautiful torsos versus anything that looked "sick" (or "irritable" for that matter). I am reminded that if we can successfully break women's bodies up into pieces we can use these "pieces" as sex objects to sell pharmaceuticals and the idea that if you take this drug your belly may even come to look like the ones pictured here!

Basow (2003) reports that the consequences of media exposure to unrealistic body images contributes to the fact that:

> the vast majority of women overestimate their body size, thinking they are at least 25 percent larger than reality. Two out of five women overestimate their body size by a larger margin, perceiving at least one body part to be at least 50 percent larger than it actually is. The more distorted the body image, the worse women feel about themselves. (p. 225)

In reality, the media images most visible in terms of the ideal body structure belong to less than 5 percent of women (Basow, 2003; Kilbourne, 1994, 1999).

These advertising images are basically tailored to make us feel inadequate and insecure, so that we will be compelled to buy their product to "fix" ourselves. Indeed a lot of money is to be made off of our poor body image by "publishers of women's magazines, cosmetic surgeons, and manufacturers of diet, exercise, and beauty products, 30 billion to be exact" (Basow, 2003, p. 226). In fact, many women resort to cosmetic surgery as the "answer." To emphasize this theme, consider the leads into one of The Learning Channel's feature programs called *A Personal Story*: "Liposuction . . . Dignity . . . Augmentation . . . Respect . . . Breast Enhancement . . . Self-Esteem. . . . " I think the message here is clear. In fact, most of these procedures also involve working to maintain that "youthful look" and now surgery is even available for "women who have given birth that will restore their 'youthful labia'" (Basow, 2003).

One might argue that women need not buy into any of this; indeed, we need to be consciously critical of the messages we take in. On the other hand, however, there are social and economic consequences for not meeting the socially mandated feminine ideal. Basow (2003) reports that women who are overweight make less money than their thinner counterparts and are perceived as "lazier, more undisciplined, less attractive, and less intelligent" (p. 224). I think most overweight women are aware of the fallout of weight discrimination and in case this applies to you, Kilbourne (1999) makes the point that women are sold ways in which to suppress "disappointment at being unappreciated" as well as "anger, resentment and hurt feelings" (p. 110). And surprisingly the best way to cope, we are sold, is through food: "He never called. So Ben [her dog] and I went out for a pint of Frusen Gladje. Ben's better looking anyway" (p. 110). Never mind the fact that our best source of support comes from our relationships and that an ad like this one perpetuates the idea that we should cope alone in the company of food.

Kilbourne (1999, 1994) also makes the point that advertising teaches us ways of relating while it simultaneously minimizes and sexualizes relationships. To illustrate this she describes an ad for frosting that begins:

> with an extreme close up of the peaks and swirls of frosting on a cake. A woman's voice passionately says, "Oh, my love." A man's voice says, "Huh?" and the woman replies, "Not you—the frosting!" With increasing excitement, she continues, "It's calling my name!" and the man replies, "Janet?" The woman cries out, "I'm yours!" as a male voiceover says, "Give in to the rich and creamy temptation of Betty Crocker frosting." As one of the peaks of the

frosting peaks, so to speak, and then droops, the woman says, in a voice rich with satisfaction, "That was great." (p. 113)

Kilbourne goes on to make the point that regarding "frosting" per se, this advertisement could arguably be harmless. Yet taking a deeper look it is evident that the relationship is minimized and that the preferred connection is with the product, not the other person. She goes on to add, "Imagine if this advertisement were for alcohol (Not you—the bourbon!). Perhaps we'd understand how sad and alienating it is" (Kilbourne, 1999, p. 113). What strikes me here so powerfully is the notion of addiction and how addiction, or any obsession for that matter, is detrimental to relationships.

The frosting ad is one example of how food is sold as "sexy" and as an object of desire, which, in this particular ad, was also satisfied. If you are asking what the problem may be with this, consider another one of Kilbourne's (1999) points particularly in reference to women: that "when food is sex, eating becomes a moral issue—and thinness becomes the equivalent of virginity. The "good girl" today is the thin girl, the one who keeps her appetite for food (and power, sex, and equality) under control" (p. 115). In essence, women are supposed to minimize their desires for the sort of things that are perceived as "self-indulgent," whereas life-giving and caretaking desires are acceptable, and assumed to be normal.

PARADOX OF DESIRE: SEXUALITY AND OBJECTIVITY

Sexualization of my daughter's behavior came early. My first memory of this happening was when she was 3 months old. An acquaintance had come over to visit, another mother, with a daughter close to Julianna's age. I was changing her diaper and in this particular moment our eyes were locked in a mutually animated connection full of smiles and giggles. The other mother commented: "Oh look, she's flirting . . . they start early don't they." I was shocked, speechless, and deeply disappointed. As Boyle, Marshall, and Robeson (2003) remind us, "Gender is not something inborn; rather it is something that exists only through our (re)creation of it on a daily basis" (p. 1326). They go on to urge us to consider the social constructionist nature of gender and that we (adults and children alike) create gender meanings as we interact socially with each other (Lorber, 2000, as cited in Boyle et al., 2003). In all fairness, my daughter did not have the opportunity to participate in any mutual meaning making, rather a particular meaning was "assigned" to her, literally, and preverbally. This was another one of those realities I was completely unprepared to deal with.

Writers on the topics of sexuality have made the point that "sexual arousal has been the most understudied topic in psychology" for decades (Buchwald, Fletcher, & Roth, 1993, p. 2). The subject of adolescent girls' sexual desire has been even more elusive. Interestingly, but not surprising, is the fact that "we have effectively desexualized girls' sexuality, substituting desire for relationship and emotional connection for sexual feelings in their bodies" (Tolman, 2002, p. 5). Girls are sent the message to "be sexy but don't be sexual, . . . don't be a prude but don't be a slut; have (or fake) orgasm to ensure that your boyfriend is not made to feel inadequate, if you want to keep him" (Tolman, 2002, p.7). These messages also include that of compulsory heterosexuality. Teenaged girls have long since learned to silence their desires for the sake of relationship, whatever the cost. "Teen magazines, movies and television contribute to the pervasive paradox: They offer advice on how to provide pleasure to boys juxtaposed with stories of sexual violation and harassment (Brumberg, 1997; Ussher, 1997; Carpenter, 1998, as cited in Tolman, 2002, p. 7).

In fact, few girls have realistic role models for exploring their "subjective" sexual desire ("objectively" this is certainly *not* the case). Madonna has been able to own her sexuality and act on her sexual desire as an exception because of her celebrity status as a pop-star icon. Yet, most young girls don't see her as a realistic role model and have unavoidably been witness to her demonization, which I believe has come, in part, due to her sexual subjectivity from both men and women. Indeed, women are "keepers of the code" and a great deal of relational aggression between girls is prompted when a girl is perceived as a slut because she has (or is thought to have) engaged in something which has damaged her reputation beyond repair (Simmons, 2002; Wiseman, 2002). Once again, being sexy is okay (even expected), but being sexual and acting on your own desire goes against the culturally sanctioned codes for femininity. In this sad vein, socially sanctioned sexual behavior becomes a source of divisiveness between girls and women, and mothers and daughters.

The mass media sends a barrage of messages to women of all ages that dictate everything from body image and sexuality to lifestyle and eating habits—and weaved within all these contexts is the dangerous and pervasive message that "you are never good enough, you can never do enough and above all else resist your sexual desires, but be sexy (for men) and if you start to feel bad about yourself there are lots of things you can do to comfort yourself like buying this product. . . ." All of these messages contribute to the unattainable "superwoman ideal" molded by the idea that we can "have it all and be all things to all people."

This myth is especially dangerous for new mothers because in our culture, the myth of the "bliss of motherhood," particularly "new motherhood," competes with the reality of life with an infant, the pain of nursing and the onset of postpartum depression, which strikes nearly 20 percent of all mothers. Additionally, a reported 80 percent of new mothers "get the blues" (Resnick, 2003, p. 262). Resnick (2003) names a sad reality: "The baby falls, and she's caught in the doctor's cool gloved hands. You fall, and nobody catches you, which seems a little more than unfair" (p. 262). To make matters worse, in our culture "women compete to see who's the best mother and who's back on the treadmill soonest. The wimps stay in bed. Real women kill themselves trying to appear perfect" (p. 265).

Before I had my daughter I had never noticed the "appearance" of new mothers until I began to prepare for my first outing to the mall with my daughter after she was born. On this particular trip I realized I had never had a reason to wonder about where all the nursing mothers went since there were none routinely visible in the mall. Suddenly, I began to feel the gazing eyes of other new mothers who would be sizing me up, to judge how I appeared to be handling the daunting new task of infant care. Shame ensued as I felt I could not compete with the image of the "new mother" who must obviously have a full-time nanny (and who must *not* be nursing) to help them achieve that fresh, wide-eyed, and blissful appearance.

On the other hand, I have an image of "Charlene," a character on the TV comedy show *Designing Women*. On one episode she had decided she would return to work when her new baby was only several months old. She went to an interview during which she fell asleep all gussied up in a red silk blouse. Finally, she wakes up horrified to discover her interviewer, a man, had been patiently waiting for her to emerge from her noisy slumber and to make matters worse, she had leaked breast milk through her blouse. Later in the episode, she complains that she is simply worthless and that she feels like a "big mom blob." Not having had any reason to identify with this scene some years back, I wonder why it stuck in my mind. Looking back, I think it was one of those honest, yet humorous images with which women really can identify that reflected a truth of the reality of women's lives. I think it also normalized the awkward demands of a woman's lactating breasts, a reality *rarely, if ever,* seen on prime-time TV.

I bet most of you have never given much thought to the "politics of nursing" and the idea that when breasts are used for the biological purpose for which they were meant (and not solely for the viewing pleasure of hungry consumers) people get offended. And indeed, women do "tend to view their bodies as objects, things to be evaluated and commented upon" making it difficult to feel comfortable in our own skin much less handling our breasts (Basow, 2003, p. 221). My sense is that nursing is offensive because women are so sexualized and objectified that many just can't get past the notion of "breast" as anything outside the realm of pornography. In reality, most of us see more breast on prime-time television than we are able to observe from a mother/infant nursing pair. Yet some find nursing pairs so offensive and distracting that nursing is banned from the grounds of most federal museums. I ask which is more distracting: a nursing pair or a screaming baby?

In a sense, you could say our culture is more obsessed with the female body than with women's interests (as deserving human beings), per se. Hutchinson (1994) warns that "obsession with the body is not the same as embodiment. We can be disembodied while being excruciatingly obsessed" (p. 155). This is precisely what happens to many women, a dilemma that is oddly paradoxical in that we are continually directed to focus on something from which we feel disembodied. In this experience:

> we are numb to our bodies, repressing dimensions of our experience that frighten us or make us vulnerable—pain, sexuality, hunger, anger, and even excitement and pleasure. . . . If, on the other hand, we were victims of physical violation or psychic violation, we may dissociate from them, writing them out of our experience of our selves. (Hutchinson, 1994, p. 155)

In essence, denial of desire and disembodiment can almost be described as developmental milestones, yet the shame of desire is something with which young girls are all too familiar.

In Lamb's (2001) book *The Secret Lives of Girls,* she spends some time talking about her research on girls' sexuality and the notion of "practice kissing." Lamb makes the point that most of her research participants reported that practice kissing between girls was "okay" until a degree of desire was noted in one of the partners. In her work, the shame associated with desire is noted in early childhood. Her adult subjects more often recalled the "other girl" as being "more into it" and mused over the difficulty adult women have at recalling or even naming their own sexual desire (Lamb, 2001, p. 34). Lamb (2001) also makes the sad point that often women equate their own sexual desire with sexual deviance. Tolman (2002) expounds on this by stating:

> Sexuality is so often thought of only in negative terms, so frequently clustered with problem behaviors such as smoking or drinking, in our minds as well as in research, that it is easy to forget that while we are not supposed to become smokers or drinkers in adolescence, we are supposed to develop a mature sense of ourselves as sexual beings by the time we have reached adulthood. Without a clear or sanctioned path, developing this sense is even harder for girls. (p. 4)

In this same book, Tolman writes of adolescent girls' first experiences with intercourse, noting that all too often the girls state that they were unclear of how it happened, it "just happened." Taking into account the denial of desire, the experiences of objectification, the disembodiment, and the notion that we are to be sexy for men, but not sexual, it is no wonder that women feel disempowered regarding their sexual needs and vulnerable to sexual assault. The notion that women aren't socialized to feel they have ownership of their own bodies is directly attributed to the sense of sexual access that others, namely men, feel entitled to.

FEAR OF SEXUAL ASSAULT

Sadly, "The fear of sexual assault that is part of the daily life of women in this country takes up a continent of psychic space. A rape culture is a culture of intimidation. It keeps women afraid of being attacked and so it keeps women confined in the range of their behavior. . . . It costs us heavily in lost initiative and in emotional energy stolen from other, more creative thoughts" (Buchwald, 1993, p. 188). In a new book, *Girls Will Be Girls*, the authors JoAnn Deak and Tereasa Barker (2002) write that in regard to sexual violence adolescent girls share their sense of helplessness with women and often stated "this is just the way it is for females" (p. 19). Deak and Barker elaborate: "Just as menstruation is a part of female life and opens you up to the risk for pregnancy, sexual assault, too is a distinct and ever present risk. Not a great situation, but an immutable way the world is set up" (p. 19).

Hill and Silver (1994) make the point that we are constantly bombarded with objectified images of women's bodies and that the mass media, and Hollywood in particular, have made sure there is no shortage of images of sexual violence. Overexposure to anything is sure to have some effect and in this case it is the desensitization of sexual violence against women. They make the point that:

The misogyny and degradation become invisible—not because they aren't present, but because we have blocked them out. The constant becomes commonplace, commonplace becomes normal, normal becomes natural; and what is seen as natural can then only be regarded as right. (p. 295)

This is another one of those insidious processes that not only normalizes sexual violence but also normalizes the expectation that a woman won't be supported (or believed) if she makes a charge of sexual assault. This reality is directly related to the underreporting of rape and sexual assault.

I believe it is difficult for women to speak openly about how they are affected by rape scenes in popular culture and it is equally difficult to quantify the damage done to women from repeated exposure to sexual violence across the life span. Buchwald (1993) believes we can see evidence of this in women's sense of helplessness, their lowered self-esteem, and in their "belief they cannot fight a predation that appears to be universal" (p.197) and indeed other writers agree (Deak and Barker, 2002). I see women tolerate these scenes in awkward silence. My style is to get up and walk out, to make a point about what is being shown, and to ask my husband quite frankly: "How can you just sit there and watch this without getting upset?" I think his apathy speaks to the damage that is done to all of us (including men, especially men) about what happens when we are repeatedly exposed to images of sexual violence.

In preparing for this section, I came across an essay I hadn't read in some time by Emile Buchwald (1993) entitled: "Raising Girls for the 21st Century." In this essay she writes of her experiences of sexual harassment, unwanted fondling, and sexual violence most women, including myself, seem to put away in a secret scrapbook that we either suppress or work to forget. One barrier to speaking about these incidents is illuminated in her essay:

I wouldn't have known how to describe these incidents to my parents without using words about parts of the body they never named or spoke of and that I sensed they didn't wish to speak of. I never mentioned these incidents to anyone because they had to do with sex, and sex in the 50's was a taboo subject, an ugly secretive activity that people were ashamed of. There was no public discussion of rape. No one said the word out loud. Each woman was alone with what had been done to her. (Buchwald, 1993, p. 185)

I don't believe that talking about sex, much less sexual violence, has become any easier for families today than in the 1950s. The reality is, if we can't talk about desire, then we can't talk about sex, which decreases the likelihood any of our daughters will share experiences of sexual violence with us. As a parent, I certainly want to be "in the loop" and speaking about it, as I speak about it now, will hopefully lay a path for my daughter to bring her fears and experiences into our relationship.

Hill and Silver (1993) make the point that the desensitization to sexual violence is more complicated for women: "Most women would rather numb themselves to block out this horrible input. To be aware of it daily may mean saying to themselves, 'I live in a world where I am degraded, systematically humiliated, and considered worthless-and where all of this is considered normal'" (p. 295). They go on to ask: "How many women are willing to allow themselves to become this aware? It might mean living every single day with unbearable pain" (Hill & Silver, 1993, p. 295). In fact, denial does give the illusion of strength and of an altered reality, one without rape. It is one of the many things we, as women, do to survive. We don't realize how much we hurt ourselves, or our daughters, by pretending it doesn't happen or minimizing it when it does. This goes for the impact of childhood sexual abuse, which is dealt with in detail in the chapter on trauma.

According to Walker (1999), "Although there is a reputable body of research on the long-term effects of childhood sexual abuse, there has been little study of the complex and psychological sequelae of dual traumatization that reinforces patterns of disconnection for African Americans and other survivors of color" which would help us to understand how both "sexual and racial violations create a dual traumatization, informed by both the individual-familial and the sociocultural history of the survivor" (p. 51). Bryant (2003) agrees and stresses that violence against women occurs in a social-cultural context that includes "race, class, ethnicity, sexual orientation, gender and disability" and that "sexual assault survivors who are members of racially and ethnically marginalized groups have to cope with multiple traumas simultaneously."

Bryant (2003) emphasizes that African Americans, Asian Americans, Native Americans, immigrant women, lesbians, and women who live in poverty all have different strengths and weaknesses related to their resilience following a sexual assault. Many of these issues are related to myths, stigmas, and stereotypes perpetrated by the media and by education and criminal justice systems. A sad reality for African American women is that the "Rapists are less likely to be convicted when the survivor is a Black woman" leaving African American women to deal with the trauma of violations related to both her assault and of "a society that sees them as unworthy of protection" (Bryant, 2003, p. 160). Thus the notions of devaluation, invisibility, and objectification take on particularly complex implications for marginalized women who are victims of sexual assault, many of which are perpetrated by stereotypes.

Among the stereotypes that women of color face, sexual stereotypes can be among the most frustrating and painful to deal with. In a recent study published in the *Psychology of Women Quarterly,* researcher Laura Abrams (2003) studied the impact of contextual variations on adolescent girls' gender identity. In her study she found that across race and class, adolescent girls shared the experience of "men's disrespect for women" in the form of sexual harassment, physical or sexual violence, rape, intimidation, abuse, and verbal harassment in their schools and neighborhoods (p. 68). Interestingly, race and class delineated the ways in which adolescent girls dealt with various stereotypes.

The African American and Latina participants reported the most frustration with sexual stereotyping that insinuated that "girls from urban communities are promiscuous, often stigmatized by labels such as 'hootchies' or 'hos'" (Abrams, 2003, p. 70). Sadly, most of the participants identified "other" girls as "hoochies" or "hos" but distanced themselves from these labels. These participants also reported experiencing stereotypes related to race and class such as "shopkeepers targeting them as thieves, or teachers holding misperceptions of them as low academic achievers" (p.70). Several of these participants blamed the media and the larger culture for perpetrating many of these stereotypes. By contrast the White middle-class participants reported they had no experience with any particular sexual stereotypes and that their primary concerns were with the issues of sexual harassment and sexual assault in general.

Thanks to many grassroots efforts around the country, women have places to go to talk and speak about their experiences of sexual violence. Yet our culture is still somewhat resistant to an open discussion on the subject of rape. This is evident in political agendas where there is more of a focus on women's reproductive rights than on the issue of stopping violence toward women—a subject that, in my opinion, deserves the same degree of passion.

CONCLUSION

In Jean Baker Miller's groundbreaking book *Towards a New Psychology of Women,* she reminds us that women in our culture are perceived as the "embodiment of weakness" (1986, p. 37). Yet, Miller stresses that because women "'know' weakness, women can cease being the 'carriers' of weakness and become the developers of a different understanding of it and of the appropriate paths out of it. Women, in undertaking their own journey, can illuminate the path for others" (p. 33). By understanding the lives of women, we can all come to better understand ourselves and work to dismantle our respective sources of oppression in order to participate more fully and authentically in growth-fostering relationships. These challenges will be met with more or a lesser sense of empowerment depending on our degree of privilege or marginalization in the larger culture. For those who face the bleakest of circumstances, remember the Israeli and Arab women of the Succat Hashalom and that incredible amounts of courage can be found in communities of resistance and in our connections with others.

Not too long ago my now 5-year-old daughter announced for the first time: "Girls are better than boys!" (in that taunting sing-song voice mastered at such a young age). When my husband and I heard this we looked at each other with some bewilderment as to where this had come from. I think on some level we were thankful she was feeling "good" about herself, yet at the same time I felt a ping of caution that she was stepping out-of-bounds. As her mother, I am keenly aware she should be socialized to be polite as well as modest (and that I am responsible for enforcing stereotypical feminine behavior as a cultural mandate). Plus, we certainly didn't want her walking around purposefully making little boys feel bad about themselves. We silently mused over our options for redirecting her. Should we tell her: "Now Julianna, that's not fair, boys are just as good as girls?" Another ping. Maybe we should leave well enough alone. We did.

We have heard this mantra many times since her initial "announcement." I am all too aware that the majority of her life experiences will teach her otherwise and that she will learn about gender privilege soon enough. In fact, every time we turn on Nickelodeon we are reminded that no matter how hard little Cindy tries, she will never be as smart as Jimmy Neutron. If you have ever wondered how and where we are taught about social order, spend a day watching cartoons with young children.

I am due to meet my second child, a son, in a matter of weeks. I have often wondered what I might do in the event he is to ever announce: "Boys are better than girls" in the same tone as his sister. Will I feel a ping? You bet I will. Will I wonder where this came from? No. Will I wonder what my options are? Yes, and I will have a hard time undoing this since his life lessons will reinforce this on a daily basis. Will I worry that such a declaration may make little girls feel bad about themselves, including his own sister? That goes without saying. Will I have a problem if his infant delights and squeals stemming from the sheer joy of connecting are framed as "flirting" as happened with my daughter? Absolutely. It is going to be an interesting journey.

REFERENCES

Abrams, L. S. (2003). Contextual variations in young women's gender identity negotiations. *Psychology of Women Quarterly*, vol. 27, pp. 64–74.

Aydt, H. & Corsaro, W. A. (2003). Differences in children's construction of gender across culture: An interpretive approach. *American Behavioral Scientist*, vol. 46(10), pp. 1306–1325.

Basow, S. A. (2003). Body image. In L. Slater, J. H. Daniel, & A. Banks (Eds.) *The complete guide to mental health for women*. Beacon Press: Boston.

Benevento, G. (February/March, 2001). Under the peace tent: Arab and Jewish Israeli women come together to build solidarity in their bitterly divided land. In *Ms.* vol. 11(2), pp. 30–32.

Boyle, E., Marshall, N., & Robeson, W. (2003). Gender at play: Fourth grade girls and boys on the playground. *American Behavioral Scientist*. Vol 46(10), pp. 1326–1345.

Brown, L. M. (2003). *Girlfighting: Betrayal and rejection among girls*. New York: New York University Press.

Brown, L. S. (2003). Women and trauma. In L. Slater, J. H. Daniels, & A. Banks (Eds.), *The complete guide to mental health for women*, pp. 134–143. Boston: Beacon Press.

Brown, L. M. & Gilligan, C. (1992). *Meeting at the crossroads: Women's psychology and girls' development*. New York: Ballantine Books.

Brown, L. M. & Jack, D. C. (2003). Anger: That most unfeminine emotion. In L. Slater, J. H. Daniel & A. Banks (Eds.), *The complete guide to mental health for women*. Boston: Beacon Press.

Brumberg, J. J. (1997). *The body project: An intimate history of American girls*. New York: Random House.

Bryant, T. (2003). Sexual abuse and rape: Thriving after sexual assault. In L. Slater, J. H. Daniels & A. E. Banks, (Eds.) *The complete guide to women's mental health*. Boston: Beacon Press.

Buchwald, E. (1993). Raising Girls for the 21st Century. In E. Buchwald, P. Fletcher, & M. Roth (Eds.) *Transforming a rape culture*. Minneapolis, MN: Milkweed Editions.

Cannon, W. B. (1932). *The wisdom of the body*. New York: Norton.

Carpenter, L. M. (1998). From girls to women: Scripts for sexuality and romance in *Seventeen* magazine, 1974–1994. *Journal of sex research*, 35(2), pp. 158–168.

Deak, J. & Barker, T (2002). *Girls will be girls: Raising confident and courageous daughters*. New York: Hyperion.

Gilligan, J. (2001). *Preventing violence*. New York: Thames and Hudson.

Glazer, I. M. (1992). Interfemale aggression and resource scarcity in a cross-cultural perspective. In K. Bjorkqvist & P. Niemela (Eds.), *Of mice and women: Aspects of female aggression* (pp. 163–172). San Diego, CA: Academic Press.

Greene, B. (1994). Diversity and difference: Race and feminist psychotherapy, in M. P. Mirkin (Ed.). *Women in context: Toward a feminist reconstruction of psychotherapy*. New York: Guilford Press.

Hartling, L. M., Rosen, W., Walker, M., Jordan, J. V. (2000). Shame and humiliation: From isolation to relational transformation. *Work in Progress, No. 88*. Wellesley, MA: Stone Center Working Paper Series.

Hill, S. & Silver, N. (1993). Civil rights antipornography legislation: Addressing the harm to women. In E. Buchwald, P. Fletcher & M. Roth (Eds.), *Transforming a rape culture*. Minneapolis, MN: Milkweed Editions.

hooks, b. (2000). *Feminist theory: From margin to center*, (2nd ed.) Cambridge, MA: South End Press Classics.

Hutchinson, M. G. (1994). Imagining ourselves whole: A feminist approach to treating body image disorders. In P. Fallon, M. A. Katzman & S. Wooley (Eds.), *Feminist perspectives on eating disorders*, pp. 152–168. New York: The Guilford Press.

Jordan, J. V., Banks, A. & Walker, M. (2003). A relational-cultural model of growth. In L. Slater, J. H. Daniels, & A. E. Banks, *The complete guide to mental health for women*, pp. 92–99. Boston: Beacon Press.

Kilbourne, J. (1994). Still killing us softly: Advertising and the obsession with thinness. In P. Fallon, M. A. Katzman, & S. C. Wooley (Eds.), *Feminist Perspectives on Eating Disorders*, pp. 395–418. New York: The Guilford Press.

Kilbourne, J. (1999). *Can't buy my love*. New York: Simon and Schuster.

Lamb, S. (2001). *The secret lives of girls*. New York: The Free Press.

Michael, A., Porche, M. V., Tolman, D. L., Spencer, R., & Rosen-Reynosa, M. (2003). To do it or not to do it is not the only question: Early adolescent girls' and boys' experiences with dating and sexuality. *Working paper No. 407*. Wellesley, MA: Stone Center for Research on Women.

Miller, J. B. (1986). *Towards a New Psychology of Women* (2nd ed.). Boston: Beacon Press.

Miller, J. B., Jordan, J. V., Kaplan, A. G., Stiver, I. P., Surrey, J. L. (1997). Some misconceptions and reconceptions of a relational approach. In J. V. Jordan (Ed.), *Women's growth in diversity: More writings from the Stone Center*. New York: The Guilford Press.

Miller, J. B. & Stiver, I. P. (1997). *The healing connection: How women form relationships in therapy and in life*. Boston: Beacon Press.

Miller, J. B. & Surrey, J. L. (1997). Rethinking women's anger: The personal and the global. In J. V. Jordan (Ed.), *Women's growth in diversity: More writings from the Stone Center*, (pp. 199–216). New York: The Guilford Press.

Repetti, R. L. & Wood, J. (1997). Effects of daily stress on mothers' and fathers' interactions with preschoolers. *Journal of Family Psychology, 11*, 90–108.

Repetti, R. L. (2000). The differential impact of chronic job stress on mother's and father's behavior with children. Manuscript in preparation.

Resnick, S. K. (2003). Postpartum depression. In L. Slater, J. H. Daniels & A. E. Banks, *The complete guide to mental health for women*, pp. 260–265. Boston: Beacon Press.

Robinson, T., & Howard-Hamilton, M. F. (2000). The convergence of race, ethnicity and gender: *Multiple identities in counseling*. Upper Saddle River, NJ: Merrill.

Saavitne, K. W. & Pearlman, L. A. (1993). The impact of internalized misogyny and violence against women on feminine identity. In E. P. Cook (Ed.) *Women, power and relationships:Implications for counseling*. Alexandria, VA: American Counseling Association.

Sanchez-Hucles, J. (2003). Intimate relationships. In L. Slater, J. H. Daniels, & A. Banks, *The complete guide to mental health for women*. Boston: Beacon Press.

Simmons, R. (2002). *Odd girl out: The hidden culture of aggression in girls.* New York: Harcourt.

Stiver, I. P. & Miller, J. B. (1997). From depression to sadness in women's psychotherapy. In J. V. Jordan (Ed.) *Women's growth in diversity: More writings from the Stone Center.* New York: The Guilford Press.

Taylor, S. E. (2002). *The tending instinct: How nurturing is essential to who we are and how we live.* New York: Times Books.

Taylor, S. E., Klein, L. C., Lewis, B. P., Gruenwald, T. L., Gurung, R. A. R., & Updegraff, J. A. (2000). Biobehavior responses to stress in females: Tend-and-befriend, not fight-or-flight. *Psychological Review, 107*(3), 411–429.l.

Thompson, B. (1994). Food, bodies, and growing up female: Childhood lessons about culture, race and class. In P. Fallon, M. A. Katzman, & S. C. Wooley (Eds.) *Feminist perspectives on eating disorders,* pp. 355–380. New York: The Guilford Press.

Tolman, D. (2002). *Dilemmas of desire.* Cambridge, MA: Harvard University Press.

Ussher, J. M. (1997). *Fantasies of femininity: reframing the boundaries of sex.* New Brunswick, N.J.: Rutgers University Press.

Ward, J. V. (1991). Racial identity formation and transformation. In C. Gilligan, N. Lyons, & T. Hammer (Eds.), *Making connections: The relational worlds of adolescent girls at Emma Willard High School,* pp. 215–232. Cambridge, MA: Harvard University Press.

Walker, M. (1999). Dual traumatization: A sociocultural perspective. In Y. Jenkins (Ed.) *Diversity in college settings: Perspectives for helping professionals.* New York: Routledge.

Wiseman, R. (2002). *Queen bees and wannabees: Helping your daughter survive cliques, gossip, boyfriends and other realities of adolescence.* New York: Crown Publishers.

Witte, T. H., & Sherman, M. F. (2002). *Silencing the self and feminist identity development. Psychological Reports, 90,* 1075–1080.

Wooley, S. (1994). The female therapist as outlaw. In P. Fallon, M. A. Katzman, & S. Wooley (Eds.), *Feminist perspectives on eating disorders,* pp. 318–338. New York: The Guilford Press.

7

Male Development and the Journey Toward Disconnection

By David Shepard, Ph.D.

REFLECTION QUESTIONS

1. *If you are male, how comfortable are you talking about your vulnerable feelings (e.g., "I am sad," "I'm afraid," "I'm feeling ashamed") with your mother? Your father? Your male friends? Your female friends? Your partner?*
2. *If you are female, imagine a male friend telling you, "I just feel scared so much of the time." How would your impressions of your friend's masculinity be affected?*
3. *Do you recall experiences of being shamed by peers during adolescence? How did these experiences affect you?*
4. *Do you believe males are naturally more violent than women? Or do you believe that men's use of violence is a learned behavior?*
5. *Do you believe men are impaired in their abilities to be intimate? Or that they define intimacy differently than women do?*
6. *What are some of the costs men have paid for learning how to be masculine in our culture? If you are a male, what costs have you paid personally?*

INTRODUCTION

The film *Field of Dreams* (Robinson, 1989) tells the story of an Iowa farmer (Kevin Costner) who is commanded by mysterious voices to build a baseball park on his cornfield. Costner obeys, and in the final scene discovers the purpose of his efforts when his long-dead father—once a baseball player—appears on the field. Costner, who had been painfully estranged from his father since adolescence, has now been given a magical opportunity to heal both of their wounds by speaking the words that needed to be said.

Both men look at each other, exchange a few superficial words, and finally, Costner says, "Dad? Do you want to have a catch?" And so they do, and everyone in the audience understands that this was the way these two men could say, "I love you." But why didn't they just say it?

I am deeply moved by this film, and the day after seeing it, feel a need to speak to my own father. We talk by phone; coincidentally, he had just seen the film as well, and was as moved as I was, but of course from different perspectives. The film touched in me feelings about being a son; he experienced feelings about being a father.

My dad and I talked about how much we liked the film, and how great the direction and acting were. And when that conversation had nowhere else to go, we both felt this long awkward pause, both of us sensing that this was the moment when, like the characters in the film, we were supposed to say to each other, "I love you."

> **My father:** "You probably want to speak to your mother, too. I'll get her. Good talking to you, Son."
>
> **Me:** "Good talking to you too, Dad."
>
> With that, he gave the phone to my mother.
>
> Just like in the movie, father and son, two generations, unable to say, "I love you."

The question is, why was it so hard? I am a professional counselor as well as a counselor educator; I, of all people, supposedly a specialist in being authentic and in touch with feelings, should have been able to say these simple words. Whatever blocked me feels so ingrained, so *old*, that when I search inside myself for the fear I experience saying out loud the most powerful words humans have for expressing a feeling of intimate connection, I can only touch something dim at the edge of conscious awareness.

As blocked as I was from saying "I love you" to my father, I am equally blocked when it comes to tears. Life presents me with numerous situations in which I want to cry; but when the tears well up, there is a stronger force shutting them down. When I explore with my male clients their inability to cry when they are feeling sadness, what I most frequently am told is, "I have a sensation that if I do cry, it will be like a dam bursting and I won't be able to stop." I recall saying the same thing to my own therapist. The fact is, we would have been able to stop, but we have stored up such a backlog of unreleased tears that we fear they would all come flooding out at once and we would lose control of ourselves.

We men pay a price for our lost opportunities, the not saying "I love you" when we yearn to connect, the not letting go with our tears when our souls are wounded. We remain in a perpetual state of yearning—yearning to feel a full and satisfying connection with those whom we cherish, yearning to feel the liberation of tears freely flowing. Too many of us are bound by constricting rules of masculine behavior that restrict our capacity to express and fulfill our needs for intimate connection.

Beginning in the 1980s, a group of theorists and researchers began exploring the processes and pressures in male development that lead to this constriction of men's range of feelings and behaviors, aspects of male psychology that impact not only a man's well-being but his partner's and family's as well. Led by such authors as Joseph Pleck, Gary Brooks, Ronald Levant, and William Pollack, the pace of research in men's studies and male development gained momentum in the 1990s and continues today unabated. These and numerous other authors are continuing to explore and elucidate male development from intrapsychic, constructivist, and feminist perspectives, inspired by the perception that there now exists a "masculinity crisis in which many feel bewildered and confused, and the pride associated with being a man is lower than at any time in the recent past" (Levant, 1996, p. 259).

This confusion stems from the contradictory messages men receive as children and adults on what constitutes positive masculine behaviors. On the one hand, our culture still supports traditional definitions of the male gender role, in which a man's self-esteem is based on being the sole breadwinner, on achievement, on having power over others, on stoicism, and on outward displays of strength no matter what difficulties life brings him. At the same time, men face increasing pressures to commit to relationships, to share vulnerable feelings and experience deeper levels of intimacy, to nurture children, and to curb aggressiveness and violence. Men are thus increasingly challenged to sort out for themselves, amid these pressures and messages, personal definitions of masculinity.

The underlying assumption in this research and in this chapter is that men must rise to this challenge, because the old rules defining masculinity have created profound disconnections for men, which involve: (a) disconnection from vulnerable feelings like sadness and fear, which are normal and appropriate parts of life; (b) disconnection from nurturing, soothing, and caregiving capacities; (c) disconnection from the vocabulary of emotions, which many men have never adequately learned; (d) disconnection from one's children, despite desires for close relationships; (e) disconnection from capacities for intimacy, and concomitantly, disconnection from those whom men love.

Bergman (1991) sums it up by stating these disconnections take men out of the "process of connection . . . the very process of growth in relationship" and in this process there is "a learning about turning away from the whole relational mode" (p. 4). The goal of this chapter is to review current thinking on the relationship between male development and these relational disconnections; and, in personal terms, to answer the question, what made it so hard to say, "I love you," to my father, or to cry when my heart is aching?

By tracing the arc of male development from early infancy through early adolescence, I will be operating from two theoretical perspectives: the first, psychoanalytic object relations thinking, which is informed by research on early childhood development; and, second, gender role socialization theory, which attempts to explain "the processes by which children and adults acquire and internalize values, attitudes and behaviors associated with masculinity, femininity, or both" (O'Neil, 1981, p. 203). The emphasis on these two frameworks reflects the theoretical underpinnings of much of the current thinking on male development. Research in early childhood male development, at least as it pertains to relational issues, is strongly rooted in psychoanalytic thinking, perhaps because of this framework's long-standing focus on the connection between the infant-caregiver matrix in the first three years of life and adult relational capacities. Adolescent and adult developmental research tends to focus on the role of socialization and the social construction of male role definitions.

Thus, in this chapter I will describe how child-rearing practices in the first three years of a boy's life set the stage for compromised capacities for intimacy and relatedness; and then show how subsequent boyhood and adolescent socialization experiences regarding appropriate male behavior codify and reinforce the unconscious belief for boys and men that intimacy is a fundamental peril. I will also discuss the developmental transitions of marriage (or committed union) and fatherhood as milestones for men to experience new levels of connection, and review the evidence on how men are succeeding at meeting this challenge. Finally, I will conclude with a discussion of the more profound results of these disconnections from self and others: depression, alcohol abuse, and at its extreme, violent behavior.

EARLY INFANCY THROUGH EARLY ADOLESCENCE

Margaret Mahler, one of the most influential developmental theorists in our country, detailed a 3-year process by which infants graduate from a symbiotic relationship with their mother to a state of separation and individuation. Based on 10 years of observing 38 children interact with their mothers, Mahler, Pine, and Bergman (1975) described a sequence in which infants gradually develop a sense of individuality after undergoing months of oscillation between clinging to mother and practicing separation. The goal of the mother, according to these authors, is to encourage this separation without making the child feel rejected, which means creating a secure bond with her child while the child tests his or her ability to safely explore the world on his or her own. This state of closeness and dependency is essential for the child to develop the requisite inner sense of security and sense of self as an autonomous individual (Mahler, Pine, & Bergman, 1975).

This developmental story, with its emphasis on separation and individuation as the capstones of healthy development, is only one lens through which to comprehend the process of human psychological growth. For example, Relational-Cultural theory, an alternative model for understanding female and male development, emphasizes that people grow through and toward connection, rather than toward separateness. Relational-Cultural theory helps us to see that the assumption that the mature self is an autonomous, individuated self, a premise underlying Mahler's and other psychoanalytic theories, reflects a particularly Eurocentric bias about human nature (Jordan, 2001).

However, Mahler's theories are historically important in understanding current thinking about male development because of recent suggestions that the separation/individuation sequence has strong negative consequences for boys (Dooley & Fedele, 1999). Presumably, according to Mahler et al.'s model of development, if parenting is good enough, both sexes begin life strongly bonded to mother, enjoying the safety and warmth of their dependency and feelings of empathic connectedness. Theorist Nancy Chodorow (1978), in a work of prodigious importance to theories of gender-specific psychological development, critiqued the assumption that girls and boys share similar experiences of bonding as well as similar processes of separation/individuation. She suggested that mothers (or the infant's primary caregiver) are pressured, by both unconscious and social forces, to push their sons out of this state of relational bonding too soon. With daughters, Chodorow argued, mothers feel comfortable keeping their little girls attached to them, perhaps because there is little social pressure to do otherwise. However, mothers or primary caregivers somehow have received the message that it is not good parenting to permit their sons to remain in this same state of attachment and dependency. The pressure is to facilitate boys' learning as soon as possible that they are different from girls and that they need to begin their journey toward becoming autonomous males. Chodorow suggested that the mother's awareness of being the opposite sex of her son may facilitate this disconnection.

Fathers play a role here, as they also encourage their sons to separate and develop identities as little men. Finally, little boys themselves, sensing the need to be different from their mothers, willingly embark on their journey of disconnection, surrendering, as Brooks and Gilbert have described it, "the emotional and sensual pleasures of attachment to their mothers" (Brooks & Gilbert, 1995, p. 253). Relational-Cultural theorists Dooley and Fedele (1999) have noted that boys are still in the concrete stage of cognitive development (alluding to Piaget's theory of cognitive development) when this separation occurs. By age 5, these authors observed, boys have experienced a concrete sense of a lost relationship with their mothers; they are emotionally on their own (Dooley & Fedele, 1999).

This hasty separation from relational bonding for boys is a critical piece in understanding how boys' development can lead to impaired capacities for connection and intimacy throughout life. Researchers in male development, and in particular, William Pollack, have argued that this loss of closeness for boys is in fact a traumatic experience of abandonment (Pollack, 1995, 1998a). The loss is profound because it occurs before little boys are emotionally prepared for such an intense degree of separation.

Pollack believes that there may be no biological or evolutionary need for little boys to be disconnected from the warm, bonded state with their mothers that little girls are permitted to enjoy throughout childhood. Older models of understanding the individuation process had suggested that the separation from mother was necessary so that boys would start looking to their fathers as models for their masculine identity (Greenson, 1968). The notion here would be that boys need to learn the male gender role, and that staying too close to mother prohibits that process. In essence, what Pollack and other developmental theorists are asking is,

Why so fast? What is the hurry for little boys to become little men? And, most importantly, what cost do boys pay for this early separation?

One price boys may pay is that later in life, as men, they may fear intimate connection. The two explanations for this phenomenon are based on the psycho-analytic notion that both childhood traumas and the yearnings of unmet needs can become repressed in the unconscious. Relational-Cultural theorist Stephen Bergman (1991) refers to this as "male relational dread" and notes that this is par-ticularly intense for heterosexual men in intimate relationships with women (p. 8). Pollack speculated that when men experience the pull toward intimacy in adult relationships, the unconscious memory of the buried pain acts as a danger signal, warning men to step away from connection. "Having experienced a sense of hurt in the real connection to their mothers," Pollack noted, "many boys, and later men, are left at risk for empathic disruptions in their affiliative connections, doomed to search endlessly . . . and yet fend women off because of their fear of retraumatization" (Pollack, 1995, p. 41).

Boys' yearnings for closeness and attachment to mother, as well as the recollec-tion of the painful loss, also become part of the unconscious. However, because boys are praised by one or both parents for their emerging independence, their toddler self-esteem becomes linked with this sense of being self-reliant. Naturally, if parents tell their little son, "What a big boy you are," when their son demon-strates independence from his mother, he will feel good about himself. His face will beam with pride. Erikson (1963) described this stage of growth when he wrote that the role of parents was to encourage the child's attempts to do things by himself while protecting him from the shame of failure as he asserts his autonomy. Hence, Erikson's terming this stage as "autonomy vs. shame and doubt" (p. 252).

However, an internal conflict is being set up; one part of the boy still desires a close, dependent, and gratifying relationship with mother, while another part is learning that big boys do not need mommy. And, of course, little boys do need their mommies (or primary caregiver), and will often have to run to them in tears to get the special kind of soothing, nurturing relief that mothers can give. Some-times, the behaviors are met with disapproval, shame, even punishment, not only from mother, but from fathers and other boys. Others may learn how our culture values independence and discourages neediness through more subtle learning processes, perhaps through caregivers' failure to correctly empathize with their sons' emotional needs for closeness and dependency. In either case, boys can learn to disavow their basic psychological need to depend on another person for love, support, and nurturance (Blazina & Watkins, 2000; Levant, 1995; Rabinowitz & Cochran, 2002).

As adults, opportunities for intimate relationship trigger this internal conflict left over from these early years. In other words, the adult male unconsciously would say to himself, "If I allow myself to get as close to a woman as I deeply wish to be, or depend upon a woman as strongly as I yearn to do, I will lose my identity as a man, which is built on a foundation of autonomy and self-reliance." True inti-macy is therefore difficult for a man because it requires him from time to time to let go of his adult inner strength, feel his vulnerabilities, and allow his partner to support him.

A second cost for the premature separation from mother is that boys may lose an invaluable learning opportunity to develop the skill of empathy, so crucial to sustaining effective relationships. As girls journey through childhood, their continued close empathic connection with mother virtually becomes an education in how to be with a person. Because boys are pressured to turn away from mother (and from the process of being in a relationship with her), they fail to absorb fundamental lessons in attending and responding to others' feeling states. According to Stephen Bergman, psychiatrist and contributing author to Relational-Cultural theory, "Not knowing how to do it, it soon becomes avoided, even more, devalued, and even its existence as a possibility denied" (Bergman, 1995, p.74). On the other hand, girls emerge from childhood "with a basis for 'empathy' in their primary definition of self in a way that boys do not. Girls emerge with a stronger basis for experiencing another's needs or feelings as one's own" (Chodorow, 1978, p. 167).

The painful separation from mother and the resulting internal conflicts can leave some boys, and adult men, with a fragile sense of self, frequently in need of bolstering by achievement, proof of masculinity, and support from intimate partners. According to Blazina and Watkins (2000), the internal conflict between needing others versus being solely self-sufficient takes a toll on the psyche by putting it under constant stress; the self "fragments," which is the psychoanalytic term for a sense of self that is vulnerable to depression and requires extensive support from others. In the context of intimate relationships, the chronic need for ego support can be experienced by partners as exhausting, leaving the partner both emotionally drained and frustrated with the relationship.

The profound loss Pollack described is also one that boys do not effectively grieve, primarily because they are discouraged by parents, peers, and other influences from expressing sad feelings, which are not compatible with masculinity. Rabinowitz and Cochran (2002) noted that boys actually tend to receive mixed messages, at times affirmation of their emotions by parents, teachers, and other children—and at other times shame or ridicule. Boys learn to psychologically defend themselves from this confusion by learning to truncate the natural mourning processes that are activated whenever intimate attachments are broken, and this lesson endures throughout life. Humans are supposed to feel loss when a strong connection with others is severed, whether by death, rejection, abandonment, separation, or interpersonal conflict. As counselors, we know that disavowed feelings of loss underlie a number of emotional disorders and relationship difficulties; helping people experience loss is one of the ways counselors facilitate healing. For men, lessons about denying loss and defending the pain through rationalization and avoidance of feelings may first be learned in this crucible of the separation experience with mother.

Bergman (1995) has put forward a compelling analysis of how men's disconnection from the mother relationship in childhood eventually leads to the adult male's drive to achieve. The developmental sequence proceeds as follows: When a little boy is pressured to turn away from his connection to his mother, he psychologically deals with this loss by asserting his "difference," his maleness. This awareness of difference leads to a tendency to compare himself with other boys, which inevitably raises the question, "Am I better or worse than the other boys?" At the same time, cultural messages tell the boy that it is important to stand out, to be superior, to exert power over others, to be special. Gradually, the boy's self-esteem becomes dependent on how he performs in competition with other boys; he needs to constantly strive to do better, to be highly competent, in order to maintain the feeling of specialness. Bergman (1995) noted that the drive toward competence in school, sports, etc., is at the expense of developing competence in relationships. With all the "doing" that boys are engaged in, there is little time for practice in being and relating. When the boy becomes a man, the striving for competence and achievement must be sustained in order to maintain his self-worth.

Bergman (1995) suggested that one negative outcome of the need to be special is that specialness and connectedness are incompatible; when a man feels special, he is feeling "better than" someone else, which means emphasizing his *individuality* rather than his *relationality* to others. I would suggest that another consequence may be that boys and men inevitably find themselves self-monitoring the state of their specialness. They can become preoccupied with comparisons of self versus others. The resulting internal monologue would run somewhat like this: "Am I special right now? I look around and see others admiring me; I've outrun three other kids in my class; the boss likes my plan better than the ones my colleagues presented. Hey, I'm feeling pretty good right now! True, to be special, I've had to be better than someone else, and my relationship with her or him might suffer. But I have succeeded." This self-monitoring process can become a habitual style of thinking, in which males engage in a continual process of checking out whether they are special (feeling OK about themselves), or the opposite (ordinary, incompetent, inferior).

Feeling better than someone else is a precarious position to maintain. When I empathize with male clients about the grind of trying to feel special, they sometimes tell me that they spend so much conscious effort to feel competent and special that they have lost confidence in their core self-worth. Instead, they feel deep down that they are "fakes" and "frauds" as people—that they do not deserve the respect and recognition they have received in life. This dilemma is one more impediment to intimacy; it can be terrifying to some men for a woman to see their inner world and deeper feelings of vulnerability. When a female partner attempts to get close by exploring with him his innermost feelings, he fears that what she will see is a weak, incompetent, fearful human being.

In this discussion, I have described the various ways in which the traumatic abandonment from mother that little boys experience may lead to impaired relational capacities in adult life. For counselors, the implication of this line of thinking is that facilitating men's connection to their repressed sadness, allowing them to grieve this critical abandonment, and acknowledging their yearnings for connection are important steps in helping men reconnect to their relational abilities. Whether the counselor makes explicit via interpretations the relationship of current relational issues to childhood development is, of course, dependent on theoretical orientation. What is important is that counselors appreciate that sadness and yearnings for closeness can be a lifelong experience for many men, and a source of the rigidity of their defenses against vulnerable feelings.

It is also important to note that the theories of Chodorow, Pollack, and others are just that—theories. They are rich in explanatory power, to be sure, but to date there is little empirical research to support or disconfirm them. In the next section, I will look at the processes by which socialization experiences also contribute to men's disconnections from self and others. If one accepts the object relations premises enumerated above, socialization can be perceived as a reinforcement of lessons learned in early childhood, part of the continuum of experiences that lead to men's difficulties with connection during adulthood. Alternately, socialization may be perceived as the primary explanation for these impaired relational capacities.

DISCONNECTION FROM THE SELF
AND THE BOY CODE

The speed at which boys begin to adopt behaviors of stereotypical masculinity during childhood development appears astonishing to many parents. A recent session I had with the parents of a 4-year-old went something like this:

She (to her husband): "Honey, I do not want to give Taylor a gun for Christmas. I know he wants it, but he's only 4 years old! I don't want him to grow up with rigid stereotypes of about what a boy is supposed to be."

He: "It's too late. Haven't you noticed that he's already making guns out of whatever cardboard he can find in the house? He's going to play with guns, anyway. We can't stop it. We have to let him be a normal boy."

She (sighing): "You're right, but it's so sad."

Gender role socialization for both sexes may take effect as early as age 2 (Ruble & Stangor, 1986). A number of factors have been described in the research on how the process of socialization occurs:

1. Boys-playing-with-boys, in school, playgrounds, locker rooms, bedrooms, and backyards, is a powerful forging experience where the rules defining masculinity are adopted and enforced (Maccoby, 1990).
2. Mothers and fathers reinforce gender role congruent behaviors, through the behaviors they respond to positively, the games they encourage, and the toys they buy (Meyer, Murphy, Cascardi, & Birns, 1991).
3. While fathers play an especially important role in encouraging their sons to behave according to gender stereotypes, both parents discourage sons from learning to express vulnerable emotions like sadness and fear (Levant, 1996).
4. Media, especially television, help form and reinforce gender role stereotypes.

5. Social learning theory proposes that boys learn through modeling, observing, and imitating the behaviors of their peers and their fathers (Bandura, 1977).
6. Cognitive psychologists have described the mental mechanisms through which gender identity formation occurs, suggesting that children remember information consistent with gender stereotyping more easily than counter-stereotypical messages (Liben & Signorella, 1993; Ruble & Stangor, 1986).

William Pollack studied hundreds of young and adolescent boys in a Harvard Medical School research project entitled "Listening to Boys' Voices." Based on his extensive observations of and interviews with boys and their parents, he termed the set of rules that define masculine-appropriate behavior for boys, the "Boy Code." According to Pollack, there are four imperatives at the heart of the Boy Code (Pollack, 1998b). These four rules were actually first articulated in 1976 by David and Brannon, as a description of the norms that define traditional adult masculinity in Western culture (David & Brannon, 1976). Pollack adapted these rules to apply to boys' behavior as well as to adult men's:

1. The Sturdy Oak. This is the injunction that boys (and men) should be stoic, to not show weakness by sharing pain or openly grieving.
2. Give'em Hell refers to the rule that boys and men should maintain a false self of extreme daring, bravado, and attraction to violence.
3. The Big Wheel emphasizes that masculinity means to achieve status, dominance, and power over others at any cost.
4. No Sissy Stuff is the injunction Pollack considers the most traumatizing "gender straightjacket." This is the rule that boys and men do not express tender feelings such as dependence, warmth, and empathy. It also serves to alienate effeminate boys by reinforcing their persecution through social injunction.

In a study written over 40 years ago, Hartley (1959) reported that boys learn to conform to the male code of behavior through repeated reprimand, embarrassment, and humiliation when they deviate from masculine role norms. The stories boys told to Pollack in his recent study confirm that shame is still the operative mechanism enforcing the Boy Code. Shame is a feeling of deep worthlessness and helplessness, "a piercing awareness of ourselves as fundamentally deficient in some vital way as a human being" (Kaufman, 1985, p.8). When a boy (or girl) feels shame, he experiences negative thoughts putting him in the worst possible light: "I am unacceptable, weak, a loser, inferior, etc." Self-consciousness increases. Rage may well up as a protective attempt to repel those who might shame him again. "Boy culture," wrote Stephen Krugman, a specialist in the relationship between male development and psychological trauma, "is competitive, insensitive, and often cruel. Being chosen last, or not at all, is a vivid memory for many men. Being picked on, afraid to fight, or forced to fight generates a welter of intense feelings, with shame at the core" (Krugman, 1995, p. 93).

Boys go to great lengths to avoid the intense pain of shame. They will mask their emotions, particularly their vulnerable ones. They will adopt a false self of toughness and bravado. Some boys isolate themselves, so afraid of shaming experiences that solitude becomes preferable. Some become bullies themselves, defending against their inner humiliation by humiliating others.

Indeed, according to researcher James Gilligan (1999), shame is at the root of much of the violence committed by males, both adolescent and adult. Boys who are persistently ridiculed by peers and parents develop an internal sense of chronic shame and a hypervigilance to the possibility of being shamed by others. Feeling respected by others can quiet the deep-seated feelings of inadequacy. Boys and men who routinely act violently are thus on a perpetual lookout to see if others are respecting them. When they feel "disrespected," whether the slight is real or imagined, they may react with violence, as a way to force respect. Gilligan points out that clearly not every male who is continually shamed becomes a violent person. Other factors play a role, such as experiencing physical abuse or emotional abuse from an authoritarian father. Still, it is the capacity to tolerate shame and contain the inner rage that shame engenders that distinguish those who react to life's slights and rejections nonviolently from those who violently act out their feelings of humiliation (Gilligan, 1999).

The power of shame and the Boy Code was brought home to me in my therapy work with Tony, who is struggling to free himself of the feelings of shame that are the core of his depression. A muscular, wiry man in his mid-20s, Tony tends to sit at the end of my couch, as though he were so wary of forming an intimate connection that he needs to sit as far from me as humanly possible in my small office. We have been working together for several months, and he is now taking the brave step of describing in detail what happened on the day he was badly beaten by a neighborhood bully in a wooded area less than 100 yards from Tony's house. Tony was 11 years old at the time. As he recalls it:

> I wiped away my tears and tried to get rid of the grass-stains on my pants so there would be no evidence. I came in the house, and my mother was there, and I remember feeling so torn. I desperately wanted to tell her what happened and let myself cry. But some part of me refused to let her see how humiliated I felt. I don't know why I was so terrified of letting my own mother see me hurt when I needed her so much. The weird thing was, I was as scared of her finding out as I was of the guy who beat me.

I asked Tony if he told his father what had happened. John looked at me as though I had asked the world's dumbest question. "My father? What would have been the point?"

Tony's victimization was a deeply traumatic experience; yet, what he most intensely recalled was the shame-based conflict of admitting his weakness and vulnerability to his parents. His need was for his mother, whom he saw as a potential source of emotional support; but revealing his trauma to her would have violated all four rules of the Boy Code. He would be admitting weakness (The Sturdy Oak); he had to maintain a façade of toughness (Give'em Hell); he could not dare to admit lack of dominance (The Big Wheel); and he would be displaying needy and vulnerable emotions (No Sissy Stuff). If the idea of talking to his mother activated an internal conflict, his thoughts of talking to his father reflected no ambivalence whatsoever. He perceived his father as an enforcer of the code, who would add only additional shame and humiliation on top of what he was already feeling.

If there is an underlying theme to the Boy Code, it is that feelings and behaviors associated with women are to be avoided, devalued, and disavowed. These socialization experiences of childhood and adolescence thus become internalized as a generalized fear of femininity; because having qualities associated with femininity is part of being human, boys learn to fear their own inner experiences. To a considerable extent, learning to be a man means learning not to be a woman.

The degree to which boys learn to devalue the feminine varies from boy to boy; cultural and class backgrounds can play a role (O'Neil, 1981). Some boys may rigidly conform to the Boy Code, manifesting in a stereotypically masculine gender identity and the display of hypermasculine behaviors, in an effort to prove his superiority as a man. Other boys appear to be more capable of resisting adherence to the rules, and may be more comfortable expressing their vulnerabilities. Does adherence to the code change as boys grow through adolescence? The research is not clear; some developmental psychologists have argued that gender role orientations become increasingly traditional from late childhood through adolescence. An alternative theory is that as boys (and girls) become older adolescents and capable of more cognitive complexity, they also become more flexible in their adherence to the rules defining masculinity and femininity (Katz & Ksansnak, 1994). Accordingly, male adolescents can become more "relationally complex" as they navigate this developmental phase.

Researchers in male development have theorized that even those who resist developing a traditional gender identity, based on being stoic, tough, competitive and dominant, may still emerge into adulthood with an unconscious fear of their feminine side. A boy may develop an internal critical voice condemning him when he feels weak and dependent; he may develop an unconscious braking mechanism that blocks the experience of tender and need-based emotions; and he may learn to fear that others will see him as weak and dependent if he displays his genuine vulnerability. According to Rabinowitz and Cochran (2002), many boys enter puberty with an underlying sadness and loss (that must, of course, be hidden from others), the result of the years of pushing aside these basic human feelings and longings. Emerging into adolescence with a coherent sense of self and a healthy, positive sense of his masculinity is a difficult challenge.

Moreover, a safe passage into puberty may be an even more formidable challenge for boys of color, growing up with experiences of racial discrimination and lack of access to many of the opportunities that come as a privilege to boys in the dominant culture. For example, for an African American who sees the doors of power in our mainstream culture closed off to him, portraying a hypermasculine cool pose may be a compensating façade, a way of demonstrating a sense of pride, strength, and power, while also showing distrust for the culture that appears to reject him (Lazur & Majors, 1995). Latino males may grow up with a cultural legacy of machismo, which can embody both positive aspects of masculinity—strength, attractiveness, dignity, and respect for others—but also exaggerated aggressiveness and arrogance (Fragoso & Kashubeck, 2000).

Similarly, Asian culture emphasizes the No Sissy Stuff rule as appropriate male behavior. Ironically, according to mens' studies researcher William Liu, whether Asian American males identify with their traditional culture, or whether they internalize mainstream American norms instead, the outcome may be a stoic, restricted, and unhealthy approach to the display of normal vulnerable emotions (Liu, 2002). In truth, however, there is very little research on how ideas of manhood differ for boys and men of different races; anything we say about male development in this chapter must be construed in light of the fact that scholarship on male development has been almost exclusively limited to the study of White males.

Nevertheless, we do hold the assumption that regardless of race or class, both genders come into this world with expansive possibilities for relational capacities, and that healthy development is about learning to balance relationality with social expectations. Boys and girls are capable of expressing tenderness, assertiveness, and competitiveness; boys and girls yearn for connection and relatedness; boys and girls have access to the full range of emotional experiences. However, out of fear of violating the Boy Code, a boy feels pressured to disconnect from his vulnerable, connection-needing, tender parts; in essence, he disconnects from himself. Writes the noted family therapist, Terrence Real, "Disconnection is masculinity" (Real, 2002, p.78). "It is not merely that boys and men are taught to disavow these human qualities (vulnerability and tenderness); they are actively taught to despise them" (Real, 2002, pp. 78–79).

DISCONNECTION FROM EMOTIONS

Hannah, a bright woman in her 20s determined to have couples counseling to turn around a struggling marriage, looks at me imploringly. "I don't know why it's so difficult to connect with him. Paul's not stupid. But he's an idiot when it comes to just telling me what he's feeling."

Paul has that deer-caught-in-the-headlights look. I ask him what he's feeling in response to Hannah's accusation.

There is a long, careful pause. "Well, I guess, anger. Maybe a little."

"Anything else?" I say.

"I don't know. Maybe. Uh, uh . . . I'm really not sure. Sadness? Does that sound right? What do you think I'm feeling, Dr. Shepard?"

Hannah rolls her eyes. "Is this what I have to accept for the rest of my marriage? Only my women friends can relate to me with feelings? You know men, Dr. Shepard. Are they incapable of feelings? Or are they just not interested in sharing them?"

The answer to Hannah's question is a critical one for counselors, because it raises fundamental questions about how we work with men on the issue of feelings;

should we accept men as having a limited capacity to express feelings? (If so, we might be more inclined to use cognitive-oriented treatment modalities.) Alternately, we may take the view that men have a compromised but potentially rich connection to feelings, and our role is to help them access these emotions. From a developmental perspective, socialization processes that begin in the second year of life and continue throughout a male's childhood development essentially disconnect many men from their capacity to feel and articulate comfortably the full range of human emotions—with the exception of anger, the one feeling men are taught is acceptable. Because this disconnection is learned, it is capable of being repaired.

Indeed, research suggests that boys start out in life more emotionally expressive than girls, displaying joy, anger, and tears with greater frequency. Male infants are also more socially oriented than girls and more likely to look up at their mothers (Weinberg, Tronick, Cohn, & Olson, 1999). But something happens to boys, beginning at around 6 months, so that these emotional capacities gradually diminish, and boys and girls end up in opposite places from where they began in life in terms of emotional expressiveness.

Ronald Levant, one of the founders of the New Psychology of Men movement, reviewed the research on infant socialization and found four phenomena that might explain this suppression of emotionality. First, when mothers interact with their baby sons, they are careful not to show too many emotions, especially sad feelings. Boys tend to cry more easily, and mothers may be reluctant to show feelings that would upset them. Second, when boys are about 1 year old, their fathers begin to take an active interest in them, and while they are comfortable roughhousing with their sons, they do not speak to them about emotions in the same way they might with daughters. Thus, boys receive less practice than girls with the vocabulary of feelings. Third, as boys get older and enter the school-age years, both parents discourage their sons' demonstrating vulnerable emotions like sadness and fear. Boys are told through numerous means that vulnerability is incompatible with masculinity. There are the obvious messages, such as statements like, "Big boys don't cry," or "Stop acting like a sissy." Socialization can work in more subtle ways as well; parents may refrain from discussing their own sadness with their sons. Or they throw their son a negative glance when the child himself displays a sad emotion, signaling to the little boy right away what is acceptable and what is not (Levant, 1995).

Pollack described this example of how parental responses to their children shape a child's belief in what feelings are consistent with male behavior. "For example, if a girl begins to cry, when she gets a bad grade at school, her parents might tend to say things such as 'Oh, you must feel awful' or 'Are you all right?' whereas if a boy begins to cry when he suffers the same fate, they might say, 'How unfair—that's ridiculous' or 'You march right in there and tell them that this just isn't right'" (Pollack, 1998b, p. 43).

Finally, peer group interaction powerfully reinforces boys' suppression of vulnerable emotions. It is in these interactions, frequently occurring in the context of sports and other games, that boys face humiliation if they dare admit they are frightened, sad, discouraged, or just plain feeling badly about themselves. It is in these interactions that boys learn to "tough it out" when they hurt and to hold back tears when they are sad. Most likely, any male reader of this chapter can recall an incident when their expressions of vulnerable emotions were met with laughter, teasing, or shaming put-downs. It is no wonder that Levant describes the process of boys learning to restrict their emotionality as "The Ordeal of Emotional Socialization" (Levant, 1995, p. 236).

Thus, throughout a boy's childhood development, the expression of emotions and using language to talk about emotion are discouraged, ultimately resulting in two possible outcomes by the time boys reach adolescence. The first is restrictive emotionality, a reluctance to express vulnerable emotions and dependency needs stemming from an internalized fear that such feelings are inconsistent with appropriate masculine behavior. It is important to stress that this fear of expressing feelings becomes deeply ingrained, so that boys and men will often hold back from displaying feelings without knowing why they are doing so.

Second, male children who are not encouraged to (or indeed are discouraged from) talk about their feelings, and who get little practice talking about feelings, may fail to develop the capacity to identify and verbalize their feelings. It is not that they do not experience sadness or joy or other feelings; it is that they have a difficult time finding the correct words for their feeling experience. In essence, these males are thus socialized to become "alexithymic," a term originally developed to describe psychiatric patients who had a markedly diminished capacity to recognize their emotions (Sifneos, 1973). What Levant calls "normative alexithymia" (Levant, 1995) may be experienced by as many as 19 percent of males (Taylor, Bagby, & Parker, 1991).

As noted earlier, the one emotion that boys are socialized to experience as acceptable is anger. While boys may be reprimanded by teachers and parents for expressing angry emotions, they are not shamed as being "sissies." Indeed, boys may be excused for angry behavior on the basis of the "boys will be boys" theory. The consequence of this particular socialization experience is not only that many boys (and men) are comfortable with their aggressive side but also that anger becomes a funnel emotion; that is, other feelings—sadness, shame, grief, fear—are expressed as anger. If anger is the one acceptable emotion, it becomes the way to express all emotions. Psychologist Gregory Smith offered a slightly different explanation for the same phenomenon: when vulnerable emotions are stirred up, males sense them as signs of weakness, an experience that then triggers anger as a response to this intrusion of unwanted emotions (Smith, 1996). Perceiving anger as a funnel emotion has critical implications for the counseling of boys and men; anger becomes understood as a cover for more tender underlying feelings that the male client cannot but needs to experience and express.

The case of Nick, a 16-year-old client, is an example both of boys' suppressed empathic capacities and emotions redirected into anger. Nick was an appealing, bright, and sensitive boy, but with a reputation for belligerence, disruptive behavior, and falling grades. Moreover, he was convinced he had ADHD, bipolar disorder, or a pathologic personality, which made it difficult for others, teachers and peers alike, to understand him and doomed him to failure. In one session, Nick asked, "What is it when I *feel* what my friends are feeling—some kind of crazy telepathy?" Nick was talking about empathy, but he'd had little in his own life, and his own empathic capacities puzzled him (Putz, 2002).

Nick's parents had had a tumultuous marriage and after they divorced when Nick was 9, his father moved across the country. Inside, he was hurt both by his father's abandonment and his mother's inability to see he needed more from her. However, without the skills and permission to voice his deeper feelings, his unarticulated and misdirected energy manifested as anger, which signaled to Nick that something was wrong with him and to his teachers that he was a troublemaker. Once Nick understood his feelings through his work in counseling, they were no longer an alien force, capable of contamination or malevolence. More vulnerable emotions could be expressed, and the belligerent posturing that was getting him in trouble subsided.

ADULT OPPORTUNITIES FOR CONNECTION

For most men in their late 20s or 30s, marriage or a nontraditional commitment to a life partner is theorized to be a normative and profound developmental event (Newman & Newman, 1995). Marriage is not only an opportunity for men to experience connection but is often a virtual class in it, as women have increasingly demanded that men respond mutually to their needs for intimacy. For the World War II generation, it was expected that if the man fulfilled his role as the provider, his wife should be content with a status quo in which her intimacy needs may or may not be met. That was the social agreement for a man's hard day at work. When divorce occurred, according to a 1948 study, the major reasons were financial nonsupport and neglect (Goode, 1969). Forty years later, women most frequently cited emotional deficiencies in the marriage as the reason for breaking up (Kitson & Sussman, 1982). Real (2002) observed that the current crisis in gender relations (women's wanting more closeness and their frustrations with men's inability to deliver) is the result of women's roles being radically transformed in recent decades while men's have not.

For men, learning to become intimate, or more relationally complex, to meet the urgent needs of their relationships can be a challenge for a number of reasons. One task is redefining the meaning of sexuality, and integrating sex and intimacy in a manner more gratifying for both partners. For many men, needs for intimacy became entangled with sexual needs during adolescent development. As teenagers, they learned to repress their needs for emotional intimacy, which seemed too associated with weakness, and substituted nonrelational, women-objectifying sexual gratification, when what they really desired was closeness or comforting (Brooks, 2001). As adults, they may find themselves having difficulty overcoming this nonrelational sense of sexuality, and must learn how to integrate sexual expression, sensuality, and emotional intimacy in order to make the kinds of connections that their partners seek.

Intimacy also means sharing feelings. Although intimacy is a difficult construct to define with precision, it is generally presumed to require self-disclosure and communicated empathy, the capacity to reflect in words what one's partner is feeling (Heller & Wood, 1998). The term self-disclosure refers to the sharing of vulnerable feelings, ranging from "I had a bad day at work" to "I am frightened that I am not good enough." To be intimate, a man must thus transcend internalized, culturally sanctioned prohibitions against self-disclosure of vulnerability and weakness. Intimacy also requires not only the willingness but the capacity to communicate feelings and inner states; as we have seen, for many men, the skills in the language of feelings were never fully learned. Similarly, men's ability to be empathic, to be attuned to a partner's inner state, may have been compromised through his gender socialization that included the devaluing of relatedness.

As a couples' counselor, I regularly observe the frustration felt by both men and women when the woman's request for a man to tell her what he is feeling is met with intellectualization, stammering, anger, or an attempt to stop the conversation completely. As mentioned earlier, Bergman (1995, 1991) termed the paralysis men feel in this situation "male relational dread"; the man becomes overwhelmed with stress at a moment when a failure to demonstrate competence in relating to his partner seems imminent. He may become flooded with negative thoughts about the impossibility of succeeding in giving his wife the kind of feeling words she is seeking; he may also experience shame over his incompetence in relating. The best he may be able to do is say, "I don't know what I'm feeling," and that feels like an inadequate response. The result is, he shuts down conversation, which leads to a disconnection. As Bergman puts it, "[E]ven though a man may desperately want connection, he again becomes an agent of disconnection" (Bergman, 1995, p. 83).

What about the argument that men and women simply view intimacy through different lenses? Wood and Inman (1993) have observed that men are intimate in their own way: sharing action-oriented activities, like taking trips or working on the house, and being physically in proximity but not communicating (e.g., she cooks; he watches TV) may be construed as equally intimate as the sharing of deep feelings. Men also tend to experience sex as a way of feeling intimately connected. All of these forms of intimacy reflect men's comfort with being in an active rather than passive mode. Perhaps, the definition of intimacy has been "feminized," and what is needed is a more inclusive understanding of the word that would equally value men's and women's subjective experiences (Heller & Wood, 1998).

Certainly, there are different styles of relating between the genders, and both need to be valued. However, I would argue that the problem with the above line of thinking is that it ignores the developmental sequence that devalues men's genuine yearning for intimacy and, through socialization, both fails to teach men relational capacities and defines those skills as unmasculine. To be sure, I have worked with many men in marital therapy who expressed extreme frustration with their partner's insistence on their relating in ways they feel are unnatural. They will say, "My wife wants me to be her girlfriend, and she needs to get it that I never will be. I'm a guy, honey. Accept it." I suspect there are many marriage therapists who espouse this view as well. However, my experience with male clients is that they openly wish they could feel closer to their wives, unless they wanted a separation, in which case they wanted to feel closer to someone else. The problem, simply, is that they do not know how.

Relational-Cultural theorists don't use the language of "intimacy," per se, rather they discuss "relationality" as a process, the goal of which is to always work toward better connection. These theorists would suggest that in order to work through relational disconnections, and to become more relationally competent, we need to become increasingly more authentic. Authenticity, in a sense, allows us to bring more and more of who we are into our relationships and ideally we are supported in this process. Male gender socialization is a process by which boys are taught to hide parts of themselves in order to meet the standards of "masculinity." As such, the process of working toward increased authenticity is about getting in touch with vulnerable feelings boys were initially taught to disconnect from and to be ashamed of (Dooley & Fedele, 1999).

Some men will use the opportunity of a committed union to develop deeper levels of connection, and the relationship can be a truly transformative experience. It is not that they become more like "girlfriends," and engage routinely in feeling talk with their partners. However, they can have moments when they learn to manage the anxiety of relational dread, and experiment with disclosing vulnerable, authentic, even painful feelings and thoughts. If their partners do not shame them for their displays of vulnerability (unfortunately, women, socialized to believe in the same constricting definitions of masculinity as their male partners, sometimes do shame them), men can discover the satisfactions of connection: a sense of not being alone in one's pain, an experience of being seen and understood, a sense of relationship as a journey mutually embarked upon.

Other men may adhere more rigidly to the Boy Code, and behave more consistently with the definition of the traditional man, characterized by stoicism, toughness, and a preference for a division of household and parenting responsibilities according to "old-fashioned" gender role norms. Finally, some men may attempt to reach out to their partners in new ways, but find that their relational dread is too overwhelming, leaving both their partners and themselves yearning for something more. It is with these men, and in these relationships that counselors can play such an important role, as the counseling relationship becomes a laboratory for experimenting with connection and intimacy.

FATHERHOOD

If ever there was a time for men to develop new skills for relating, connection, and intimacy, the experience of fathering should be the ideal opportunity. For many men in young-to-middle adulthood, becoming a father is a significant developmental transition. Yet, even as new fathers adore their children, joyfully play with them, and share in their partners' child-rearing responsibilities, gender role socialization may still restrict their full engagement in the process.

Why do men become fathers? Rabinowitz and Cochran (1994) cited several reasons: the desire to keep his name alive for future generations; the desire to transmit his values to his children; the desire to share with his loved one the process of parenting; the longing for the experience of personal growth that fatherhood entails. In many cases, becoming a father is not a planned part of life. As a developmental task, the decision to become a father can be an opportunity for the man to examine his relationship with his own father, reflecting on the values he has adopted or rejected and ultimately coming to his own definition of good fathering. It is at this point that some men reflect on the ways in which their own fathers gave or failed to give love and emotional sustenance; feelings of abandonment can come up, as well as the determination to give their new child what they themselves did not receive. Parenting thus becomes a developmental opportunity for inner healing.

Some of my male clients who have embarked on this journey have cited the desire to experience feelings of love as their primary motivation. They talk about wanting to feel something they sense women frequently experience but has eluded them so far in their lives. There is a sense of hope and excitement that the baby will allow them to express a buried part of themselves, a part that can connect wholly and unhesitantly with another human being.

Clearly, our culture is far more open to fathers playing nurturing roles than it was even two decades ago. This expanded definition of fathering is not only indicative of an increasing cultural sensitivity to the father's desire to be a more involved parent; it is also a reflection of changing economic and social conditions. Some women must work for the family to survive economically, and some women choose to work for both economic and self-aspiration reasons; in either case, men are expected to assume more child-raising responsibilities. The shift from the model of the father as the sole breadwinner to a more engaged parenting partner is reflected in media images of fathering. For example, in situation comedies in the 1950s and 60s (such as *Father Knows Best, Leave it to Beaver, The Dick Van Dyke Show*) the prevailing image was the dad who comes home from work, hugs his kids, provides pearls of fatherly wisdom, and administers firm discipline while soft-hearted mom winces. Today, television fathers share parenting responsibilities (though are often inept at it) and appear to spend as much time at home as their wives.

Thus, for the contemporary father, the developmental task is not adjusting to his role as the breadwinner, but learning to manage the tension between work and the yearning to be close to his child. This tension may be linked with both internal and external pressures. For example, internalized traditional male gender roles may equate successful fathering with providing economic security for his family; it is a man's own father who is still the working psychological model for what it means to be a parent. Moreover, the culture continues to support traditional role divisions, in which the father spends more time at work and the mother at home, regardless of the father's desires to be more involved in child-rearing. Also, some men may discover that it is easier to feel competent at work than it is to take care of a child.

Research confirms the difficulty of this struggle: Men may complain about longing to feel more connected to their children; yet, whether because of economic realities or their own choices, they are performing only one-third of the tasks associated with child care (Pleck, 1997). Developmental researchers have argued that if men are to fully engage in the parenting process, as many men wish to, the definition of fathering needs to change first. This new version of fathering would mean our society would view the emotional bond between father and children as importantly as the bond between mother and children. Ideally, the roles of father and mother, in terms of child-raising and economic providing, would become interchangeable; mothers would be seen as being as competent as men in the world of work, as fathers would be understood to be potentially as competent as women in the world of caretaking (Silverstein, Auerbach, & Levant, 2002).

MALE DEVELOPMENT AND VULNERABILITY TO DEPRESSION AND ADDICTION

Up to this point in this chapter, I have suggested that normal developmental sequences in men's lives can lead to various disconnections—from self, from feeling states, and from intimate or potentially intimate partners. Developmental researchers are suggesting that disconnection from their own inner pain can be for some men a serious psychological problem, threatening not only their capacities for intimacy but their own inner well-being. Indeed, it has been argued that because of the early childhood developmental sequence of premature separation from their mothers, all men are vulnerable to depression (Pollack, 1999). This experience, as noted earlier, is a profound loss, which in an ideal world would be grieved, primarily through expressing the emotion of sadness. Herein lies the rub; little boys are discouraged from expressing sadness and crying, through the various injunctions from parents and peers that real boys stay tough. From a psychoanalytic perspective, grief that is never expressed, and therefore never resolved, becomes unconscious; as boys grow into men, this repressed sadness persists as a kind of white noise of pain in the back of a man's brain.

There are two consequences to this repressed grief: first, it may contribute to men's avoidance of sad emotions in general. They may unconsciously fear that feeling sadness in the present may tap into and then unleash the dormant sadness left over from the painful separation experience in childhood. As one client told me, "If I started to feel the sadness that has always been there, I am afraid it would all come rushing up, and I would be overwhelmed with pain."

Second, the repressed grief acts like a thorn in a man's psyche, leaving him vulnerable to a true depression when faced with critical loss at some point in life (such as the end of a relationship; death of a loved one, loss of a job). Obviously, not every man becomes depressed when rejected by a partner or when laid off at work. However, the argument goes, the culture's discouragement of boys' and men's feelings of sadness leaves some men particularly at risk when they experience a significant loss; the *appropriate* painful feelings will simultaneously trigger the emergence of the unconscious sadness, resulting in a state of depression.

For men who do experience depression, regardless of the trigger (which may, of course, be unrelated to childhood separation wounds), their gender role socialization against admitting and displaying vulnerable feelings heightens the danger that their distress will go both unnoticed and untreated. In what has been called masked or covert depression (Real, 1998; Cochran & Rabinowitz, 2000), some depressed men may hide their depression by refusing to admit they feel badly inside, because they think of their depression as a sign of weakness and are therefore too ashamed to tell anyone. Some men may mask their pain by resorting to bravado, making sure that everyone gets that they are OK. Some depressed men may not realize that their inner pain *is* depression, mislabeling it as stress or as an ordinary sense of being "down"; or they are so out of touch with their emotions (the alexithymia discussed earlier) that they are not aware of their sadness.

The following clinical vignette illustrates this dilemma for men: Steven, a 35-year-old lawyer, came to me for counseling because he has not been able to get over his breakup with his girlfriend of six months. He can't stop thinking about her, even when he is trying to concentrate on a legal case.

For the first several sessions, Steven talks about how bad the breakup has made him feel—nothing pleases him, he doesn't know what to do with his life, he cannot imagine being happy again. I also find out more about why his girlfriend, Sarah, left him. Steven would frequently withdraw into a shell where Sarah could not reach him. He would spend hours watching television, or drink several glasses of wine after dinner and have nothing to say for the rest of the evening. Steven's capacity to develop an intimate relationship with Sarah was highly impaired.

I ask him if he knows that he is describing symptoms of depression. He says he has no idea. "I suppose it is possible, but it never occurred to me. How would I know?" Stephen is not in denial. I am sure that Steven does not know that he is depressed. Because he is disconnected from his pain he does not know how sad he is and how that sadness permeates his body and his personality. Soon, the subject of our work becomes his depression, and he is relating to me heartbreaking memories of a painful childhood. It is my job to help him reconnect the sadness with his experience and transform the feeling into words; this is the beginning of healing and the first step toward authentic relating.

The cultural injunctions against admitting pain unfortunately means that some depressed men may not tell their intimate partners about their feelings, and therefore lose an opportunity for soothing and support (Shepard, 2002). They are also afraid to seek counseling or report their symptoms to their physician (Good, Dell, & Mintz, 1989). Indeed, there is some evidence that the rules defining masculinity may play a role in causing depression. Some researchers have demonstrated that the yearning to talk about feelings versus the fear of showing weakness and femininity creates an inner turmoil that may be linked with depression (Good & Mintz, 1990; Shepard 2002). These men thus experience a double whammy: the rules of masculinity may have both helped cause the depression and, paradoxically, prevent them from seeking help for it.

Trapped by the Boy Code learned in boyhood, some depressed men may channel their pain into symptoms designed to keep their depression under wraps by displaying angry feelings instead of sad ones, self-medicating with alcohol, and/or withdrawing into themselves. In all three cases, the effect is to disconnect from their intimate partners, leaving their partners puzzled, hurt, and frustrated, and the men themselves feeling isolated and alone in their pain.

Abuse of alcohol (or other substances) can be an especially damaging pain-masking strategy, because of its impact on a man's health, the changes it can engender in his personality, its addictive power, and its destructive impact on his intimate relationships. Male socialization appears to play a role in encouraging alcohol use. We know, for example, that men abuse alcohol with greater severity than do women. They begin drinking at an earlier age and they drink more frequently, to the extent that there are three to four times as many problem male drinkers as women (Lemle & Mishkind, 1989). For generations, popular culture has connected alcohol use with the gender role rules related to toughness, competitiveness, and dominance. The nexus of televised sports and beer commercials is one obvious example. When lite beer was first introduced to the marketplace, television ads featured prominent male athletes proudly drinking the stuff, as though to dispel any suggestion that "lite" meant "feminine" (Katz, APA presentation). Commercials also emphasize alcoholism as a young adult male ritual. As I write this chapter, one beer company is pushing its brew by portraying a young married man being lured from his domesticity by an obviously unattached, more wild friend; together, they "party all night," and the married friend returns home bleary-eyed to his waiting and angry wife. The message is clear: Intimate relationships are a ball-and-chain on freedom and pleasure, and a night out drinking represents the man's prerogative to have fun. The few empirical studies on the relationship of alcohol abuse to male socialization confirm that adherence to the traditional male role is related to drinking (Blazina & Watkins, 1996; McCreary, Newcomb, & Sadava, 1999). The data demonstrate that the more rigidly men adhere to the notion that their masculinity is tied to having status in society, to restricting their emotions and avoiding displays of vulnerability, the more likely they are to use alcohol. Shame also may play a role in male alcohol abuse. Because men learned to fear shame so intensely during childhood and adolescence, they may resort to alcohol to dissolve the shame sensations that occur when they feel vulnerable.

Understanding the developmental forces that can underlie a man's depression and lead to masking strategies like alcohol use is highly relevant to how counselors conceptualize and treat depression in male clients. Counselors need to be aware that facilitating the expression of feelings prematurely may heighten a man's internal conflicts about appearing weak, and consequently strengthen defenses against feelings, instead of loosening them. It may be helpful first to explore with him the relationship between feelings and beliefs about masculinity. Only after a man's fears of feeling and/or appearing unmasculine are addressed can the healing unburdening of emotions occur.

MALE DEVELOPMENT AND VIOLENCE

We cannot discuss the impact of socialization pressures on male development without commenting on its relation to male violence—toward women and other men. As Kimmel and Mahler (2003) observed in their study on school shootings, the powerful fact from which we cannot escape is that it is males who commit these acts of violence. And therefore, the question that begs to be asked is, what is it about becoming a male that leads to boys killing students, or to the fact that violent crimes by adults are overwhelmingly committed by male adults? When we look once again at the Boy Code, we can see how socialization can contribute to men's use of violence as a problem-solving strategy. Men's studies researcher Chris Kilmartin noted that each rule in the code sanctions aggression (Kilmartin, 1994). "The Sturdy Oak" rule that boys do not show weakness means that when a boy feels emotional pain, it is preferable to act tough than to reveal sadness or fear. The "Give 'em Hell" injunction tells boys that masculinity *means* aggressiveness and bravado. Being "The Big Wheel" means having dominance and power over others, which can be achieved through physically hurting others. Finally, "The No Sissy Stuff" injunction, the rule that boys should at all costs avoid appearing feminine, means that boys cannot back down from a fight. The words, "No Sissy Stuff" were used by authors David and Brannon to describe this rule some three decades ago (David & Brannon, 1976).

Today, the rule might be better described as "no fag stuff," "no gay stuff," and certainly "no girly stuff." There is no more shaming put-down than challenging a boy's heterosexuality, and one way boys can ensure that there is no doubt about their sexual orientation is to act aggressively. This phenomenon clearly delineates the etiology of "homophobia" and some of the causes of violence against gay men and women. Aggression, in response to having one's masculinity and heterosexuality (and in our culture these are one in the same) is an act of "disrespect." In an effort to restore and maintain one's place in the world men restore their sense of power and control through the use of violence (Gillligan, 1999).

Male violence toward women may also be strongly related to socialization and adherence to the traditional masculine ideals of stoicism, toughness, and power over others. The most recent statistics suggest that an estimated 2 million women are severely abused by men every year (Conrad & Morrow, 2000). The propensity for so many men to batter their wives and girlfriends is truly a manifestation of the "dark side of masculinity" (Brooks & Silverstein, 1995, p. 280). Understanding how socializing forces in a man's development relate to physical aggression toward women is neither an excuse for battering, nor a rejection of the feminist perception that battering is an inevitable outcome of a society that supports men in an oppressive, one-up position over women.

Indeed, implicit in a socialization explanation is a critique of a definition of masculinity that begins as the Boy Code and for too many men becomes a rigid gender role "straightjacket" throughout their adult lives. Additionally, any explanation of how socialization impacts male violence must include the role of shame, which enforces conformity to the rules and can trigger in men's psyches both self-hatred and rage. This relationship between feeling shame or humiliation and violence has been clearly supported in research (Gilligan, 1999).

At the simplest explanatory level, some men are violent toward women because the culture has taught them that violence is an acceptable method of maintaining power and control. When men perceive their female partners as threatening that power, they may become violent as a way of restoring their power (Dutton & Browning, 1988). However, more recent research has offered a more complex explanation, suggesting that male batterers are not only conforming to what they perceive as acceptable rules of behavior but they are also especially vulnerable to loss of self-esteem. When women gain power in a relationship, some men may experience a wound to their masculine pride, and use violence to restore it (O'Neil and Harway, 1997). To be sure, many men will feel some sense of shame when their female partner threatens their power and dominance. Most do not batter, either channeling their anger in less destructive ways (yelling, sulking, storming out the door, etc.) or taming their hurt feelings by recognizing shame as an irrational response to a situation that is in reality not a threat to the self. According to Gilligan (1999), most men are prevented from acting violently because they feel "guilt and remorse over hurting someone else; empathy, love and concern for others" (p. 37). What makes batterers different is their willingness to rid themselves of their shame-feelings through violence, regaining their sense of lost strength by aggressive demonstrations of their power over women. In short, for some men, rage is empowering.

Jennings and Murphy (2000) suggested there are several developmental phenomena in men's socialization experiences, all occurring in the context of male-male relationships, that can account for violence toward women. For example, men are socialized to constrict their emotions, and are "haunted by the persistent threat of humiliation," similar to that which they experienced with male mentors and peers in childhood and adolescence (Jennings & Murphy, 2000, p. 26). When intimate relationships call for the expression of feelings, they react with angry defensiveness. Also, batterers tended to have grown up as loners, fearing that their need for connection to other men signified weakness, or worse, "feminine feelings." As adults, these men tend to become dependent on their female partners, their sole source of nurturance and intimacy. Terrified of losing this dependent connection, they can become excessively jealous and enraged if their partners threaten to leave them, especially if it is for someone else. If the woman and man get into an argument, the batterer expects she will back down, just as he was able to bully his boyhood peers into submission. The woman's failure to do so is experienced as further humiliation, escalating his impulse toward aggression and violence. The common theme in all of these explanations is that the batterer is both identified with rigid rules of masculinity and simultaneously terrified of being humiliated if he does not live up to them during confrontations with female partners. It is as though there were a hidden chorus of males inside his mind, ready to taunt him and strip him of his masculinity if he should ever feel or demonstrate powerlessness with a woman.

While there is plenty of data demonstrating a connection between shame and violence (Gilligan, 1999), empirical research exploring the relationship between adherence to male role norms and violence toward women is still in an early stage. Data suggest such a relationship exists, but also, importantly, indicate that other factors, unrelated to masculinity concepts, may play a greater role in accounting for battering (Jackupck, Lisak, & Roemer, 2002). Nevertheless, we should not be surprised that so long as the culture teaches men to devalue and fear the feminine—their internal feminine qualities or women themselves—men will act out violently.

CONCLUSION

I have discussed in this chapter developmental forces in a man's life that make connections to both self and others problematic. Drawing from theories and recent research in male development, I have described ways in which men may become cut off from aspects of their own psychic life and from optimal relationships with partners. I began by highlighting the role of a normative early developmental trauma for boys, in which the unhealed wound from a premature separation from mother leads both to unconscious fears of intimacy and a chronic, unconscious sadness. This separation, and the cultural emphasis on little boys becoming self-reliant, also diminishes opportunities for learning empathy, acceptance of dependency needs, and the vocabulary of emotions—important components for healthy intimate relationships. I also described the Boy Code (Pollack, 1998b) as a set of rules constricting men's healthy development, rules enforced by peer and parental shaming of boys when they violate the code. I discussed how intimate partnerships and fatherhood can be opportunities for new experiences of connection, though men must struggle to overcome gender role socialization to take full advantage of these opportunities. Finally, I reviewed the ways in which both early and later developmental sequences that devalue men's inner feminine aspects leave some men vulnerable to depression, alcohol abuse, and, in some cases, acting out shame through violence toward women. Throughout the chapter, I have noted the relevance of developmental issues to counseling male clients.

I began the chapter with a personal anecdote about the way in which the film, *Field of Dreams*, reflected my father's and my desire to connect and our mutual difficulty saying, "I love you." In the years following the movie's release, over 25,000 people a year have been visiting the actual baseball field in the Midwest where the film was shot. On a daily basis, between 50 and 100 people wander about the field, some playing catch, some silently lost in their own thoughts. Scholars have attempted to make sense of this phenomenon, interviewing the field's guests as to why they came and searching for common themes in their explanations. The words of two interviewees in one study have stood out for me: "You know how everybody likes to be touristy and stuff like that, but I think everybody has some idea in their minds of something that they're searching for and hoping that they'll find it out here." And another: "You come here and you become young again" (Aden, Rahoi, & Beck, 1995, p. 373). I cannot help wondering if some of the men who visit this field are searching for a room in the memory of their youth where reside unrequited yearnings for connection. They cannot put a finger on what went wrong, but something did, something prevented them from experiencing a complete, mutual, and soothing relatedness with a family member or friend. If only they had not been bound by the rules of boyhood, if only they had been freed from the worry that they were not good enough, competent enough, or better than, they might have felt the peace of just being close.

REFERENCES

Aden, R. C., Rahoi, R. L., & Beck, C. S. (1995). "Dreams are born on places like this": The process of interpretive community formation at the Field of Dreams site. *Communication Quarterly, 43*, 368–380.

Bandura, A. (1977). *Social learning theory*. Englewood Cliffs, NJ: Prentice Hall.

Bergman, S. J. (1991). Men's psychological development: A relational perspective. *Work in Progress, No. 48*. Wellesley, MA: Stone Center Working Paper Series.

Bergman, S. J. (1995). Men's psychological development: A relational perspective. In R. F. Levant & W. S. Pollack (Eds.), *A new psychology of men*, pp. 68–90. New York: Basic Books.

Blazina, C., & Watkins Jr., C. E. (1996). Masculine gender role conflict: Effects on college men's psychological well-being, chemical substance abuse, and attitudes toward help-seeking. *Journal of Counseling Psychology, 43*, 461–465.

Blazina, C., & Watkins Jr., C. E. (2000). Separation/individuation, parental attachment, and male gender role conflict: Attitudes toward the feminine and the fragile masculine self. *Psychology of Men & Masculinity, 1*, 126–132.

Brooks, G. R. (2001). Challenging dominant discourses of male (hetero)sexuality: The clinical implications of new voices about male sexuality. In P. J. Kleinplatz (Ed.), *New directions in sex therapy: Innovations and alternatives*. Philadelphia: Brunner-Routledge.

Brooks, G. R., & Gilbert, L. A. (1995). Men in families: Old constraints, new possibilities. In R. F. Levant & W. S. Pollack (Eds.) *A new psychology of men*, pp. 252–279. New York: Basic Books.

Brooks, G. R., & Silverstein, L. B. (1995). Understanding the dark side of masculinity: An interactive systems model. In R. F. Levant & W.S. Pollack (Eds.), *A new psy-chology of men*, pp. 280–336. New York: Basic Books.

Chodorow, N. (1978). *The reproduction of mothering*. Berkeley: University of California Press.

Cochran, S. V., & Rabinowitz, F. E. (2000). *Men and depression: Clinical and empirical perspectives*. San Diego, CA: Academic Press.

Conrad, S. D., & Morrow, S. T. (2000). Borderline personality organization, dissociation, and willingness to use force in intimate relationships. *Psychology of Men & Masculinity, 1*, 37–48.

David, D., & Brannon, R. (1976). *The Forty-nine percent majority: The male sex role*. Reading, MA: Addison-Wesley.

Dooley, C., & Fedele, N. (1999). Mothers and sons: Raising relational boys. *Work in progress, No. 84*. Wellesley, MA: Stone Center Working Paper Series.

Dutton, D. B., & Browning, J. J. (1988). Concern for power, fear of intimacy, and aversive stimuli for wife assault. In G. Hotalint, D. Finkelhor, J. T. Kirkpatrick, & M. A. Strauss (Eds.), *Family abuse and its consequences: New directions in research* (pp. 163–175). Thousand Oaks, CA: Sage.

Erikson, E. H. (1963). *Childhood and society*. New York: W.W. Norton & Company.

Fragoso, J. M. & Kashubeck, S. (2000). Machismo, gender role conflict, and mental health in Mexican American men. *Psychology of Men & Masculinity, 1*, 87–97.

Gilligan, J. (1999). *Preventing violence*. New York: Thames & Hudson.

Good, G., Dell, D. M., & Mintz, L. M. (1989). Male role and gender conflict: Relations to help seeking in men. *Journal of Counseling Psychology, 36*, 295–300.

Good, G., & Mintz, L. (1990). Gender role conflict and depression in college men: Evidence for compounded risk. *Journal of Counseling and Development, 75*, 44–49.

Goode, W. J. (1969). *Women in divorce*. New York: Free Press.

Greenson, R. (1968). Dis-identifying from mother: Its special importance for the boy. *International Journal of Psychoanalysis, 49*, 370–374.

Hartley, R. E. (1959). Sex role pressures and the socialization of the male child. *Psychological Reports, 5*, 457–468.

Heller, P. E., & Wood, B. (1998). The process of intimacy: Similarity, understanding and gender. *Journal of Marital and Family Therapy, 24*, 273–288.

Jakupcak, M., Lisak, D., & Roemer, L. (2002). The role of masculine ideology and masculine gender role stress in men's perpetration of relationship violence. *Psychology of Men & Masculinity, 3*, 97–106.

Jennings, J. L., & Murphy, C. M. (2000). Male-male dimensions of male-female battering: A new look at domestic violence. *Psychology of Men & Masculinity, 1*, 21–29.

Jordan, J.V. (2001). A relational-cultural model: Healing through mutual empathy. *Bulletin of the Menninger Clinic, 65, 1*, 92–104.

Katz, J. (2002, Chicago). For crying out loud: Images of male vulnerability post-9/11. Paper presented at the meeting of the American Psychological Association, Chicago, Il.

Katz, P. A., & Ksansnak, K. R. (1994). Developmental aspects of gender role flexibility and traditionality in middle childhood and adolescence. *Developmental Psychology, 30*, 272–282.

Kaufman, G. (1985). *Shame: The power of caring*. Rochester, VT: Schenkman Books, Inc.

Kilmartin, C.T. (1994). *The masculine self*. New York: Macmillan.

Kimmel, M. S. & Mahler, M. (2003). Adolescent masculinity, homophobia, and violence: Random school shootings, 1982–2001. *American Behavioral Scientist, 46*, 1439–1458.

Kitson, G. C., & Sussman, M. B. (1982). Marital complaints, demographic characteristics, and symptoms of mental distress in divorce. *Journal of Marriage and the Family, 44*, 87–101.

Krugman, S. (1995). In R. F. Levant & W. S. Pollack (Eds.), *A new psychology of men*, (pp. 91–128). New York: Basic Books

Lazur, R. F., & Majors, R. (1995). Men of color: Ethnocultural variations of male gender role strain. In R. F. Levant & W. S. Pollack (Eds.), *A new psychology of men*, (pp. 337–458). New York: Basic Books.

Lemle, R., & Mishkind, M. E. (1989). Alcohol and masculinity. *Journal of Substance Abuse Treatment, 6*, 213–222.

Levant, R. F. (1995). Toward a reconstruction of masculinity. In R. F. Levant & W. S. Pollack (Eds.), *A new psychology of men*, (pp. 229–251). New York: Basic Books.

Levant, R. F. (1996). The new psychology of men. *Professional Psychology: Research and Practice, 27*, 259–265.

Liben, L. S., & Signorella, M. L. (1993). Gender-schematic processing in children: The role of initial interpretations of stimuli. *Developmental Psychology, 29*, 141–149.

Liu, W. M. (2002). Exploring the lives of Asian American men: Racial identity, male role norms, gender role conflict, and prejudicial attitudes. *Psychology of Men & Masculinity, 3*, 107–118.

Maccoby, E. E. (1990). Gender and relationships: A developmental account. *American Psychologist, 45*, 513–520.

Mahler, M. S., Pine, F., & Bergman, A. (1975). *The psychological birth of the human infant*. New York: Basic Books.

McReary, D. R., Newcomb, M.D., & Sadava, S. W. (1999). The male role, alcohol use, and alcohol problems: A structural modeling examination in adult women and men. *Journal of Counseling Psychology, 46*, 109–124.

Meyer, S., Murphy, C. M., Cascardi, M., & Birns, B. (1991). Gender and relationships: Beyond the peer group. *American Psychologist, 46*, 537–537.

Newman, B. M., & Newman, P. R. (1995). *Development through life: A psychosocial approach (6th ed.)*. Pacific Grove, CA: Brooks/Cole Publishing Company.

O'Neil, J. M. (1981). Patterns of gender role conflict and strain: Sexism and fear of femininity in men's lives. *Personnel and Guidance Journal, 60*, 203–210.

O'Neil, J. M., & Harway, M. (1997). Multivariate model of explaining men's violence toward women. *Violence Against Women, 3*, 182–203.

Pleck, J. H. (1997). Paternal involvement: Levels, sources, and consequences. In M. E. Lamb (Ed.), *The role of the father in child development (3rd ed.)*, pp. 66–103. New York: Wiley.

Pollack, W. S. (1995). No man is an island: Toward a new psychoanalytic psychology of men. In R.F. Levant & W.S. Pollack (Eds.), *A new psychology of men*, (pp. 33–67). New York: Basic Books.

Pollack, W. S. (1998a). The trauma of Oedipus: Toward a new psychoanalytic psychotherapy for men. In W. S. Pollack & R. F. Levant (Eds.), A *new psychotherapy for men* (pp. 13–34). New York: Wiley.

Pollack, W. S. (1998b). *Real boys: Rescuing our sons from the myths of boyhood.* New York: Henry Holt and Company.

Pollack, W. S. (1999, Boston). Real boys, real men, real depression: New models of diagnosis and treatment. Paper presented at the meeting of the American Psychological Association, Boston, MA.

Putz, V. Personal communication, November 21, 2002.

Rabinowitz, F. E., & Cochran, S. V. (2002). Deepening psychotherapy with men, Washington, DC: American Psychological Association.

Rabinowitz, F. E., & Cochran, S. V. (1994). *Man alive: A primer of men's issues.* Pacific Grove, CA: Brooks/Cole Publishing Company.

Real, T. (1998). *I don't want to talk about it: Overcoming the secret legacy of male depression.* New York: Fireside.

Real, T. (2002). *How can I get through to you? Reconnecting men and women.* New York: Scribner.

Ruble, D.N., & Stangor, C. (1986). Stalking the elusive schema: Insights from developmental and social-psychological analyses of gender schemas. *Social Cognition, 4,* 227–261.

Robinson, P. A. (Writer/Director). (1989). *Field of Dreams* [Motion Picture]. United States: Universal.

Shepard, D. S. (2002). A negative state of mind: Patterns of depressive symptoms among men with high gender role conflict. *Psychology of Men & Masculinity, 1,* 3–8.

Sifneos, P. E. (1973). The prevalence of 'alexithymic' characteristics in psychosomatic patients. *Psychotherapy and Psychosomatics, 22,* 255–262.

Silverstein, L. B., Auerbach, C. F., Levant, R. F. (2002). Contemporary fathers reconstructing masculinity: Clinical implications of gender role strain. *Professional Psychology: Research and Practice, 33,* 361–369.

Smith, G. (1996). Dichotomies in the making of men. In McLean, C., Carey, M., & White, C. (Eds.), *Men's ways of being*, (pp. 29–50). Boulder, CO: Westview Press.

Taylor, G. J., Bagby, R. M., & Parker, J. D. A. (1991). The alexithymia construct. *Psychosomatics, 32,* 153–164.

Weinberg, K. M., Tronick, E. Z., Cohn, J. F., & Olson, K. L. (1999). Gender differences in emotional expressivity and self-regulation during early infancy. *Developmental Psychology, 35,* 175–188.

Wood, J., & Inman C. (1993). In a different mode: Masculine styles of communicating closeness. *Journal of Applied Communication Research, 21,* 279–319.

8

Coming Out and Living Out Across the Life Span

by Stacee Reicherzer, M.A.

REFLECTION QUESTIONS

1. *Consider each of the following terms: bisexual . . . gay . . . lesbian . . . transgender. What thoughts and images come to mind? On what do you base these thoughts and images? If you have known someone who is BGLT, in what ways have your perceptions changed?*

2. *If you are heterosexual, reflect back on the period during which you identified your attractions to members of the opposite sex. When you began to experience your sexuality, how did you feel about yourself? Was it awkward and embarrassing?*

3. *Now imagine that just as naturally and intensely as you developed your attraction to the opposite sex, you developed an equal attraction to members of your own sex. Added to your adolescent feelings of awkwardness are feelings of shame and confusion because everything you hear about bisexual, lesbian, and gay people is negative—and is even referred to as "perverse," "unnatural," or otherwise worthy of condemnation by God and by society.*

4. *Imagine you developed an exclusive attraction to members of your own sex. At 13, what would you do? How would you feel about yourself? How would you feel about God?*

5. *Think about being a woman or a man. Aside from your physical form, how do you know you are a woman or a man? Imagine what it would be to have this sense of your own gender but be in the body of the opposite sex. What feelings come to mind? How hard would you work to appear as that gender? Would you wear the clothes and play the roles of that gender or the gender that feels more natural to you?*

INTRODUCTION

I introduce this chapter by acknowledging ambivalence and difficulty because I am a gender-reassigned woman. Because I live and "pass" as a heterosexual woman in the non–BGLT world, meaning that I can go virtually anywhere and not receive the mistreatment that others of my BGLT community would experience, disclosing that I am transgender means giving up some of the privilege that goes along with being "straight" in a heterosexist world. At the same time, it is also empowering because it acknowledges a distinction that is a part of me just as it has been a part of a few proud people that have existed throughout history. That I would feel hesitant to divulge this is evidence of how deeply pervading shame can be, a shame that is socially sanctioned by the outright hatred and discrimination that exists against people who are transgender identified as well as bisexual, gay, or lesbian.

The bisexual, gay, lesbian, and transgender communities together form a rich cultural mosaic. With some estimations placing the number of "out" BGLT Americans at 10 percent of the population, it is probably of little surprise that such diversity exists among the millions of Americans who self-identify in one or more of these groups.[1] Considering that there are many other people who, for various reasons, may emotionally experience but not profess a same-sex attraction or an ambiguous or transgender identity, it can be inferred that the percentage of Americans within this group is significantly higher.

The experiences of so many are difficult to capture in one chapter. The focus of this chapter then will be of common experiences that are largely shared by members of these communities, such as coming out, BGLT adolescent experiences, coupling, parenting, and aging. In addition, special sections will focus on bisexual and transgender issues.

An important focus of this chapter will be the role of shame and oppression in shaping the lives and relationships of BGLTs. Legal bans on same-sex marriages, lack of legal protection for lesbian and gay parents, hate crimes committed against BGLT individuals, family and social rejection due to BGLT status, and the volumes of other atrocities committed against these individuals are significant social constructs of BGLT experiences. Dr. Judith Jordan describes this further in her powerful social commentary, "Relational Development: Therapeutic Implications of Empathy and Shame" (1989):

> A powerful social function of shaming people is to silence them. This is an insidious, pervasive mode of oppression, in many ways more effective than physical oppression. In a supposedly egalitarian society, shaming becomes a potent, indirect exercise of dominance to subdue certain expressions of truth. By creating silence, doubt, isolation, and hence immobilization, i.e., shame, the dominant social group (in this case white, middle class, heterosexual males) assures that its reality becomes *the reality*. This dynamic has dictated the social experience of most marginalized groups, be they women, blacks, lesbians, gays, Hispanics, or the physically challenged, whose voices have for too long been unheeded and whose reality has thus been denied. (p. 7)

Carter Heyward (1989), an Episcopalian priest who struggled to overcome shame and accept herself as a lesbian, describes this shaming oppression as "heterosexism" in that it is to homophobia what sexism is to misogyny and racism is to racial bigotry and hatred (p. 5). Exploring and challenging one's own heterosexism is a critical step in personal development, especially for BGLTs in the helping professions. The insidious nature of heterosexism has been its message that BGLT people are somehow second-class citizens, oddities who deserve unequal treatment. Shame has created internalized homophobia/biphobia/transphobia[2] (described later in the "Self Acceptance" section) as well as a hostile or fractious nature to others in the BGLT communities. I can remember when one popular gay "men's bar" in the town I grew up in refused to admit me because of my transgender status. Additionally, I have been challenged for using a women's restroom in a lesbian bar. Ironically, this has never happened to me in a straight bar or anywhere else outside the gay and lesbian communities!

Understanding heterosexism as a social construct in the lives of BGLT individuals is imperative. As Gartrell (1984) states of her clinical work with lesbian clients, "simply being a lesbian is not the only prerequisite for understanding homophobia. One must also make a concerted effort to be educated about the personal, political, social, and cultural ramifications of homophobia" (p. 5).

It is my experience that despite its incidence of fractions within and between individuals, by and large BGLTs function to form communities that care for their members and mobilize around important political and social issues. While BGLT individuals follow similar popular migration shifts due to vocational and other life choices as heterosexuals do, BGLT communities often center in large urban areas. This is due to the relative safety and BGLT specific services available within a neighborhood that is BGLT friendly. One can sense the community-mindedness in such a neighborhood, particularly if attending a drag show–community fund-raiser or sipping espresso with the regulars at a local coffee bar.

The helping professions have been at the forefront in advancing awareness of these communities' needs. In 1973, the seventh printing of the *Diagnostic and Statistical Manual of Mental Disorders* (American Psychiatric Association, 1980) was revised to reflect the removal of homosexuality as a form of mental illness. The American Psychiatric Association has decried the efforts of clinicians and laypersons to convert a person from homosexual to heterosexual as unsafe and unethical, a view that has also been upheld by the American Psychological Association and other human service professional organizations. These organizations also parent associations that seek to provide culturally competent services to the BGLT client groups by advancing the awareness of the helping professions they serve.

Most helping professionals will encounter BGLT clients at some point in their career. While the role of heterosexism is certainly a salient issue in the lives of BGLTs, it may be experienced to a greater or lesser degree depending on a number of factors, including the individual's ethnicity, gender identity, socioeconomic status, degree of physical ability, physical appearance, and age.

Each individual presents a unique narrative that must be explored without preconceptions. Even so, I believe it is important to offer some common terminologies that I use as well as those used by others who experience BGLT lives in contemporary Western culture. These terminologies were developed by Carroll, Gilroy, and Ryan (2002). None of these has a clean definition that is commonly used by all members of any one community. The lack of solid definition illustrates the limits of linguistic and cultural awareness in gender identity.

GENDER A set of cultural norms and mores whereby humans identify as "men" and "women," or "transgender." Gender refers to that which a society deems masculine or feminine.

GENDER IDENTITY An individual's self-identification as a man, woman, transgender, gender variant, or other identity category.

GENDER VARIANT Individuals who stray from socially accepted gender roles in a given culture. This word may be used in conjunction with other collective designations, such as gender variant gay men and lesbians.

INTERSEX Otherwise referred to as hermaphrodites, individuals termed intersex are born with some combination of ambiguous genitalia. The intersex movement endeavors to stop the practice of pediatric surgery and hormone treatments that attempt to normalize infants into the dominant "male" and "female" roles.

SEXUAL ORIENTATION This term refers to the gender(s) to whom a person feels attracted. Examples of sexual orientation include homosexual, bisexual, and heterosexual. Transgender, transsexual, and gender variant people may identify with any sexual orientation, and their sexual orientation may or may not change during their gender transition.

TRANSGENDER A range of behaviors, expressions, and identifications that challenge the pervasive binary gender system in a given culture. It is an umbrella term that is used to describe any of a variety of cross-gender identities.

TRANSSEXUAL A person who is dissociated with his/her birth gender and wishes to use hormones and in some cases, gender reassignment surgery, as a way to correct his/her body to the gender he/she recognizes as his/her own. Once surgery is complete, they often designate gender according to their surgical reassignments: female to males (FTMs) designate as males, and male to females (MTFs) designate as females. (Carroll, Gilroy, and Ryan, 2002)

It is the intent of this chapter to give voice to that which for too long has been silenced, by sharing with helping professionals some of the realities of BGLT development. It is not all-inclusive, and I advise that in order to better understand BGLT individuals and clients one should read professional and other contemporary literature on such topics as laws that ban same-sex marriages and adoptions, the impact of AIDS in the gay community, teen suicide, BGLT parenting, and other important social issues.

SHAME: THE TRAUMA THAT
SILENCES US

Many of us knew from young ages that we felt different from others of our own sex about our sexuality and/or gender. In many cases, if anything related to BGLT people was ever mentioned when we were growing up, it was very negative and cruel. Imagine the experience of hearing about this mysterious circumstance called homosexuality and realizing that this is who *you* are. Who could you tell that you were scared as hell because no matter how hard you tried, you just didn't like the opposite sex in the same way you liked others of your own sex? Who could you tell that you wanted to be a girl and not a boy, or vice versa? When you're 11 years old and your caregivers and everyone you know to be authority figures say that this thing you hold as your deepest and scariest secret is bad and deserving of condemnation, not just by humankind but by God the Creator, what would you feel? When you're 16 and in your isolation, you look for someone, anyone, whom you can trust to share this with and whom you think will make you feel okay about yourself, you tell your best friend that you're gay and the next day "FAG" is written across your locker and five guys you've never met before follow you to your bus stop and beat the shit out of you. What would these experiences do for you? How would this trauma shape your life?

BGLT people have heard religious leaders describe them as abominations that are a threat to families and to America's social fabric. Gay men have been scapegoated for the spread of AIDS. Same-sex love has been called "sodomy," a term rooted in the biblical story of the sinning towns of Sodom and Gomorrah. Condemnation by religious institutions, particularly when these institutions are the same ones in which people have been raised, can create tremendous fear and shame. BGLT people growing up in such institutions and hearing words condemning a part of what they are very often tend to internalize this as self-hatred.

Condemnation of gay people does not occur exclusively in religious institutions. Indeed, political platforms have been built on it! (Witness the number of people at the 1992 Republican Convention who held printed signs that stated "Family Values Forever. Gay Rights Never!") The military ban on lesbian and gay people, the ban on same-sex marriages, the lack of state and national legal protection of BGLT employment rights, all bespeak the government-sanctioned discrimination of gay people.

Even in noninstitutional settings, discrimination and harassment take place. Hate crimes committed against gay people are not uncommon, and rather than protecting people against such atrocities, police are sometimes the perpetrators (Berrill, 1992). BGLT people in many cities must carefully consider whether they can enter a public establishment such as a restaurant or convenience store based on whether or not such a place is "friendly." After all, who wants to be called names and risk being assaulted just for having dinner?

Same-sex couples in the United States do not have the right to legally marry, although this issue is currently being reexamined in a number of states, resulting in some modifications to existing same-sex marriage bans. Depending on where they live, they may not be able to even hold hands in public or otherwise display their affection in any way that denotes anything other than platonic intimacy without the potential of being harassed or even physically harmed. The frustration of not being able to publicly share an innocent embrace or kiss with one's loved one is a foreign concept to most heterosexuals.

Another difficulty in acknowledging oneself as BGLT is the belief that committing to a same-sex partnership means losing the right to parent. In her study of lesbians and generativity, Slater (1995) illustrates this:

> [T]he social and legal barriers to lesbians retaining custody of their children is formidable. As it was for all previous generations of lesbians, coming out for many is still synonymous with agreeing not to raise children. As a result, these women lose their primary generative link the moment they adopt a lesbian identity, even if they are only in early adulthood. (p. 5)

The realization of a same-sex preference or transgender identity as a component or components of one's identity can be very often difficult to accept at first. To not only identify but to accept oneself as BGLT, embracing one's truth, is a major accomplishment to be lauded. It means that despite all of the oppression and cruelty that is heaped upon BGLT people, the individual holds a deeper value: honesty of one's true nature and a willingness to live according to this truth. As Slater (1995) describes of lesbian development, "Lesbians and other additionally oppressed women in particular must struggle to enhance, not inhibit, attention to their own personal development" (p. 4). This may be said for all BGLTs. In the process of coming out of the closet, coming out to oneself is the first courageous step.

SELF-ACCEPTANCE

Self-acceptance is often a very difficult challenge for gay and lesbian people, as the previous section illustrates. It often begins with realizing a lack of sexual interest in members of the opposite sex at a time when one's own peers are developing these sexual interests. At the same time, the person begins to feel attracted to people of the same gender. Although the realization of same-sex attraction may take some time to occur, there is a gradual realization that indeed one holds a same-sex attraction that is as present as the opposite-sex attractions that are held by one's heterosexual peers. Accepting one's sexual preference is the first step to coming out as gay or lesbian (the bisexual and transgender processes are a bit different and are explored in their respective sections).

In exploring acceptance of sexuality or gender identity, it is necessary to examine what these aspects of self mean for each of us. Gender identity is the way we experience our selves as a gendered person both internally and externally. A person has an understanding of gender and how s/he experiences it, thus an understanding of self in relation to members of both genders emerges. Western culture, in its failure to embrace gender ambiguity and less gender-polarized roles, tends to create a message that a person can be only one gender or the other. Sexuality is the way we experience ourselves as relational beings with the people with whom we choose to couple (or join in greater numbers than couples in the cases of some bisexuals who identify as polyamorous). That it is called "sexuality" demonstrates the limited understanding our language affords us in describing how we experience relationships. Marriages and unions are about more than sex. They are about constructing lives together that are based on interactive roles of sharing. This encompasses not just sex but leisure, emotional intimacy, sharing responsibilities to each other and often to a shared residence, and creating a life-giving family experience.

Accepting oneself as BGLT is therefore very challenging. While one may have the realization of attraction to members of the same sex or a nontraditional experience of gender identity, acceptance of this fact comes at a higher price. It is not easy to embrace something that will inevitably lead to some form of mistreatment. Simply put, American society in many ways treats BGLT people poorly.

BGLT ADOLESCENT DEVELOPMENT

Gender identity and sexual orientation begin to develop during adolescence (Newman & Newman, 2003). Identifying a same-sex attraction or a cross-gender orientation at this development stage is challenging for a number of reasons. It's important to note that there is still considerable professional debate on the genesis of both same-sex attractions and cross-gender identities. Despite a growing body of research that suggests biological origins for both, sadly, there are still many who assert that being BGLT is a choice (though probably few who are BGLT have ever said this!). The intent of this chapter is not to highlight such debate. Through my experiences in knowing hundreds of BGLT individuals personally, professionally, and clinically, as well as through acknowledging my own identity as a transgender woman, I have concluded that it is a matter of accepting one's nature rather than a choice. Reflecting on my adolescence, I can honestly say that I never chose to be emotionally and physically assaulted, sometimes daily, for my gender-atypical behavior. My gender identity was simply a part of me that was ever present and impossible to hide.

Adolescence is an age period of heightened self-consciousness. A primary developmental task is to identify a peer group that serves to affirm one's identity. Unless the BGLT adolescent has a particularly BGLT-friendly family or peer group, this difference may result in feelings of isolation. This is particularly challenging because adolescents generally have very limited access to community resources that could provide responsible and BGLT-affirming information.

BGLT issues are often negatively socialized in families and among teenagers. "Fag" is the most widely heard epithet in schools today. While the tern "fag" is often used at people who are not actually gay, its intent to demoralize people who are even thought to be gay should not be overlooked. Additionally, the word "gay" is often used in adolescent parlance to denote something that is thought to be stupid, boring, or otherwise unacceptable. The language we use to construct meaning of our world is powerful in how it shapes our identities. If the adolescent has only heard of BGLT people called "fags" or "dykes" and described as sexual deviants, the difficulty of feeling different becomes even more painful because the person now feels different *and* reviled. While no one else may be aware of how the BGLT adolescent feels, the adolescent may experience a self-shaming dialogue: "If they only knew, they'd hate me and probably try to kill me," or "Dad will for sure throw me out of the house!"

Heyward (1989) describes heterosexism as holding in place "deeply personal institutions such as the self-loathing of homosexual youths and the hatred of such youths by their peers" (p. 3). This segues to the problem of identifying same-sex attractions or a cross-gender identity. The emotional difficulty of accepting one-self as a castigated member of society is difficult and painful. Experiencing this castigation firsthand can be physically dangerous. Violence against BGLT youth is seldom publicized. This probably has much to do with the shame BGLT adoles-cents may experience. They may be afraid to tell their parents they are being harassed, even making up elaborate stories to explain evidence of physical assault on their persons. Even when they do know the reason their child is being victim-ized, families often are too shocked and embarrassed at the fact that their child is BGLT and do not want to publicize their children's experiences. At a time when other students are focused on dating, football games, and their homework, the BGLT adolescent is often experiencing great physical and emotional danger in the school system.

It should be noted that not every BGLT person's sexual orientation or gender identity is evident. Further, many non-BGLT adolescents support their BGLT peers and accept them (this tends to vary by region and social fabric at a particular school). Fortunately, many cities have gay and lesbian teen hotlines and support groups. Unfortunately, not all BGLT youth have access to such services, which are primarily in larger cities with sizeable gay and lesbian communities. Also, these are sometimes targeted as gay or lesbian youth services, and may not address the addi-tional complexity of either bisexuality or transgenderality.

To be ostracized at such an age can lead to depression, isolation, high inci-dence of drop out from secondary schools, significant needs for affirmation that result in risky sexual behavior, and heightened risks of suicide attempts. McBee and Rogers (1997) estimate that gay and lesbian youths may be two to three times more likely than heterosexual youths to attempt suicide. They also identify suicide risk factors, some of which are not limited to gay and lesbian youth: previous sui-cide attempts, substance abuse, family dysfunction, identity confusion, social ties, and feelings of social inequity.

Families can also have a strong influence on the adolescent's work toward self-acceptance. The adolescent needs, like all adolescents, to know that she or he is loved and accepted as a unique individual. Family rejection at any stage can be emotionally devastating, but at adolescence it can be significantly damaging. Alternatively, families who support a BGLT youth and wish to ensure his or her well-being can be powerful advocating agents who intervene with the school to ensure the youth's safety. Parents and Friends of Lesbians and Gays (PFLAG) is a support network that has emerged as part of this effort. Such efforts not only demonstrate to the adolescent the family's love and support of her or him but also teach the adolescent important lessons about self-advocacy that are likely to be necessary throughout her or his life.

WHAT IT MEANS TO BE "OUT"

> We cannot simply up and leave heterosexism behind until we leave this world via death. Coming out refers not to leaving behind us the structures of oppression, but rather exiting from "closets" of psychospiritual, physical, and political bondage. A closet is a lonely, cramped place in which to hide . . . a place of disconnection and disembodiment in which, because we are out of touch with one another, we are out of touch with ourselves. (Heyward, 1989, p. 4)

The need to be "out" is often perplexing for heterosexual people to comprehend. If one will experience oppression as a result of being openly BGLT, why would *anyone* who could pass for straight and traditionally gender-expressed choose to be out? Pope (1995) provides us answers to this: (a) personal reasons such as honesty, integration of one's sexuality into every aspect of one's life, recognition of who the individual is as a person, and the need for support from those around them; (b) professional, political, or societal reasons such as providing a role model for other gay males and lesbians, desensitizing one's coworkers and oneself toward the issue, and eliminating any fear of blackmail; (c) practical reasons such as obtaining health benefits for one's partner, allowing one's partner to attend business or social events, and to avoid slips of the tongue and embarrassment when it inevitably comes out. The most important reason, Pope states, is the full integration of every aspect of the self into one fully functional human being.

Carter Heyward (1989) had this to say of her own coming out,

> Coming out pushes me further, a day at a time, into a realization that I don't want to pass, not really [as heterosexual]. This has been the most liberating, creative, and painful lesson of my life—learned in an educational matrix shaped by friends and lovers and enemies, by students and teachers and therapists and compañeras and all sorts and conditions of other creatures. (p. 7)

Perhaps the most succinct reason is provided by Weinberg's (in Gartrell, 1984) contention that the amount of psychological damage in lying is directly proportional to the amount of self-contempt that motivates the lie. Lying about being BGL legitimizes the shame that is taught by a heterosexist culture. Remaining in the closet discredits a normal form of human sexuality and enforces the notion that there is something wrong with being BGL.

The term "out" has even more meanings. It may mean expressing oneself as a transgender rather than the sex a person appears. It may also mean disclosing bisexuality after previously proclaiming oneself gay, lesbian, or straight. Coming out describes an honest and authentic proclamation of an important part of one's identity that is shared as a means of allowing the individual to live a more authentic life.

It should also be stated that for many being out is an aspirational goal, but not necessarily a very realistic one. The need to remain in the closet for fear of losing one's job or familial and social status are all important considerations as people make decisions whether and how much to disclose about their sexuality.

Acceptance of self for BGLT individuals tends to be rather challenging because of all the negative socialization Western culture constructs about anything other than heterosexuality and traditional gender roles. Sharing this experience of self with family and friends is perhaps even more daunting, for it means facing the possibility of rejection. Fear of rejection is a relatively common occurrence for people who interview for jobs, ask others for a date, and engage in other daily aspects of life. This fear is often a cognitive distortion, an amplification of potential risk. For BGLT individuals, a fear of coming out is not based on distorted beliefs. Rejection of BGLT people by their families is often a reality and is something many of us are witness to our entire lives.

To be rejected for something as central to one's nature as sexuality or gender identity may be difficult to grasp for many heterosexuals, who experience socially sanctioned gender roles. It would mean having to forge a completely new support network. In many cases it means losing a sense of home. Where does someone who is rejected by family go for holidays or to celebrate his or her birthday? To be cut off, to be told either directly or indirectly that one does not belong, would be extremely devastating. The risk of coming out to one's family is possibly one of the most defining moments of BGLT life.

BGLT people often come out to trusted friends before they come out to family members. These friends may be other BGLT people who are already out. There are support groups in many communities to help people who are coming out make decisions on how and when to come out to people in their lives. This illustrates community mindedness that comes from the hard work of many socially conscious BGLT individuals.

It is important for gay and lesbian people to identify and confront their own feelings of homophobia before coming out. Gartrell (1984) states that for lesbian and gay clients in counseling the process of coming out should begin from a position of as much strength and pride as possible. If a client has recently had a relationship loss, it may be an inappropriate time to come out to one's parents. Confronting typical parental concerns that a child will have a more difficult life by being gay or lesbian at a time when one is already experiencing the pain of a breakup will lead to unnecessary self-doubt.

There is no one right way to come out. Many factors influence the decision. How do different members of the family perceive homosexuality? Has the family had exposure to real BGLT life as opposed to media-slanted antigay propaganda? Are there any close friends to the family who are BGLT friendly who could help ease the process? It is often helpful to have useful information and phone numbers on community resources such as Parents and Friends of Lesbians and Gays (PFLAG).

In addition to coming out to one's family, a working BGLT person must decide if it is safe to come out to an employer. Would it mean risking one's job? If a BGLT chooses not to come out at work, he or she often has an "alibi" planned to inevitable questions about one's marital status or children. What does one say when a well-meaning mother wants to set up a closeted lesbian coworker with her son? Some BGLTs must construct elaborate stories about their personal lives in order to skillfully rebut these situations as a means of survival.

Coming out is a complex and difficult process that is never really "complete" because in a heterosexist culture, people who are BGLT are very often assumed to be heterosexual. Every time a BGLT person encounters a new person or setting s/he must make a decision on whether to be out when innocuous questions about marriage or family status are asked. For transgenders who "pass" as our reassigned genders, whether to be out occurs every time we're asked on a date. We are also often forced out when we have to do anything concerning birth records or that otherwise reveals our birth name and birth sex: applying for passports, marriage licenses, driver's licenses, and so forth.

While heterosexuals are safe to openly discuss their partner and family situations in virtually any context, this is much more restricted for BGLT people, even those who are out. Even so, there are many ways to experience being out. Some are open in their social settings, workplace, and with families of origin. Others are open to selected friends and family members while electing not to disclose to employers or other institutions. Still others maintain privately BGLT lives that are not shared with anyone in their family of origin. Ideally, the world should be safe enough for everyone to be open about being BGLT. This is not the case, however, and individual BGLT people must make their own decisions about how and to whom they choose to disclose their sexual or gender identities.

RELATIONSHIP CHALLENGES

If you are in a heterosexual relationship, imagine what it would be like to constantly have to pretend that your spouse or partner is nothing beyond a platonic friend. How would you act? Would you guard the laughter you share so as not to appear too intimate? Would you publicly hug, kiss, or hold hands? Probably not! In any case, you would most likely feel very self-conscious about how you as a couple looked and acted and would work very hard to try not to draw attention to yourselves. If you are in a heterosexual relationship, try it for a day with your significant other. Try for a single dinner date to act only as two platonic friends. Does it sound ridiculous and frustrating? Unless they are in a particularly predominantly BGLT-friendly area, this is the daily reality of same-sex coupling.

Given the history of shame that BGLTs have had to endure, there is perhaps a greater danger of shame-based behaviors surfacing in BGLT relationships than in heterosexual relations. As Jordan (1989) suggests, "People who are shame prone are used to taking the blame for relational failures. Therapy can offer an invaluable opportunity to heal the long range wounds of shame" (p. 7) by acknowledging the reality of heterosexist shaming.

Perhaps a mixed blessing for BGLT relationships is the lack of rigid gender roles that dictate the behaviors of heterosexual men and women. Men and women who move into partnerships generally have some idea of what coupling looks like based on their experiences in their families of origin. Same-sex couples, for better or worse, often do not have the same cultural norms governing specific behaviors for each partner. This leaves couples more freedom, with less of a blueprint from which to construct their relationship.

To be BGLT means that one cannot simply rely on conventional wisdom about partner selection and family making. It requires one to challenge the most basic notions one was raised to believe about his or her role in relationships. The concept of families as patriarchal and heterosexual institutions has created the notion that BGLT families are somewhat of an oxymoron. This is not the case, as progressively more BGLT couples are electing to have children (Johnson & O'Connor, 2002). For a discussion on same-sex parenting, see Chapter 13 on "Familial and Relational Transitions Across the Life Span."

BISEXUALITY

If gay and lesbian people are underrepresented in contemporary literature, bisexuality is virtually nonexistent. In a study entitled *Counseling Bisexual Clients* (Smiley, 1997) the author identified that out of seven counselor education textbooks, on average, a 400-page book devoted under 1 percent of the text, less than two pages, to gay and lesbian issues. Only one textbook made any specific reference to bisexuality, in a paragraph about 100 words long. My research on the topic five years later is equally disappointing.

As a sexual identity and preference, bisexuality has historically been oppressed not only by heterosexism but by many in the gay and lesbian communities as well. Bisexual people are often inappropriately labeled as confused or partially out of the closet. It would seem that just as homophobia is beget by heterosexism, biphobia is beget by both heterosexism and homosexism. The very qualities that activists for gay and lesbian rights indict in heterosexism may also be understood to be behind the oppression of bisexuality. As gay and lesbian people are misunderstood and labeled by heterosexists, so are bisexual people misunderstood by both heterosexists and homosexists. For this reason, it is more difficult for bisexual people to find a supportive community.

Bisexuality is described by Christopher James (1996) as "the sexual or intensely emotional, although not necessarily concurrent or equal, attraction of an individual to members of more than one gender" (p. 218). Making this term complex is its use as a means of describing a variety of sexual attractions. People often describe themselves as bisexual, homosexual, or heterosexual according to changing life circumstances. For some people, bisexuality is a self-identifier used as a stage in the process of coming out as gay or lesbian and is later abandoned. Alternatively, the self-identified bisexual may later abandon the description and become in-volved in exclusively heterosexual relationships. For many others, it is a term used intermittently and alternatively with homosexual and heterosexual to describe a current life circumstance in relation to another person. A growing number of bisexual–identified people, however, are embracing the description and recogniz-ing it as their sexual identity.

Many bisexual women, men, and transgenders continue to identify with gay culture because it forges a political identity that is difficult to ignore.[3] In spite of this, the bisexual community is often mistreated by non-bisexuals, who often label them as confused. Another criticism that bisexual men and women hear from some people in the gay and lesbian communities is that they don't want to come out of the closet all the way, thus losing all "rights" associated with life in the het-erosexual world. It is further stated by many gay and lesbian people that bisexual-ity is a myth and that it weakens gay and lesbian political causes.

What precisely then is a bisexual identity? No one answer will suffice. It is many things to many people. According to a study performed by Rust (1996), many people change the way they describe their sexual preference and identity as their life circumstances change. These changes can reflect relationship changes or social and political changes. One of Rust's participants, for example, previously described herself as a lesbian but changed this designation when she began feeling an attraction to a man. The attraction later developed into a sexual relationship. She adopted a bisexual identity to more adequately describe her former relation-ship with a woman and her new relationship with a man.

In other circumstances, a single bisexual relationship does not lead to a com-plete identity change. In some cultures, bisexual and homosexual behavior may be structured according to active and passive sex roles. According to Fox (2000) and Williams (1988) this is typical of some Mediterranean and Latin American coun-tries, where bisexuality is normative for a great part of the male population. The active partner has relations with both sexes, yet identifies as heterosexual. The receptive male partner has relations primarily or exclusively with other men and is seen as homosexual.

Many respondents to Rust's 1996 study stated that they identify as gay or les-bian, because, as stated above, they share the gay and lesbian political affiliation. This has been especially true of women, who did not want to lose the lesbian iden-tity until they thought they could feel connected to a bisexual movement and political force via a bisexual identity. Some are now discovering that a bisexual identity indeed does connect them. This is giving rise to many former self-declared lesbian and gay people now proclaiming their bisexuality.

Fox (2000) identified several typologies of bisexuality that are based on the extent and timing of past and present heterosexual and homosexual behavior. He cites Klein's study in 1978 that differentiated transitional, historical, sequential, and concurrent bisexuality. According to the study, transitional bisexuality repre-sents a stage in the process of coming out as gay or lesbian, or alternatively, a gay or lesbian identity may be a step in coming out as bisexual. Historical bisexuality refers to individuals who presently identify as homosexual or heterosexual but have previously experienced relationships that were contrary to this current designation.

Some individuals have had relationships with both men and women, but with only one person at a given time (sequential bisexuality), while others have relationships with both men and women during the same period (concurrent bisexuality). This last, concurrent bisexuality, is often referred to as "polyamorous" in bisexual parlance.

Fox (2000) also uses M. W. Ross's description of circumstantial bisexuality. A person may be hiding, experimenting with, or transitioning to a gay or lesbian (defense bisexuality). When a society provides no alternative to heterosexual marriage, gay or lesbian relations may take place away from home (married bisexuality). There may be a cultural expectation of some form of same-sex relations, as in some African tribes as well as Melanesia (ritual bisexuality). Some people do not use gender as a criterion for sexual attractions (equal bisexuality). Finally, the male who performs the non-receptive role in same-sex intercourse is considered heterosexual ("Latin" bisexuality).

There are clearly many instances in which one might describe her/his self as "bisexual" depending on past or present relations, culture, contexts, and political ties. As with all forms of sexual identity and preference, the term "bisexual" is a concrete term that is used to describe something fluid and abstract. As our language evolves to describe the great complexity of human experience, it is likely that people's self-description will also change to more accurately reflect their personhood.

TRANSGENDERS AND THE GENDER CONTINUUM

Just as sexuality exists on a continuum in which individuals identify themselves, gender identity may also be seen as a separate but parallel continuum. A variety of gender expressions have been demonstrated across many cultures throughout human history. These may have included occasional cross-dressing or opposite gender role-playing to hormonal and surgical reassignment of morphological gender. Because people's gender identities have been expressed in a number of ways, there is no definitive transgender or transsexual identity. Expression of gender identity, and indeed how transgenders experience themselves, may be understood as largely a social construction.

Leslie Feinberg (1996), a female to male transgender author with a nonpolarized gender identity, describes the social construction of a transgender identity further:

> All together, our many communities challenge all sex and gender borders and restrictions. The glue that cements these diverse communities together is the defense of the right of each individual to define themselves. . . . Transgender people traverse, bridge, or blur the boundary of the gender expression they were assigned at birth. . . . I am transgender and I have shaped myself surgically and hormonally twice in my life. I reserve the right to do it again. (p. xi)

To emphasize the social construction of transgender identities, I offer as a comparison the tradition of the two-spirits in Native American cultures. Many North American tribes, including but not limited to the Lakota, Zuni, Papago, Yaqui, Kodiak, Cheyenne, Omaha, and Navajo, considered the androgynous two-spirits of their respective tribes to be sacred because they were believed to be created as the embodiments of male and female duality. As such, two-spirits were treated with great respect and often held the role of shaman, or healer for the sick. They were generally males who from early childhood were identified for their special "gift" of androgynous or effeminate behavior. There is also evidence of girls who were raised to be warriors, generally when the child demonstrated natural proclivities of the male tribal role. For these, I use Walter Williams (1988) descriptor, "amazon."

When the parents identified such a child, they would begin the process of seeing to it that the child was trained in the tribe's tradition of the two-spirit; in the case of boys this generally included dressing them in girls' clothing and teaching them how to perform the tribal functions of women as well as those uniquely assigned to two-spirits. The tribes accepted their two-spirits and embraced them for their unique gifts, which were seen as blessings of great fortune by the spirit (Williams, 1988). Similar traditions exist in indigenous tribes throughout the world.

While the two-spirits lived as women and the amazons lived as men, it was understood that in neither case were these truly men nor women: they were considered their own separate but equal genders. They were also not what we think of in contemporary Western culture as "transsexuals" because there was no transition between sexes.

Contemporary Western culture offers a different view of gender identity from that of the two-spirit/amazon tradition. Gender is considered an "either/or" proposition, rather than the "both/and" that appears in the aforementioned tribal cultures. Additionally, contemporary Western culture has constructed a narrow range of behaviors that are considered acceptable for either gender. The diagnosis of Gender Identity Disorder is described as a strong identification with one's opposite gender and a "persistent discomfort or strong sense of inappropriateness" (APA, 2000, p. 581) with one's morphological sex. This view of gender tends to influence the socialization of people who are nonpolarized on the gender continuum. To emphasize the role of culture in how we construct identities, the differences in the socialization of gender identity between indigenous cultures and that of ours are contrasted.

While the conditions in the tribes cited above promote a child's growth into her/his gender role, contemporary Western culture oppresses it. Poor treatment of BGLT youth by our contemporary culture has been cited earlier in this chapter. Akman & Fisher (2002) describe the problems of transgender youth in our culture as being particularly complicated because of the hostility of other adolescents in response to their atypical gender behavior. As two-spirits and amazons have transitioned into adulthood, their roles in the tribe afford them great respect. Their wisdom and guidance are sought by other members of the tribe. In contemporary Western culture, transgenders are often assaulted, discriminated against, and subject to perhaps a higher degree of the maltreatment than the general BGLT communities experience. Finally, while two-spirits/amazons are given the respect of their tribes and accepted as shamans and hunters, respectively, transgenders in contemporary Western culture are viewed as having a "disorder" (APA, 2000).

It should be noted that Native American two-spirit culture has largely been suppressed or in some cases hidden due to the influence of the dominant culture over the lives of Native Americans. This suggests again the pervading role of shame, and illustrates how shame works to discredit and dissolve nondominant cultural practices.

Professionals within the Western medical community have provided a model designed to aide health care professionals in assisting Gender Identity Disorder (GID) patients with diagnoses and care. The Harry Benjamin International Gender Dysphoria Association's Standards of Care for Gender Identity Disorder provides for a number of medical and mental health issues, including the recommended requirements for a patient to receive gender-reassignment surgery. Currently, this includes two mental health professionals' identification of Gender Identity Disorder and recommendations for surgery (Meyer, Bockting, Cohen-Kettenis, Coleman, Diceglie, Devor; 2001). Further requirements include living in one's opposite gender for one year, and six months of opposite-sex hormone treatment.

Many people who would otherwise be classified as having varying degrees of Gender Identity Disorder are denouncing the classification of their biopsychosocial state as a disorder, instead pronouncing transgender, gender variant, or any of a variety of cross-gender nomenclatures as their gender classification, rather than male or female (Carroll et al., 2002). Many transgenders are at odds with the Standards of Care for Gender Identity Disorder (Meyer et al., 2001), which are seen as assigning behaviors to one gender or the other. Further, the use of mental health professionals as enforcers of this binary system is also at odds with an understanding people have of themselves as existing in a nonpolarized place on the gender continuum. According to Carroll (2002), The International Bill of Gender Rights that was established at the 1993 Conference on Transgender Law and Employment Policy included the right to "freedom from psychiatric diagnosis and treatment" (p. 133).

Expression of gender can be significantly enhanced or hindered by the community in which the transgender lives. If the individual is able to pass as the opposite gender, or alternatively, if she or he lives in a trans-friendly community, gender expression may be made easier.

An important note for counselors is the difficulty transgenders often experience in their access to human services. Domestic violence shelters, for example, are traditionally established for women and have been known to deny access to transgenders/transsexuals. Similarly, inpatient rehabilitation services and homeless shelters, which are sometimes dorm style with separate sections for men and women, often refuse transsexual/transgender people admittance.

There is often a mistaken assumption that all transgenders have heterosexual orientation (that MTFs are attracted to males and FTMs are attracted to females). Rosario (1996) identifies a case of a FTM who wanted to be a gay male. The description of such cases as rare by the DSM-IV (1994) provoked, as Rosario describes, skepticism for the psychiatrist. States Rosario:

> The paradoxically designated "heterosexual transsexual"—the female to gay male or male to lesbian, poses an enormous problem to this heterosexual paradigm. As several psychiatrists perplexedly exclaimed to me, 'If a woman is attracted to men, as normal, why would she want to become male in order to be a homosexual?' This violates the imaginary balance of sexual psychopathologies, which weighs the stigma of transsexualism against that of homosexuality. This hegemonic heterosexual logic seems to be that the female-to-straight-male transsexual is gaining the normalcy of heterosexuality at the price of her breasts and genitals, while the FTGM [female to gay male] is striving for a dysfunctional goal of double social opprobrium. (pp. 39–40)

In this case, the transsexual felt a strong desire to be male and dissociate with her female gender (gender identity). At the same time, she felt attracted to males (sexual orientation). She wanted to both be a male and have intimacy with other males. She identified herself as a gay male. Thus, the psychiatrists who assumed that she would be sacrificing her breasts and female genitalia when she felt attracted to men were unclear of the difference between sexual orientation and gender identity. There was for this patient no sacrifice because she did not identify as a woman.

The second bias of the other psychiatrists in Rosario's study is the heterosexist assumption that it is somehow better to be a heterosexual woman than a gay male. While certainly gay males may experience more oppression and mistreatment than heterosexual women, the notion is that the FTGM should ignore her identification with the male gender in order to fit into heterosexist cultural norms. This bias by people in the helping professions is further illustrated by Namaste (1996):

> At gender identity clinics, transsexuals are encouraged to lie about their transsexual status. They are to define themselves as men or women, not transsexual men and women . . . they are to conceive of themselves as heterosexuals, since psychiatry cannot even begin to acknowledge male-to-female transsexual lesbians and female-to-male transsexual gay men. (p. 193)

Well-intentioned helping professionals may try to encourage transsexual and transgender people to adhere to gender norms, including appropriate "heterosexual" relations. No matter the intention, this ignores the rich complexity of human development and the freedom of individuals to self-determine their identity. It places the emphasis on conditional acceptance of the individual according to norms that may or may not fit rather than allowing the client to construct her/his own narrative.

AGING AND THE BGLT POPULATIONS

The diversity of experiences among BGLT individuals, who span numerous cultures and ethnicities and who all have unique and separate family situations, makes it difficult to generalize such a broad topic as aging. For the purpose of this section, I describe general characteristics and developmental occurrences that are based on some of the research on the subject of aging and sexual orientation.

As BGLT individuals progress in life, new developmental challenges occur. Kimmel (2002) demonstrates this:

> Think of the activity you most enjoy doing, whether it is painting, playing golf, hunting, cooking, traveling, or playing with your grandchildren . . .
>
> Now think of the person or people you love the most and how important they are to you . . .
>
> Now imagine that you fell and broke a hip and wound up in a hospital and then in a nursing home for rehabilitation—hopefully for only a few weeks until you can return home.
>
> However, in that nursing home you cannot let *anyone* know your favorite activity, or the person or people you love the most. You cannot mention anything about them. (Kimmel, 2002)

This description provides the experience of many elderly BGLT individuals, who do not wish to infer their sexual orientation by disclosing information about activities or loved ones. Kimmel closes this description by adding:

> Finally, imagine that you are going home with a home health aide whom you do not know. As you think about the way you left your home, you recall all the photos of your loved persons on the shelf and all of the signs of your favorite activity scattered around the house. How are you going to explain them to your new aide? (pp. 17–18)

To understand this occurrence from the experience of elderly BGLT individuals, it is important to trace the historical roots of the BGLT movements during the 20th century. Many men and women with same sex attractions who served in the armed forces throughout the 20th century were discharged if their sexual orientation became known. In many cases, this prevented access to GI benefits. In 1953 President Eisenhower issued Executive Order #10450, which encouraged dismissing homosexuals from government jobs. This executive order coincided with the anticommunist crusade of Senator McCarthy (Kimmel, 2002). Gerassi (1968) and Loughery (1998) describe in Kimmel the police harassment of 1,500 men in Boise, Idaho, who were interrogated about their sexuality, including ten prison sentences for men who were involved in gay relationships. An investigation at the University of Florida in 1958–59 led to the witch hunt of homosexual faculty. Sixteen faculty members, including several who were published and two who were Fulbright scholars, were terminated. All were later evaluated as capable or even outstanding faculty (Kimmel, 2002).

Berger (1982) describes the experience of gay men in a study for his work, *Gay and Gray*:

> Those who "came out" after World War II and had access to large metropolitan areas were able to frequent the few bars which catered to homosexuals. But

these places were hidden away and were often difficult for the novice to find; once he got there, the atmosphere was less than relaxed. Dancing, touching, or any display of affection was not permitted, and drinks were watered down and expensive. Patrons were likely to give first names only and were often unwilling to meet outside the gay bar. Police raids were common. Friendship cliques met regularly, but it was often difficult to break into these. And if the individual held a professional or other responsible job, he might not want to join the clique whose members were too "obvious," lest his employment be endangered. He might also be hiding his involvement from family members.

A particular suburban house became a meeting place for a group of homosexual men. Invitations were solely by word of mouth; the host carefully screened each potential guest before informing him of the location, lest he be a police informer. Guests were instructed to arrive one at a time, in order to avoid attracting the neighborhood's attention. Guests who traveled to the area in small groups would split up before approaching the house. (p. 188)

During the 1950s, the Daughters of Bilitis and the Mattachine Society were formed by individuals who were willing to announce their sexual orientation and wished to reach out to others (Beeler & Herdt, 1998). Kimmel (2002) cites Loughery's account of the Mattachine Society's conference in 1959, which, when it was reported in the *Denver Post* on September 4–6 led to the arrest of the local organizer. His home was searched and he lost his job during the 60 days he was held in jail.

The police raid of the Stonewall Inn Bar in New York City's Greenwich Village was a turning point in BGLT history. At 1:00 A.M. on June 28, 1969, the police raided the bar, which was privately owned. The crowd fought back. The event received a great deal of press, and the next night, 1,000 people flocked to the site. Several days of rioting and protest followed. One week after the event, a gay march commemorated the uprising at the Stonewall Inn. One year later, a pride march in New York and several other cities was held. On the 30th anniversary of the event, the site of the Stonewall Rebellion was listed on the National Register of Historic Places.

In 1963, Betty Friedan's revolutionary work, *The Feminine Mystique*, called attention to women's oppression in American culture. Feminism, the women's movement, was born. During the 1970s, feminism gained momentum, and following the events of the Stonewall Rebellion, many women in the feminist movement began coming out as lesbians.[4] For lesbians, feminism was an especially powerful and necessary political call to action that went hand in hand with their identity as lesbians: it represented overcoming male political oppression as well as male control over women as sex objects. It represented a woman's rights to choose not only her education, career, and household status, but also the people she would love and nurture in relationships.

In the 1980s, the AIDS epidemic rapidly spread in the gay male community, leading to the stigmatization of AIDS as a gay disease. Religious fundamentalists trumpeted AIDS as God's condemnation of gay people and homophobic hatred was fueled.

The history of the BGLT movements, and the influence these experiences had on the people who have lived through them creates tremendous diversity in the elderly population. Many BGLTs have been active and out from the beginning of the rights movements. Others may have lived portions of their lives closeted in response to these events, even marrying and having children, choosing later to come out selectively to trusted family and/or friends. For still others, the historical events have been peripheral to their lives, whether they were leading BGLT lives or not, not entirely aware of the events that were occurring of historical significance to the BGLT movements.

The events of the BGLT movements have influenced the American culture, and thus the context in which BGLTs may come out to families. Coming out today is different than it was in previous decades. Consideration of how the person's sexuality will be perceived by one's family is a great consideration in the decision of how and when to come out. As BGLT awareness evolves in American culture, the perception of safety in coming out to one's family is likely to evolve.

Coming out for older BGLTs has significant differences than for those who are younger. As Beeler and Herdt explain (1998), younger BGLTs may still be dependent upon their families of origin for emotional and financial support, including their housing. Older people generally are more financially and emotionally independent and have often established support networks that serve as families of choice. In many cases, they may be providing support for their families of origin.

Beeler and Herdt (1998) also point to the difference in reaction to disclosure of one's nondominant sexual orientation or gender identity between families of older and younger BGLTs. Parents of younger BGLTs will likely consider how their children's sexual orientation will create problems in the children's lives in the future: discrimination, hate crime victimization, AIDS risks. They may also experience their own feelings of despair over the loss of dreams they had for their children, such as marrying an opposite-sex partner and having children. They may have concerns over how the family will be perceived by others, and blame themselves or the other parent for the child's sexuality.

For older BGLTs, disclosure to parents brings less concern over the future. Certainly, the threat of hate crimes and other social problems still exists, however the family has the person's history as a basis of comparison, and is likely to be less concerned over this potential than parents of younger BGLTs who do not have this. Additionally, parents of older BGLTs have had a longer period of time to adapt the idea of their children never becoming parents. A single male child of 55, for example, has less pressure to marry and have children than one who is 25.

Instead, parents of older BGLTs question and explore the past. As Beeler and Herdt (1998) describe, families may feel that the years previous to coming out were based on a fraudulent or deceitful relationship with the newly out person. This may be especially true of the BGLT who is married or has children from a current or previous heterosexual relationship. Alternatively, families of older BGLTs may not be entirely surprised, perhaps always suspecting that a single uncle who has lived with the same male roommate for 20 years is gay.

In a 2001 study performed by Grossman, D'Augelli, and Herschberger (cited by Kimmel, 2002) of a sample of 416 BGLT participants over 60 years old, 84 percent said their mental health was good or excellent, 14 percent said it was fair, and only 2 percent reported it to be poor. Their self-rating of mental health related to income and inversely related to sexual orientation. Living with a domestic partner was associated with higher self-esteem and greater personal satisfaction. Eight percent reported being depressed because of their sexual orientation; 10 percent said they consider suicide sometimes; 13 percent said they had made a suicide attempt, most often between the ages of 22 and 59. Comparing this statistic to McBee and Rogers's (1995) study of suicide risks among gay and lesbian adolescents (described earlier in the adolescent section of the chapter), which was found to be associated with concerns over acceptance, social ties, family dysfunction, and feelings of social inequity, one can conclude that older BGLTs have greater coping abilities made possible through their lifetime of accomplishments and social ties that have been established.

It is important to remember that while all BGLTs experience heterosexism as a culturally repressive element, older BGLTs experience the added burden of ageism. To understand the effects of aging in the BGLT population, Kimmel (2002) compares and contrasts ageism and heterosexism:

There are numerous similarities between aging and minority sexual orientation, which may be summarized as follows. Both old age and minority sexual orientation:

1. Have been the focus of an active search for biological origin, and possible cure, despite the fact that both are normal human characteristics.
2. Evoke irrational fear and avoidance in some people, who tend to avoid close contact and physical touching with both groups.
3. Evoke confusion with associated conditions: ageing with senility or death; sexual orientation with gender identity or promiscuity.
4. Operate as a master status that obviates other relevant social positions and characteristics.
5. Are perceived as being best to avoid if possible; they are both dealt with by "Don't ask, don't tell" policies.
6. Are characterized more in terms of their perceived disadvantages than their advantages; losses are thought to exceed gains, strengths are seen only as compensations for weakness.
7. Are discriminatory views—ageism and heterosexism—that emphasize the importance of fertility and propagation as normative for everyone.
8. Are conferred a special status in some cultures, in which the individuals may be seen as having special powers resulting from their minority status.

In contrast, there are four clear differences between ageism and heterosexism:

1. Most people hope to become old one day; few hope to become a sexual minority.
2. No one blames the individual's choice, or his or her mother, for becoming old.

3. Families openly acknowledge and celebrate becoming older; few families celebrate their children coming out as lesbian, gay, bisexual, or transgender.

4. Churches and moral guardians do not urge older persons to avoid acting old, but they often urge sexual minorities to avoid acting on their erotic or romantic attractions. (p. 31)

Clearly, the historical oppression of BGLT people and the subsequent rights movements have had significant influence on individuals as well as for future generations of BGLTs. The development of social awareness of BGLT issues has also changed the context for coming out. There is great plurality in the lives and experiences of older BGLTs, which influence their beliefs about their sexual orientation and how or whether they choose to present it.

CONCLUSION

There are a large number of variables that influence the lives of BGLT individuals and communities. It is important to consider the influence of social oppression on individual development and examine what could be different if this oppression would not exist. It is also relevant to examine the institutions that hold such oppression in place, and consider how these institutions have been used throughout history to deny rights of women, people of nondominant ethnicities, people of nondominant religions, and all of the rest of humanity whose names, faces, or physical abilities or attributes did not match those of the people in power.

The experiences of BGLT people do not have to include shame, abuse, or ostracism. Ask yourselves what necessitates these three dimensions, and whether they are worth the price of BGLT teen suicide, drug abuse, or even the development of low self-worth. What possibilities might exist for people if they did not receive extreme social oppression?

I ask you to consider how humanity could be made better by extinguishing social oppression. I ask you to reflect on the many gifted and talented individuals who have flourished and built societies as shackles have been removed, and consider how much more becomes possible when BGLTs would no longer be oppressed. I invite you to work with me in advancing a more enlightened age where, rather than denigrating individual differences, we celebrate both distinctions and similarities. Let us build communities that transcend social or political boundaries and instead experience the humanness of who and what we are. Considering ourselves as relational beings, we can begin to take responsibility for the experiences of all humanity, thus advancing principles of love and acceptance.

NOTES

1. The term "BGLT" will be used to describe the collective populations of bisexual, gay, lesbian, and transgender people. The reason for this is because there is no one designation that reflects the rich diversity of these groups. To use "gay" to

describe all BGLT populations, when "gay" is a designation adopted by gay males, would undermine the rich distinctiveness of the lesbian experience and would be completely inaccurate when describing bisexual people and nonhomosexual transgender people. Also, rather than using the more common "GLBT" or "LGBT," which assume political bias, I instead alphabetize my designations.

2. "Biphobia" and "transphobia" are words I have coined to describe the unique experiences of oppression for bisexual and transgender people. The experiences of these groups are described in their respective sections.

3. For more information on bisexuality within the feminist movement, read Udis-Keller, 1996.

4. I do not mean to imply that all feminists are or were lesbians, nor that all lesbians are or were feminists. Indeed, equal rights for women and overcoming historical sexism is a battle shared by women and men across all forms of sexual orientation and identity. The political coincidence of the women's movement with the beginning of the BGLT rights movements led to the natural cross-fertilization of lesbian rights and feminism. Consequently, many lesbians were and are active feminists.

REFERENCES

Akman, J. & Fisher, B. (2002). Normal Development in Sexual Minority Youth. In Oldham, J. M. & Riba, M. B. (Series Eds.); B. E. Jones & M. J. Hill (Vol. Eds.) *Review of psychiatry: Vol. 21. Mental health issues in lesbian, gay, bisexual, and transgender communities* (pp. 1–13). Washington DC: American Psychiatric Publishing.

American Psychiatric Association. (1980). *Diagnostic and statistical manual of mental disorders* (3rd ed.). Washington, DC: Author.

American Psychiatric Association (1994). *Diagnostic and statistical manual of mental disorders* (4th ed.) Washington, DC: Author.

American Psychiatric Association (2000). *Diagnostic and statistical manual of mental disorders* (4th ed., Text Revision). Washington DC: Author.

Beeler, J. & Herdt, G. (1998). Older gay men and lesbians in families. In D'Augelli, A. R. & Patterson, C. J. (Eds.). *Lesbian, gay, and bisexual identities in families. Psychological perspectives* (pp.177–196). New York: Oxford University Press.

Berger, R. M. (1982). *Gay and gray*. Urbana: University of Illinois Press.

Berrill, K. T. (1992). Anti-gay violence and victimization in the United States: An overview. In Herek, G. & Berrill, K. T. (Eds.). *Hate Crimes. Confronting violence against lesbians and gay men* (pp.19–45). Thousand Oaks, CA: Sage.

Carroll, L., Gilroy, P. J., & Ryan J. (2002*). Counseling transgendered, transsexual, and gender-variant clients. *Journal of Counseling & Development*, 80, 131–139.

Feinberg, L. (1996). *Transgender warriors: Making history from Joan of Arc to RuPaul*. Boston: Beacon Press.

Fox, R. C. (2000). Bisexuality in perspective. A review of theory and research. In B. Greene & G. L. Croom (Eds.), *Psychological perspective on lesbian and gay issues: Vol. 5. Education, research, and practice in lesbian, gay, bisexual, and transgendered psychology. A resource manual* (pp. 161–206). Thousand Oaks, CA: Sage.

Friedan, B. (1963). *The feminine mystique*. Norton Press: New York.

Gartrell, N. (1984). Issues in psychotherapy with lesbian women. *Work in Progress, No. 10*. Wellesley, MA: Stone Center Working Paper Series.

Halberstam, J. (1998). *Female Masculinity*. Durham, NC: Duke University.

Heyward, C. (1989). Coming out and relational empowerment: A lesbian feminist theological perspective. *Work in Progress, No. 38*. Wellesley, MA: Stone Center Working Paper Series.

James, C. (1996). Denying complexity: the dismissal and appropriation of bisexuality in queer, lesbian, and gay theory. In B. Beemyn & M. Eliason (Eds.), *Queer studies. A lesbian, gay, bisexual, and transgender anthology* (pp. 217–241). New York: New York University Press.

Johnson, S. & O'Connor, E. (2002). *The gay baby boom. The psychology of gay parenthood.* New York: New York University Press.

Jordan, J. V. (1989). Relational development: therapeutic implications of empathy and shame. *Work in progress, No. 39.* Wellesley, MA: Stone Center Working Paper Series.

Kimmel, D. J. (2002). Aging and sexual orientation. In Oldham, J. M. & Riba, M. B. (Series Eds.); B. E. Jones & M. J. Hill (Vol. Eds.) *Review of psychiatry: Vol. 21. Mental health issues in lesbian, gay, bisexual, and transgender communities* (pp. 17–36). Washington DC: American Psychiatric Publishing.

McBee, S. M. & Rogers, James R. (1997). Identifying risk factors for gay and lesbian suicidal behavior: Implications for mental health counselors. *Journal of Mental Health Counseling,* 19 (143–155).

Meyer, W.; Bockting, L., Cohen-Kettenis, P.; Coleman, E.; Diceglie, D.; Devor, H.; et al. (2001, January–March). Harry Benjamin International Gender Dysphoria Associations' the standards of care for gender-identity disorders (6th version). *International Journal of Transgenderism,* 5,1. Retrieved January 17, 2003, from http://www.symposion.com/ijt/soc_2001/index.htm.

Namaste, K. (1996). "Tragic misreadings": Queer theory's erasure of transgender subjectivity. In Beemyn, B. & Eliason, M. (Eds). *Queer Studies. A lesbian, gay, bisexual, and transgender anthology* (pp.183–203). New York: New York University Press.

Newman, B., & Newman, P. (2003). *Development through the life cycle. A psychosocial approach* (8th ed.). Belmont, CA: Wadsworth/Thomson Learning.

Pope, M. (1995). The "salad bowl" is big enough for us all: an argument for the inclusion of lesbians and gay men in any definition of multiculturalism. *Journal of Counseling and Development,* 73, (301–304).

Rosario II, V. A. (1996). Trans (homo) sexuality? Double inversion, psychiatric confusion, and hetero-hegemony. In B. Beemyn & M. Eliason (Eds.), *Queer studies. A lesbian, gay, bisexual, and transgender anthology* (pp. 35–51). New York: New York University Press.

Rust, P. C. (1996). Sexual identity and bisexual identities: The struggle for self-description in a changing sexual landscape. In B. Beemyn & M. Eliason (Eds.), *Queer studies. A lesbian, gay, bisexual, and transgender anthology* (pp. 64–86). New York: New York University Press.

Slater, S. (1995). Lesbians and generativity: Not everyone waits for midlife. *Work in Progress, No. 72.* Wellesley, MA: Stone Center Working Paper Series.

Slater, S. (1998). *The lesbian family life cycle.* Urbana: University of Illinois Press.

Smiley, Elizabeth B. (1997). Counseling bisexual clients. *Journal of Mental Health Counseling,* 19, 373–382.

Udis-Kessler, A. (1996). Identity/politics: Historical sources of the bisexual movement. In B. Beemyn & M. Eliason (Eds.), *Queer studies. A lesbian, gay, bisexual, and transgender anthology* (pp. 52–63). New York: New York University Press.

Williams, W. (1988). *The spirit and the flesh: Sexual diversity in American Indian culture.* Boston: Beacon Press.

9

The Developmental Impact of Trauma

By Amy Banks, M.D.

REFLECTION QUESTIONS

1. *What are the characteristics of a healthy, nontraumatized man? Woman?*
2. *Choose a relationship in your life that is supportive. What are the qualities of that relationship that make it work successfully?*
3. *How do you define trauma in a person's life? Have you experienced a trauma? If so, what did you notice happening?*
4. *If a young child has been abused and develops symptoms of post-traumatic stress disorder (PTSD) can he or she heal?*
5. *In what ways can oppression and "power over" relationships be traumatizing?*

INTRODUCTION

The Relational-Cultural model of development, as described by Jean Baker Miller, Irene Stiver, Judy Jordan, and Jan Surrey, at the Stone Center, Wellesley College, states that growth and development occur within relationship and toward relationship (Jordan et al., 1991). At its extreme the theory would hold that every human interaction during an individual's life, whether healthy, abusive, or somewhere in between, will make a mark on the future relational capacity of the individual.

In an ideal relational world, an infant would be born into a community of people who understood that the need for respect, empathy, and love are as profound as the basic need for food and shelter. In this ideal world the child would be surrounded by "growth-fostering relationships" (Miller and Stiver, 1997). These relationships are characterized by mutual empathy, authenticity, and mutual empowerment. These relationships would form powerful templates within both his or her mind and body. Within these relationships the children would develop their own relational capacities and broaden their own relational world. He or she would move from an immature dependence on direct caretakers to a wide mature mutuality with the human community.

Unfortunately, the human world is not ideal. In the United States alone, one out of every three girls and one out of every seven boys will be sexually abused before they reach the age of 18 (Finkelhor & Browne, 1984). There are almost 1 million substantiated cases of child maltreatment in the United States each year. There are countless others that are not reported. Given the number of cultures that are organized with "power over" dynamics, which objectify individuals and devalue relationships, it is unlikely that the "ideal" world will ever exist. Research into the neurobiology of early life stress and trauma is beginning to show us what the impact of abuse is on the brain and the body. We can now see concretely the potentially devastating biological ramifications of trauma. These biological changes alter behavior and can have a dramatic impact on relational development.

Most Western psychological models of development theorize that a child is born dependent and that the goal of parenting and socialization is to make him or her independent and autonomous. These models view development as a linear movement away from "mother" or primary caretaker (Mahler, Pine, & Bergman, 1975). When there is adequate parenting, a child will internalize enough self-esteem and self-confidence from his or her primary relationship to move through life with ease. One goal of these "separation models" is to create a person who is psychologically independent. In these models, separation from mother is crucial to the child individuating and becoming "his own person" (Mahler, et. al 1975).

Most of these models do not fully take into consideration the relational and cultural context of most traumas. Without the appropriate context, the "psycho-pathology of trauma" is something that is localized within the traumatized individual, as an individual failure rather than a failure of relationship or of community and society. As an example, you could think of a gay or lesbian teenager growing up with the constant message that his/her lifestyle or behavior is unacceptable, immoral. He or she will grow up with images of widely publicized hate crimes against gays and lesbians. He or she may begin to feel fearful, anxious, withdrawn around people, fearing he or she is going to be judged or worse, physically hurt because of who he or she is. You could look at this individual out of context and believe him or her to be unfriendly, a loner, perhaps even paranoid. Without an understanding of how these behaviors began and what purpose they are serving to keep the individual safe in an abusive culture, the locus of blame may wrongly be placed on the gay or lesbian client rather than on the homophobic society.

SOCIAL AND CULTURAL IMPLICATIONS

While individual traumas are devastating, it is important to also look at the ingrained social traumas and stressors that occur in our culture every day. Teicher et al. (2002) has studied and named the effects of chronic stress on children. If we look out at our culture what chronic stressors do we see? Certainly there are the obvious childhood physical and sexual abuse and neglect, domestic violence, and assault. But there are also more subtle, but perhaps equally important forces at play in our society. Racism and other forms of oppression are widespread today.

While most people can readily see the power abuse in childhood abuse or even in domestic violence, it is important to recognize the widespread use of power over others to maintain a patriarchy. In this system, there are dominant and subordinate populations of people. The dominant group makes the rules that all will follow. The subordinate group is expected to follow these rules and often to take care of the daily needs of the dominant group. Any movement by the subordinate group to move out of this situation is met with shaming strategies and disempowering maneuvers, aimed at keeping the status quo in place (Miller, 1976). However, being in this type of invisible, undervalued, and paralyzed state is untenable and stressful. It is possible that some of the most basic dynamics in our culture contribute to the chronic stress in individual lives and the extraordinarily high levels of major stress-related conditions—substance abuse, post-traumatic stress disorder (PTSD), depression.

Janie Ward (2000) discusses racism and its changing face in her book, *The Skin We're In*. Racism in the new millennium may look different; it does not always have the white sheets of the "Klan," but there is a silence that Ward describes as "destructive, even deadly." She talks about the climate of denial around racism today—that White people are suspicious of those who claim it exists, and Black parents are sure attacks will happen but do not know when or how. This anticipation of assault leaves the minority population with a sense of hypervigilance, much the way a woman would respond in a situation of domestic violence. In Ward's research on Black parents and their children she has found that Black parents worry that with earlier and more exposure to White culture and White sense of superiority, their children will internalize inferiority before they develop the tools to combat it.

A father from Janie Ward's research describes the relational impact of racism: "[T]he survival that they have to worry about is psychic, I think. People messing with your mind, trying to convince you that you're not as good as you are, or trying to convince you somehow that you don't deserve what you've gotten." In this type of interaction—one I would describe as "power over"—the more powerful one has the ability to deny the reality of the individual who is beneath him or her in the pecking order. This is a dynamic that plays out continually with marginalized populations in our culture. We have already seen that within the relational model of development, in order to have healthy relationship, it is imperative to represent your feelings and thoughts authentically in a relationship. When this is not accepted, or worse is denied or denigrated, then the individual within the relationship will often go underground and keep much of himself out of the relationship. Essentially, he moves into a place of isolation (Miller and Stiver, 1997). By definition this is a place of stress. We are social animals and we do not grow nor do we thrive in isolation.

Fortunately, there are many writers and parents who come from the margins who are focusing on helping their children to understand and manage the oppressive forces in their lives. They are willing to talk with their children about oppression. They are trying to teach them strategies to manage the silence, to manage the politics of oppression. As we have learned earlier, what is important for children in times of chronic stress or trauma is who is helping. If there are trusted friends or adults who are acting responsibly and helping to protect the child and to give him or her unconditional love and acceptance, this can do much to buffer the stress.

Another cultural situation which may contribute to early life stress is the gender-straightjacketing for both girls and boys. Carol Gilligan (1982) has written poignantly about the dilemma of young girls as they mature and become young women. With the introduction of opposite-sex attraction and romantic involvement they begin to lose their voice: they shouldn't be too smart, too in control, too tomboyish. They begin to leave large parts of themselves and their competencies out of relationship in order to be in relationship.

Bill Pollack (1998) and others are now writing and researching on the gender pressure on boys. He argues that little boys, even in the best of situations, are shamed out of voice by the age of 4–5. This is done through the enforcement of the "Boy Code," which is a set of expected behaviors aimed at helping the young boy to "toughen up" and begin his journey to manhood. This is also powerfully reinforced by an abnormally early and severe separation from mother. As Pollack writes, "I have come to believe that this forcing of early separation [from mother] is so acutely hurtful to boys that it can only be called a trauma—an emotional blow of devastating proportions" (p. 12).

Pollack describes the imperatives of the Boy Code for men and boys: (a) being the "sturdy oak,"—one who is stoic, stable, and independent; (b) having a false sense of self with much daring, bravado, and attraction to violence; (c) achieving status, dominance, and power; and (d) avoiding at all costs expressing the "sissy stuff," those urges that are seen as feminine, such as dependence, warmth, and empathy. However, in the long run, all of these aspects of the Boy Code affect the boy's ability to be in connection, and leave him in condemned isolation. It is this condemned isolation that is experienced as stressful and that gets the whole biological cascade rolling—the end result can be catastrophic and it can be permanent. Again, Pollack speaks powerfully about the effect of the code, "Let's imagine what the experience of ruptured connection must actually feel like for a little boy. Let's imagine the sense of loss a boy must feel as he is prodded to separate from the most cherished, admired and loved person in his life, the shame and embarrassment he often encounters whenever he is asked to 'act like a man,' but doesn't" (p. 29). On some level might this not be the emotional "neglect" that Teicher (2002) has studied in his biological research on boys and stress?

RELATIONAL-CULTURAL THEORY AND RELATIONAL DEVELOPMENT

Relational-Cultural theory places "growth-fostering relationships" at the heart of human development. A healthy relational matrix is both the cause and the outcome of growth and development. A growth-fostering relationship, as described by Miller and Stiver (1997), is one in which both individuals feel a sense of zest, have more ability to act and do act, feel a greater sense of self-worth, have a greater clarity about themselves and the relationship and have a greater desire to be in relationship. A growth-fostering relationship is one in which empathy is experienced freely and without shame; a relationship in which two individuals are growing and developing together. It is also one in which there is mutuality. This does not mean a balance of power, but rather a commitment by both parties to be impacted by one another (Jordan, 1991). In its healthiest form, the mother-infant relationship may be our first mutual relationship. In a non-abusive home, the infant, without having experienced shame or humiliation, searches for a mother's face, and her reactions (Schore, 2003). He can see his experience mirrored in hers; she can see him change and respond to her responsiveness. The child does not understand words at this point, but there is a presence, a quality of being, a commitment to reaching out for one another. Despite the differences in roles and maturational levels between mother and infant, these early interactions are mutually impactful.

Relational-Cultural theory suggests that the internalized templates of previous relational experiences, relational images, are carried into new relationships (Miller and Stiver, 1997). As human beings grow and develop our relational world becomes increasingly complex. Inevitably, disconnections happen in developing relationships not only when there is a failure of empathy or lack of understanding but also when the felt sense of old, more destructive relational images and the relational expectations they generate are brought into these new relationships. Essentially, each new relationship is not simply two people interacting, but rather the imprint of two relational worlds weaving a tangled web of connections and disconnections. For those who have had a number of less than ideal relationships in their lives (in this culture, this includes many), this can help explain why the beginning of relationships can create such feelings of vulnerability for most people.

Jordan et al. (1991) have described relational development in terms of an individual's relational competence, his or her ability to form and recognize growth-fostering relationships. Jordan and Dooley (2000) have described the skills involved in relational competence: Engagement, empathy, mutual empathy, authenticity, mutuality, diversity, and mutual empowerment can all be thought of as "markers of good connection" (p. 20).

"Engagement" refers to a quality of being present and caring about the relationship. This includes both verbal and nonverbal messages about how much you value and respect the other person's world. It involves actively participating in the growth of each person in the relationship or interaction.

"Empathy" can be thought of as a way of understanding and joining another person's world through feelings. Perhaps you can recall a time when a friend shared with you a difficult life experience, a breakup for instance. Having been through a breakup of an important relationship yourself, you felt as though you understood his/her experience. You could join them in their place of pain; you could imagine what they were going through. This is empathy. Empathy allows both people to feel more connected, more validated, and more understood in their relational experiences. Empathy always involves understanding the context of the experience.

"Mutual empathy" can be experienced between two people when an empathic response is shared or made visible in the relationship. In the example above, when your friend speaks of the breakup of an important relationship, you can understand his/her experience by relating to an experience you have had. But unless you share with that person your growing understanding of what he/she is going through, the full potential of mutual empathy will be missed. Mutual empathy means being in the experience together, feeling with each other and being moved by each other and seeing that the other is moved. Letting the other person know that you are being moved by his/her experience is powerful. It communicates that the other person matters, that he/she is being understood. The experience of "mattering," of having an impact on the other person, increases one's sense of relational competence.

"Authenticity" is the honest expression of one's needs and feelings within a relationship. When authenticity is practiced within the context of a "growth-fostering relationship," it is done with keen attention paid to the possible impact on the other person and on the relationship.

"Mutuality" refers to a sense of mutual respect within the relationship. It involves both parties sharing their feelings and being part of a growing connection together. Mutuality involves bilateral responsiveness and initiative. Affecting another person as well as being open to being affected or impacted by the other require a sense of "safe enough" vulnerability. If you can remember a time you were in a non-mutual relationship, one in which you had little power, it is likely that it felt very difficult to share any of your feelings or thoughts. You may have assumed that you would not have been listened to. Power differences often interfere with mutuality, although this does not have to be the case. Even in some unequal power relationships, such as parent-child relationships, an emphasis on empowerment and mutuality can be encouraged. When parents share their feelings and invite their child into their experience, it moves the relationship toward greater mutuality.

An awareness of "diversity" and differences within relationships is a key marker for good connection. This awareness involves learning about, appreciating, valuing and being changed by difference within a relationship. This can be a large task as it means appreciating the vast number of ways that individuals and groups of peoples can be different as well as understanding the distorting influence of power on the experience of diversity. Learning from one another's differences deepens relationships and honoring diversity results in mutual empowerment.

"Mutual empowerment" and empowerment are both forms of power that exist within healthy relationship. This power is not focused on winning as the ultimate goal, but rather is meant to enhance the sense of strength and courage of each person and to contribute to the formation of a larger, empowering community. This form of power embraces difference so that we can learn from each other. When we embrace diversity and value difference, mutual empowerment is the inevitable outcome.

These relational markers may be used as guides as I discuss healthy relational development. But how, when, and where do we learn these skills? Are they with us at birth or are they taught to us throughout our lives? What happens to our growing sense of relational competence when and if it is impacted by trauma? These are questions that we will explore in this chapter.

In order to understand and appreciate how normal relational development is impacted by trauma, it is important to think about what "normal" relational development looks like. Relationship starts at conception. The first 8–9 months of a child's life were once thought to be spent safe and protected in the mother's womb. The physical bond can be profound. There was a time when the people believed that a baby was protected from the stresses of the world until being born. Research now tells us that the condition of the mother during pregnancy can have a huge impact on the developing fetus and on the eventual relationship that the mother and baby will develop. For example, Watson et al. (1999) studied the effects of prenatal exposure to a large trauma. By looking at babies born to women who had survived a large earthquake in China, they discovered that children who were in their second trimester of development during the earthquake ended up with significantly higher rates of depression in adolescence than did children in other stages of prenatal development. Imagine, an event that happened before you were even born could impact your emotional and relational life more than a decade later.

Within a competitive "power over" culture the relational skills described by Jordan and Dooley (2000) are not taught or valued, so you may not see them in even your closest colleagues or friends. However, if you look at young children what is very clear is how, from our earliest moments, we are programmed to connect. A closer look at a few major developmental theorists can provide a window into how relational competence may develop.

According to Daniel Stern (1985), the first year of life is filled with a growing sense of relational wonder. He has documented the complex interaction between mother and infant as early as the first few days of life. The mother and child exchange responsive communications through facial expressions, touch, and eventually sound. The child is building the capacity to impact people and to be impacted in relationship. He/She is growing and developing very much within the context of relationship. This is the basic foundation of trust in relationships. In an unresponsive relationship, an infant will initially protest, and if this does not lead to engagement with a caretaker the infant will withdraw and become despondent.

Despite a growing body of evidence supporting the value of responsiveness to infants and children, some parenting texts counsel parents to "get their child under control or to get the child on a schedule." One clear example of this is when parents "ferberize" their child's sleep schedule. In order to develop a child's capacity to go to sleep on his/her own, Ferber (1985) suggests that you should let the infant cry for incrementally longer periods of time. Over time, the child learns that when he/she cries out for comfort as he/she is going to sleep, he/she will not be responded to. Eventually he/she stops crying. Yes, the child learns to sleep on his/her own, but at what relational cost? For children raised in this manner, even if done within a loving context, there can be early disconnections in primary relationships. The unreasonable behavioral expectations placed on infants and young children may cause children to disconnect from their own experience as it is labeled wrong and ignored, punished, or behaviorally shaped. It can also lead to the expectation of disconnection rather than responsiveness in future relationships. As I have already discussed, this process is evident in boys in this culture who may actually be traumatized by the "normal" gender socialization process (Pollack, 1998).

The husband and wife team of Martha and William Sears have written about "attachment parenting" in their series of books and articles on parenting (Sears and Sears, 1995). Their writings focus on the importance of connection and responsiveness in raising children. As they reviewed the studies done on attachment they drew the conclusion that "how we become who we are is rooted in the parent-child connection in the first few years of life" (p. 15).

Attachment parenting starts with being open to the cues and needs of the baby as opposed to focusing on getting the child on a set schedule or under control. Sears and Sears argue that when a parent knows his/her child, the parent will have a better chance of understanding what the child is asking for. The child, in turn, will have an inner sense of "feeling right" and will be "less impulsive, less angry and less likely to misbehave" (p. 18).

The six features of attachment parenting that the Sears' outline are focused on enhancing the quality of presence with an infant or child. They make the following concrete recommendations aimed to increase attachment between parent and child: (1) responding to the baby's cues; (2) breastfeeding; (3) wearing the baby, which literally means having the baby attached to your body as often as possible throughout the course of the day; (4) spending time playing with the baby, which gives the parent and child a chance to know each other in a much more in-depth way; (5) sharing sleep, and (6) being a facilitator (a person who helps the child learn how to conduct himself in the world). Not all of these techniques are possible for all families, an attitudinal shift from controlling a child to listening to a child is possible for everyone.

Sears and Sears describe a "growth-fostering" parenting relationship. One in which a child can begin to form an initial relational template for future relationships. Because of the way that preverbal and early toddler children learn and remember experiences, this is often memory that will be stored in their bodies for life. As Teicher et al. (2002) comment, "[T]his is a time of heavy pruning of the nervous system—a friendly environment will form the brain in a direction toward people and relationships, an abusive environment or an ignoring environment will cause the brain to form and develop in an alternative way, the individual will be preparing itself to manage a hostile environment throughout their lives" (p. 414). This development in response to early life stress may help an individual in terms of crises, but the brain, locked in a state of chronic hyperarousal and hypervigilance is not well equipped to function in the context of safe relationships.

DEVELOPMENTAL STAGES, RELATIONSHIPS, AND TRAUMA

The parent or primary caretaker acts as the guide in early relational exploration, helping the child to negotiate life's stressors. Too much stimulation, related to abuse, trauma, and neglect, from the world around him/her will overwhelm his/her developing nervous system and be experienced as stressful and aversive; but too little stimulation is also stressful. In fact, studies have shown that an infant's regulatory functions, including his or her stress-response systems, are dependent upon his or her relationships with the primary caregivers (Polan and Hofer, 1999; Kraemer, 1992).

There has been limited writing on early development from a Relational-Cultural perspective. So I am left to translate observations from other theorists into a Relational-Cultural framework. When one is observing human behavior the observations are never objective; they are shaped by the subjective belief system and the values of the observer. The developmental theories of Margaret Mahler and Piaget have shaped much of the historical dialogue around childhood development. Mahler's developmental theory focused on the concepts of separation and individuation, which she described happening in discrete developmental stages throughout childhood. Piaget describes similar stages of development with a primary focus on a child's cognitive development. I will review the highlights of these theories and suggest a Relational-Cultural alternative view for each developmental stage.

INFANCY AND EARLY CHILDHOOD

Margaret Mahler et al. (1975) described differentiation as the first stage of a child separating from his/her primary caregiver. This stage occurred between 5 and 10 months of age. Mahler believed that during this window of time the infant begins to differentiate himself/herself from another "object." In Mahler's description, differentiation from the object (mother) is the beginning of separation.

In Mahler's model, separation is followed by a period of "practicing" from the ages of 10 to 16 months. During this period, Mahler postulates that the young child is practicing the separation from object (mother) by physically moving away from her, and believes that the child is able to do this if the mother is present to provide security for the child. Rapprochement occurs between 16 to 24 months and Mahler describes this as a time when the child is more aware of his/her mother and has a desire for the mother to share in his/her new skills and experiences. Finally, Mahler describes the final stage of separation and individuation as "consolidation and object constancy," which happens during the second and third years of life. In this stage the child has the ability to hold onto objects when they are not present. The end result of Mahler's theory is that the child has achieved a "definitive individuality." This occurs when the child is developmentally able to imagine the object when it is not directly in front of him/her.

Piaget describes the first two years in terms of cognitive development (Kaplan and Sadock, 1991). Much of his work overlaps with the stages described by Mahler. He emphasizes that most infants are born in an engageable state. The newborn infant's motor and sensory reflexes are literally programmed to interact and to change in the context of his/her environment, in the context of his/her relationship to the world around him/her.

During early infancy (the first 5 months), Piaget's model focuses on the coordination of the senses that the child will use his entire life to help him engage in the world. By 5 to 9 months the child begins to seek out "new stimuli" in the environment. He/she begins to anticipate the consequences of his/her own behavior and to act purposefully to change the world around him/her. Piaget identifies this as the beginning of intentional behavior. In Piaget's model of development, "object permanency" usually occurs by the end of the first year of life. This is a full year before Mahler felt it occurred. Cognitively, this means the child is able to "hold on to relationship" and to begin to realize that others are different.

Piaget describes the second year of life as one of great cognitive advancement. He felt that a child's environment was "mastered" through assimilation (incorporating new stimuli into his/her cognitive world) and accommodation, which is the modification of behavior in response to the new stimuli.

Certainly, these first 2 to 3 years of development for a child are crucial on many different levels. Within the context of a "growth-fostering" relationship this can be a time when a child is learning about "the world response to his actions and behaviors." If the relational response is empathic and in tune, he begins to lay down very early, primitive relational images of the world as being benign, connected, consistent, and helpful.

The child in a growth-fostering relationship will begin to move out and seek new experiences. He is understanding himself better and is producing novel behaviors. By the end of his second year he is capable of abstract thought and some reasoning. This is also a time when a child's vocabulary is expanding greatly. He is much more able to relate his world to those around him.

With these relational observations in mind, let's consider what happens when these same developmental observations are viewed through the lens of growth in connection rather than the lens of separation and individuation. If the child is in a growth-fostering relationship, then the same observations can be explained much differently. With a relational focus, the goal of development shifts from the ability to function autonomously to the development of a complex relational capacity. What is then impressive about these early years is not how proudly independent young children become but rather how complexly related they are at a very young age.

I have already mentioned the work of Daniel Stern on the intense, reciprocal relatedness between mother and infant in the first few months of life. Those first early imprints of relationship are crucial. Though the infant does not yet understand the intellectual concepts of human, family, or mother, he certainly knows the comfort that can occur from "mother" when he is tired, wet, or hungry. He feels the physical relief of a hug. Though this may not be stored as a visual relational image, it is the early precursor to relational images. It is an internal statement of his relationship with his world. The expectation is that "when I am distressed and cry out for help, something in the world can comfort me."

Certainly the ability to differentiate self from other within a relationship, as described by Mahler and Piaget, is an essential stage in developing relational capacity. The young person needs to recognize that there is another person in the relationship. So, I would agree that differentiation can be a key component to development within the first year—however, the importance is not in "moving away" from mother. The importance is in the beginning of clarity about self and the other in the relationship (Jordan, 1987). You cannot respect and honor another's feelings and thoughts if you are not able to recognize where you stop and they begin. Adult relational competency requires both parties to be clear about themselves and the relationship—they must be able to know and represent their own feelings and needs in a relationship and to be able to take in the representation of the other person in the relationship. Differentiation begins this process.

The practicing stage that Mahler describes may be viewed relationally as the toddler moving out into his world and expanding her connections. Mahler highlights the child's desire to move away from mother, but Mahler also admits that the child is able to move freely out into the world only when mother is present for security. It is essentially the relatedness that fuels the child's desire to move out into the world and look for more connections, not her desire to separate. Don't we all move more freely in the world as adults when we have an effective and supportive network of relationships to sustain us? The goal is not to separate from mother, but a desire to seek other safe relationships; in a sense, to find more mothers. As Jean Baker Miller (1997) comments, in a growth-fostering relationship, there is a desire to find more healthy relationship. Yes, the child is practicing, but she is practicing relating not separating.

Mahler's rapprochement stage is a time of more obvious desire for mother, a need for mother to share in the skills and experiences of the child. This observation is, of course, predicated on the previous description of the child's need to move away from mother in the practicing stage. However, even Mahler admits that the child does not actually move away from mother—therefore, why describe it as a rapprochement. At this age, between 1 and 2 years old, the child has more verbal physical skills to share his world with his mother. As he is able to represent himself more fully and clearly in the relationship and can understand more of what mother is doing and saying, there will be a shift in interest and intensity between the two. I would not see this as a moving back into connection, but rather as a child integrating more of his developing skills into his connections with the important people in the world.

The third year of life marks the period of "object constancy." Relationally, this could be thought of as a time when the child is able to hold on to relational images in a more abstract way. The relationship does not need to be immediately in front of him, or impacting him in the moment. He can carry it with him. Mahler indicates that this is clear evidence of the achievement of individuality. I would argue that this is one of the most crucial developmental tasks of *relating*. The ability to carry relationships around with you at all times. To form mental relational images, which, if healthy, will fuel an individual's desire for connections throughout her life.

EARLY CHILDHOOD

The ages of 3 to 6 tend to be a time that many children begin expanding their relational worlds outside their immediate family. Freud described this period as the phallic-Oedipal phase (from 3 to 4 years) and focuses on the child's aggressive competitive feelings being raised within the context of his relationships. His focus on a child's genital excitement and his fear of retaliation by the same-sex parent for his fantasy desires for his opposite-sex parent seem archaic to many practicing now; however, these views regarding Oedipal fantasies still permeate our culture.

Erik Erickson's (1963) developmental theories also focused on aggressive drives during this period of time. In his model, the ages from 3 to 5 years were the time of Initiative vs. Guilt. In his model, initiative describes the child's struggle to move forward with his/her own thoughts and desires. Erickson assumes there are strong aggressive drives that lead the child to struggle with feelings of guilt. His ability to manage these guilty feelings and to keep "initiating" behavior is what defines a successful passage into the next developmental stage. Like Freud, Erickson is focused on aggressive feelings not only toward the opposite-sex parent (if there is one), but also toward siblings.

Relational-Cultural theory would look at this phase much differently. This is a period of time when a child may be exposed to differences outside of the family for the first time. The child begins to interact with children his/her own age, and is learning the invaluable relational competencies of sharing and negotiating individual needs. The child has a growing ability to communicate within relationships and must learn and adapt as his/her communication skills become more defined. Intellectually, the child lives in a rich fantasy world and will explore relationships, both healthy and abusive, in his/her play.

LATE CHILDHOOD

The ages of 6 to 12 once were termed the "latency phase" by Freud, who believed children at this stage were less focused on genitals. He felt that the child had "resolved" his Oedipal complex and was now able to channel his sexual drives into more socially acceptable goals such as schoolwork or sports. Freud felt that it was during the latency phase that a child developed a superego responsible for moral and ethical thinking. It was this superego that would eventually help tame the wild "id" that rested inside. Freud's theory of development was based on human drives such as sexuality and aggression; it is also a model, which postulates that development is a process where an individual develops internal resources to cope with the external world.

Piaget described this period of development as the concrete or operational phase; it is a period focused on logical (cause and effect) thinking. Piaget felt this was a period of time when a child could take on another point of view, as well as partial and whole relationships. It is clearly a time of important abstract thinking.

Relationally, much fine-tuning of relational competence is done from the age of 7 on. This is a time when the child begins to take the mother's point of view and has a complicated picture of the world. He is able to know people around him with more depth. It may be at this stage that engagement begins to be recognizable as described by Jordan and Dooley (2000). The child is able to take in, respect, and value another person's world; he may truly care about the relationship in a more complex way

While Freud focused on this as a latent time within relationships because of the de-emphasis on sexuality, this is actually a rich relational time for a child as same-age and same-sex peers become a major influencing force. As this stage progresses, children begin to feel that relationships outside of the family are as important or more important than within the family (Kaplan and Sadock, 1991). It is therefore a time of expanding relational awareness. With a full range of emotions available, with empathy, love, and compassion solidified, the child is learning to interact with peers on many different levels and is capable of mutuality.

ADOLESCENCE

Piaget describes the period of 11 years old on as the period of abstract thought. It is a time when young adults are gaining the capacity to use two systems of reference simultaneously. They develop the capacity for deductive reasoning.

For Freud this is the "genital phase," the final stage of psychosexual development that begins with a child's capacity for orgasm but does extend to the individual's growing ability to be in intimate relationships. Once again, both of these theorists focus on the developing internal world of the child with very little attention to how the child is integrated within the context of a larger relational world.

Relationally, the teenage years are a time of great change and confusion. The teenage body is overrun with hormones as the child begins the final surge into adulthood. This is a period where relationships can become sexual and the young person must learn how to negotiate a physical relationship for the first time. There may be peer pressure to belong to the "in" crowd. It may be difficult for teenagers to enter into relationships with a clear sense of who they are—there may be great pressures to belong to the group. Carol Gilligan (1982) talks about this period as being a time when young women lose their voice. Heterosexual girls get the message that boys are interested in them sexually, but also that being assertive and strong is not what boys want. Many teenage girls leave out parts of their experience, and they literally disconnect in order to stay in connection. Stiver (Miller and Stiver, 1997) calls this leaving of large parts of oneself out of relationship in order to be in connection the "relational paradox." It is likely that the same disconnection from internal experience happens with boys as well as they learn how to become men. Becoming a man in this culture often means losing virginity with a girl (Jordan, 1987).

Again, when in the context of a growth-fostering relationship, children can use these developing skills to understand the complicated relational world they live in. They can be taught and they have the capacity to observe two systems existing in their world. They may have the capacity to understand a "power over" system and the effects of it on their world and particularly on relationships. They may also be able to learn and value a system that is based on connection, on respect, on empathy. It is the ages from 7 till young adulthood that children hone their relational skills and use their new cognitive and emotional strengths to develop deep relational competencies even in the face of a largely "nonrelational" world.

RELATIONAL IMPACT OF TRAUMA

If one believes the major task of human development is the acquisition of relational competencies over the life span, it is not difficult to see how devastating the effects of trauma, abuse, and neglect are to relational development in any individual. Given the extent of violence in our culture it is likely that a child will experience some traumatic event during his/her childhood (Solomon and Davidson, 1997). The quality of relationships in the aftermath of the trauma has a significant impact on how the child copes with the trauma. Finkelhor and Browne (1984) and McFarlane (1988) have documented that secure attachment bonds serve as primary defenses against childhood-trauma-induced psychopathology in children who have been exposed to severe stressors. In fact, secure attachment bonds are seen as the most important determinant of long-term recovery.

The concept of trauma covers a wide range of life experiences. The definition of trauma according to the DSM-IV (1994) diagnosis for post-traumatic stress disorder is "having or witnessing a life threatening experience." Lazarus (1985) has described stress as being experienced when "an individual is confronted with a situation which is appraised as personally threatening and for which adequate coping resources are unavailable" (p. 399). These two definitions can include nonhuman-related traumas such as natural disasters or accidents, as well as any human-related abuse such as war experience, rape, domestic violence, assault, or childhood sexual or physical abuse, or neglect. While all of these traumas can lead to long-lasting relational difficulties, it is more likely to happen when the violence or trauma is committed in the context of human relationship. It is even more likely to happen if the violence is at the hands of a known person such as a parent, a close relative, a sibling, or a spouse. The stronger the relationship, the more intense the betrayal, and the more difficulty the person has in integrating what has happened into his/her life.

What do relationships look like to someone who has a long-standing history of trauma? While many areas of a person's life are affected by violence or other traumas (physical health, religious beliefs, cognitive functioning) the destruction of relationships can be the biggest obstacle to healing. Many survivors of trauma are aware of intense longings for connection, but these longings are accompanied by other equally intense feeling—the terror of being hurt again, rage at past perpetrators or traumatic events in life, or even despair at feeling so isolated.

Many, but not all, traumatized individuals will develop chronic post-traumatic stress disorder. This condition represents a shift in the person's biological ability to cope with stress. Individuals with PTSD often "reexperience" the trauma through nightmares, flashbacks, or intrusive memories of the event. This reexperiencing can happen not only when something stressful is happening but also when the person is trying to increase his/her intimacy in relationships. Particularly for those who have been victimized by another person, the fear of being vulnerable can be overwhelming. The contradictory messages about relationships—that they long for one, but are terrified of them—can make relationships very confusing. The survivor may respond to the perceived threat by fleeing the relationship, staying and fighting constantly, or by feeling frozen and powerless in the relationship. With little experience with healthy relationships, many survivors find themselves in disrespectful, event violent relationships again and again (Banks, 2001).

Coping strategies such as substance abuse, eating disorders, or self-mutilation develop as a way of shifting the altered brain chemistry. However, these strategies only further isolate the individual, leaving him/her in a state of condemned isolation (Miller and Stiver, 1997).

A closer look at early childhood abuse may help to understand how trauma can produce long-lasting trouble in relationships. Healthy bonding in infancy is believed to be a mutual process. Both the caretaker and infant read and respond to each other's facial signals. What does the infant experience when the face staring back at him/her is not positively responsive but is blank, unresponsive or worse, actually mean or antagonistic? When abuse happens in early infancy the earliest relational templates are based on the child's isolation and inability to have someone care for his/her basic needs. When abuse happens in infancy the child is maximally vulnerable—unable to comfort or care for himself/herself or to regulate overwhelming feelings. The infant's stress response system is dependent on the relationship with his/her primary caregiver (Polan and Hofer, 1999; Kraemer, 1992). To an infant, the primary relationship with a caretaker is his world, so that when that caretaker is abusive, the whole world is abusive. This primary, negative relational image distorts the infant's experience of every relationship from that moment on.

Severe abuse or trauma during early childhood, before the child is 3 years old, is likely to leave more severe relational scars. The young child is exploring the world, integrating new information. Complex verbal and cognitive ability develops later in childhood; therefore a young child is unable to understand an abusive experience. The abuse may become encoded in a more global way such as "relationships or people are dangerous." The relational image is "when I am feeling small, needy, or vulnerable, I am in danger from people around me 'or' relationships will not comfort me, they will hurt me." Since a young child does not have other means (other than relationship) to comfort himself, the child is left with a profound dilemma: to remain hurt and overwhelmed alone or to risk moving out into relationship to seek comfort. When the abuse happens within the context of the primary caregiving relationship, ironically, the child will return to that very same relationship for comfort. This seemingly contradictory behavior is of course understandable because of the primacy of the child's first bonding experience.

These earliest preverbal relational images can be very confusing. They often are experienced in later childhood or by adults as overwhelming affect (terror, rage, or grief) unattached to an event, in response to connection or closeness. Of course, the reverse can be true—if the earliest relationships are supportive and nurturing, the relational image of connection can be one of exhilaration, joy, or a deep sense of safety in intimate relationships. The main trouble with early childhood abuse is that the relational images, which rule adult connections, are not "remembered" visually in any way that makes sense for the adult person.

When abuse happens during the ages of 2 to 3 the child is more aware of the people and relationships around him. This is the time when he is expanding his relational world, when he is developing more abstract thoughts, when he is developing more ability to hold onto specific relational images. If the trauma is at the hands of another human, even though there have been many healthy relationships, he now has a clear image of relationships that are scary and hurtful. As we will see later in talking about the biology of early life stress, these traumatic images return with more volume than a non-traumatic relational image. His newly acquired abstract thought may now be a problem because he can concretely imagine more hurtful relationships in the world. His relational world may become very confusing with competing relational images. His engagement may be unpredictable; one minute he may feel safe and relaxed in the context of a trusted relationship, and then the next moment he may be overwhelmed by an image of danger and get scared or angry. He may also feel more needy and demanding in the relationship or he may become despondent and withdrawn.

THE BIOLOGY OF LIFE STRESS AND EARLY TRAUMA

Thus far I have talked about trauma and abuse as if it were simply an intrapsychic or a relational problem. What is becoming increasingly clear through the research on the biology of trauma is that the relational devastation is caused by very specific biological changes that happen in the brain and body of individuals exposed to trauma and early life stress. These biological changes result in traumatic relational images being encoded more firmly and with more volume than healthy relational images. Research into the biology of early-life stress and PTSD is growing rapidly. The information presented here will be only part of a rapidly expanding puzzle. What is clear is that chronic stress and trauma can leave a long-lasting imprint on the brain and body development and functioning.

Before discussing the impact of trauma, it is worth considering how the brain develops in the absence of trauma. The brain was long believed to be one organ in the human body that is not able to regenerate. However, as we are able to look more closely at this complex organ we begin to see a much more dynamic picture. Before birth the developing brain has two to three times the number of immature, developing neurons than will be found in the adult brain (Sidman and Rakic, 1973). However, more than 50 percent of these are eliminated even before we die (Rakic and Zecevic, 2000). The level of corticoid steroids (stress hormones) plays a critical role in the process of cell death (Gould et al.,1991).

After birth, the brain continues to change in dramatic ways. From birth to age 5, the brain triples in mass as the neurons become myelinated. Myelination refers to the addition of large areas of fat on the neurons which help in speeding the flow of information throughout the brain. This myelination during the early years of life explains the dramatic increase in the complexity of behaviors between an infant and a 5-year-old. The motor areas of the brain, like the prefrontal cortex tend to myelinate before other areas of the brain (Benes et al., 1994). This is why children can walk and run before they develop more abstract thought or complex verbal skills. There may also be a gender difference in myelination with girls myelinating earlier than boys (Benes et al., 1994).

Jacobson (1973) describes a process of neuronal modification by selective depletion. This happens to the brain throughout later childhood. In the first few years, in addition to myelination, there is also a marked expansion in the numbers of axons and dendrites and an increase in the connections and the communications between nerves. However, in response to environmental stresses, the final shape of the brain is "pruned" down with the elimination of some of the axons and dendrites. This pruning is happening during childhood—the active reshaping of the brain, which helps explain why trauma and chronic stress during childhood can have such severe and possibly irreversible effects on brain development (Teicher et al., 2002).

Given this information about the developing childhood brain, what do we know about the effects of early life stress on a child's brain? We do know that childhood abuse and trauma are associated with many axis I diagnoses, including major depression, post-traumatic stress disorder, and substance abuse in adults. Famularo et al. (1996) studied a group of severely maltreated children who were taken out of the home because of abuse. Thirty-five percent met the criteria for PTSD. Widom (1999) found the lifetime prevalence rate for PTSD to be 36.5 percent when looking at a group of victims of substantiated abuse and neglect. Interestingly, Kiser et al. (1997) reported that abused children and adolescents who did not develop PTSD exhibited more anxiety, depression, and externalizing behaviors than those who developed PTSD, of which the three major symptoms are reexperiencing, avoidance/numbing, and hyperarousal. It raises the question of whether the development of PTSD is in some ways protective to an individual initially. In general population studies of women who have been abused, the rates of depression ranged from 13 to 64 percent (Weiss, Longhurst and Mazure, 1999). Studies have shown anywhere from 59 to 98.9 percent of individuals diagnosed with post-traumatic stress disorder have at least one other axis I diagnosis (Kessler et al., 1995, Centers for Disease Control, 1988; Kulka et al., 1990; Cashman et al. 1995).

As adults, the cerebral cortex and the corpus collosum are intricately related. The corpus collosum is an important area of the brain that carries messages between the two halves of the cerebral cortex. Teicher et al. (1997) reports a marked reduction in the middle part of the corpus collosum in child psychiatric inpatients with a history of substantiated abuse or neglect. Additionally, DeBellis et al. (1999) reports that the most prominent finding in children with a history of abuse or neglect was a reduction in the size of the corpus collosum. Boys and girls differed in the type of abuse that caused the reduced size. For boys, neglect was more apt to cause a decrease in the corpus collosum, while for girls sexual abuse was correlated with a significant decrease in corpus collosum size (Teicher et al., 2002).

One significance of the decrease in corpus collosum size seems to be related to the decrease in communication between the two halves of the cerebral cortex. In general, the left side of the cerebral cortex is associated with logical, rational thinking, while the right side is more associated with affective, creative thinking. A decrease in communication between the two halves of the brain can result in memory being altered and disintegrated. When children who have been traumatized recall a neutral memory it comes from the left half of the brain, whereas, traumatic memories, with more affect attached, come from the right side of the brain (Schiffer et al., 1995). In relational terms, it may be that traumatic relational images are encoded in the right side of the brain, disconnected from safe relational images stored in the left side of the brain.

Given the number of changes seen in the brain of children exposed to stress and the large number of children who are exposed to some level of early life stress, trauma, abuse, or neglect, Teicher et al. (2002) are now arguing that childhood trauma may push the brain into an alternative line of growth. This may not be a bad thing. After all, it prepares the individual well for being alert to all further stressors. The system works under chronic stress; however it becomes problematic when you are in the middle of a safe relationship and the danger signals are continually firing.

THE NEUROBIOLOGY OF PTSD

The research on adults who have been traumatized is far greater than on children. In fact, there is a long history of interest in post-traumatic stress disorder particularly in combat veterans. In the last 20 years, research into PTSD has also included studies of adults who have been exposed to childhood abuse, domestic violence, natural disasters, or terrorism. A brief look at the stress response system may be helpful as a road map to the review of findings.

An individual's stress response has two components: the sympathetic nervous system and the hypothalamic-pituitary-adrenal (HPA) axis. Sympathetic nerves run throughout our bodies and brains and when stimulated, release norepinephrine (adrenaline), which prepares our bodies to fight or flee the stress. Adrenaline diverts blood flow and precious energy away from nonessential organs like the stomach and toward large muscles and the heart to prepare the body to meet the stressful situation.

Essential systems throughout the human body usually are balanced by another system that helps to modulate the response. The sympathetic stress response described above is balanced, in part, by the hypothalamic-pituitary-adrenal axis. The hypothalamus and the pituitary gland are both located within the brain and are responsible for the production and release of many hormones and chemicals that help regulate our bodies. One of these hormones, cortisol, an "antistress" hormone, is particularly important in balancing the stress response system.

Within the HPA axis, cortisol-releasing factor (CRF) is made and released from the hypothalamus. CRF travels to the pituitary resulting in the release of adrenocorticotropin hormone (ACTH), which, in turn, travels out of the brain to the adrenal glands, causing the release of cortisol. Cortisol feeds back to the sympathetic nervous system, turning off the release of norepinephrine. This is one way that the stress response is self-regulated.

Individuals who develop PTSD in response to stress have a number of alterations in this stress response system. There is an exaggerated sympathetic response to any ongoing reminders of trauma (Southwick et al, 1997). The HPA axis itself shows a number of changes. Yehuda (2000) has studied the HPA axis extensively and has found that individuals with chronic PTSD have a decreased level of cortisol rather than an elevated level at the time of the trauma (Yehuda et al., 1998; Resnick et al., 1995). Therefore, during periods of further stress, the norepinephrine released by the sympathetic nervous system is only mildly opposed by the cortisol. Van der Kolk (1988) has theorized that these high levels of norepinephrine may be directly related to symptoms of hyperarousal such as agitation, sleep impairments, nightmares, and flashbacks.

One of the most troubling symptoms of PTSD is the intrusive reexperiencing of traumatic material in the form of intrusive memories, nightmares, and flashbacks. Pitman (1989) has argued that the high levels of stress hormones at the time of the acute stress may result in memories of the traumatic event being "overconsolidated," or remembered in an exaggerated state. This may explain how traumatic relational images are recalled with more intense feeling than relational images of safe connection.

The animal model of "inescapable shock" is one that can be helpful in understanding some of the underlying biological changes seen in PTSD. In this model, a rat is repeatedly exposed to a mild shock without the ability to escape. Initially, the rat tries to avoid the shock, but when avoidance is impossible, the rat develops a "learned helplessness" and is frozen, no longer even attempting to avoid the shock. Researchers have used this model to help explain the biology of post-traumatic stress disorder.

Thoughts, feelings, and movements occur within the human body only when nerve cells communicate with one another. Within the brain, this communication happens across a synapse (the space between two nerve endings). An electrical impulse travels down the arm of one nerve stimulating the release of neurotransmitters into the intersynaptic space (see Figure 1). These neurotransmitters (i.e., norepinephrine, serotonin, dopamine, endogenous opiods) travel across the space

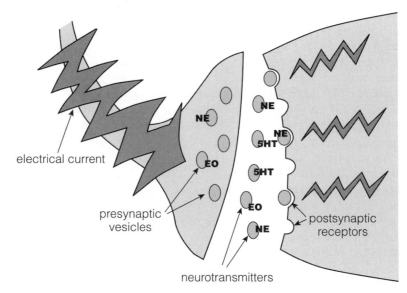

Figure 1. Synapse

From Amy Banks (2001). PTSD: Post-traumatic stress disorder: Relationships and brain chemistry. *Project Report #8. Stone Center Publications.* Wellesley, MA.

and find a postsynaptic receptor on the next nerve. When the neurotransmitter connects with the receptor, the message is carried forward.

Organisms prefer to have a relatively even flow of neurotransmitters across this intersynaptic space. One way that this flow is regulated is by shifting the number and sensitivity of the postsynaptic receptors. During periods of time when there is an overflow of neurotransmitters pouring into the space, the body will decrease the number and sensitivity of the postsynaptic receptors. This "down-regulation" of the receptors makes it more difficult for the neurotransmitter to find a receptor site to bind to, therefore less of the "chemical message" gets relayed (Figure 2). Alternatively, if too few neurotransmitters are being produced or released into the intersynaptic space, the postsynaptic receptors will increase in number and sensitivity, making it easier for the few neurotransmitters to find a receptor to bind to. This process of "up-regulation" maximizes the chances of message transmission (Figure 3).

In the model of inescapable shock, the rat is under chronic stress. Over a period of time the rat brain first tries to deal with the large outpouring of neurotransmitters by down-regulating the receptors. However, as the stress continues and the neurotransmitters are being released in large numbers, the transmitters become depleted. The production of neurotransmitters in the body literally cannot keep up with the demand for release in the face of chronic stress. The brain then "reads" this state as one of neurotransmitter depletion, and the body responds by "up-regulating" the postsynaptic receptors. Over time, the rats under

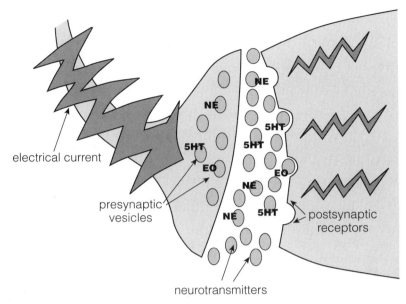

Figure 2. Down-Regulation

From Amy Banks (2001). PTSD: Post-traumatic stress disorder: Relationships and brain chemistry. *Project Report #8. Stone Center Publications*. Wellesley, MA.

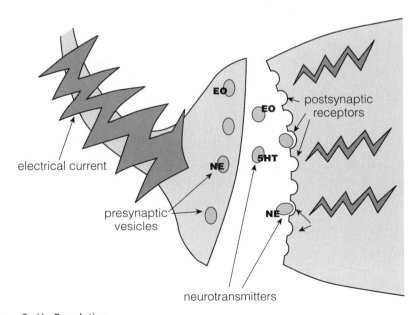

Figure 3. Up-Regulation

From Amy Banks (2001). PTSD: Post-traumatic stress disorder: Relationships and brain chemistry. *Project Report #8. Stone Center Publications*. Wellesley, MA.

chronic stress lose their ability to up-regulate and down-regulate the postsynaptic receptors. The receptors get "stuck" in an up-regulated state. When the body replenishes its supply of neurotransmitters and releases them into this up-regulated system, the messages are "amplified." As the large number of neurotransmitters is released most are able to find a receptor site to bind to (Figure 4). This amplified state of reactions is thought to mimic the reactivity of post-traumatic stress disorder.

Researchers are now able to look directly at the brains of individuals who have been traumatized. CT scans or MRIs can provide detailed cross-sectional pictures of the brain. Additionally, functional imaging studies such as PET scans or SPECT scans are used to video the brain in action.

Adults who have PTSD have been consistently found to have a decrease in hippocampal volume. Bremner et al. (1995) and Gurvits et al. (1996) have documented this finding in combat veterans with PTSD while Bremner et al. (1999) and Stein et al. (1997) have documented a decrease in hippocampal volume in individuals with a history of sexual abuse and PTSD.

The hippocampus is one area of the brain heavily involved in learning and memory (Pinchus, 1978). It plays a particularly crucial role in fear and stress in that it interprets the signals from the amygdala. The amygdala sends out the danger signals and the hippocampus tells the individual which signals are dangerous in which setting. The hippocampus puts a context to the warning. If one assumes that a decrease in volume of the hippocampus translates to a decrease in function then the person with PTSD and a decreased hippocampal volume would have a

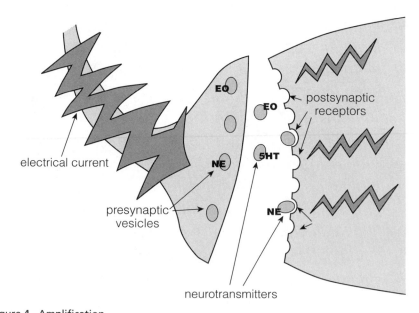

Figure 4. Amplification

From Amy Banks (2001). PTSD: Post-traumatic stress disorder: Relationships and brain chemistry. *Project Report #8. Stone Center Publications*. Wellesley, MA.

difficult time distinguishing what is truly dangerous. Relationally this could explain why individuals with PTSD have a difficult time feeling safe in relationships. They may even choose abusive partners or friends over and over again.

The few functional imaging studies are done with slightly different techniques, thus it is difficult to make broad statements about what they can tell us. However, there are some intriguing patterns that are emerging. A number of studies have documented an overactive or irritable amygdala in response to recalling a traumatic event (Rauch et al., 1996; Shin et al., 1997; Liberzon et al., 1999). The amygdala is the area of the brain that is sending out danger signals. It is the "fire alarm" of the stress response system. To further this analogy, in a person with PTSD, this fire alarm is being triggered not only when the house is on fire, but also whenever someone is cooking toast. If you apply this to relationships, it helps to explain how any movement in connection either toward more closeness or away from connection is perceived as dangerous.

The medial prefrontal cortex (MPFC) is one area of the brain known to have regulatory, inhibitory input into the amygdala (Sesack et al., 1989). Within the MPFC, subregions have specific effects. For example, the orbitofrontal cortex promotes the extinction of fear conditioning while the anterior cingulate inhibits the acquisition of fear conditioning and promotes the extinction of fear conditioning (Morgan and Ledoux, 1995). A number of studies have shown a decrease in functioning in various MPFC areas during traumatic memories (Bremner et al. 1999; Bremner et al., 1999; Shin et al., 1997; Shin et al., 1999). Essentially this would mean that during traumatic reexperiencing, not only is the amygdala overactive but the area of the brain meant to keep the amygdala in check is not working as well. Again, the individual is left to overinterpret stimuli as dangerous whether it be sudden noise or relational movement.

Jean Baker Miller has talked about "feeling-thoughts" as being essential currency in relationship (Miller and Stiver, 1997). It is the ability to represent one's experience in a relationship with cognition and affect together. Both are needed to fully represent one's self in relationship. For victims of trauma and abuse, particularly those who develop chronic post-traumatic stress disorder, feelings and thoughts become uncoupled. The entire stress response system is dysregulated such that the warning signals are turned up full blast, while the area of the brain (the MPFC) that is supposed to help regulate the warning signals is not working well. This combination of changes makes it almost impossible for a person with PTSD to stay steady and present long enough to develop a trusting relationship. Disconnection is the norm, but it is important to remember that the disconnection is not only a psychological survival mechanism but it is biologically driven and not volitionally controlled.

The world is, at times, an unpredictable and dangerous place. Even if we could eliminate violence and oppression, trauma and stress would still exist in the form of accidents, natural disasters, death, and loss. Given the inevitability of trauma and stress, it is essential that we understand the wide reaching impact of trauma on our brains, bodies, and eventually relationships. However, it may be even more important to further explore the mystery of connection.

CASE STUDY

Wendy is a 38-year-old lesbian woman who was raised in a strict Irish Catholic home. She sought counseling a year ago after experiencing her first panic attack while riding on a bus. This attack happened about a month after her father had died and she felt it was related to her grief. She adored him despite how he controlled her every move as a child. He disciplined his children with a very heavy hand and belt.

Still, she believed she would get over her grief with time. She told no one. She had learned well that you do not share vulnerability or pain with others. Over the next few months the panic attacks came more frequently and Wendy began to fear leaving the house. She withdrew from friends and would go to work only. She had recently started dating a woman she had met at a party, but suddenly, she could not return the phone calls. The retreat into isolation did not help. In fact, her symptoms got worse. She began having nightmares, often of running from her father as a child, or protecting her little sister by standing in his way. She would wake up in a cold sweat and not be able to return to sleep. The images of his beatings began to invade her daytime thoughts. She literally thought she was going crazy. Finally, after one extreme flashback where she felt she was back in her childhood home, she called a therapist.

The therapist talked about Relational-Cultural therapy and how it could help. She asked Wendy far more questions about other people than she was comfortable answering. She felt she was betraying her family, her friends, even the church. However, she felt it was her job to answer everything she was asked from someone who had more power than she did. She was surprised when this therapist recommended a medication consultation. Wendy had little faith in medicine; certainly it was nothing one took for emotional weakness. After much prodding she went. The psychiatrist, a woman who understood childhood abuse, diagnosed her with post-traumatic stress disorder and recommended she take Prozac to help stop the anxiety, the fear, and the panic symptoms. Wendy was desperate, so she agreed.

After about six weeks on the medication, Wendy began to notice a significant decrease in her anxiety and panic attacks. She began to feel she could call some of her friends. Her therapist encouraged her to tell those closest to her what had happened. This was difficult, though she seemed genuinely surprised when her closest friend began to weep when she was telling about an incident of abuse she had experienced as a child. It was hard to go to the appointments, it still felt wrong to talk so much about herself. However, she was getting a different take on her life. She was finding the therapy interesting but also still painful.

Once the more severe symptoms were stabilized with medications, the therapist began to describe the "work" they would need to do to heal. It seemed to be a much bigger task than she had signed up for. For starters, she now realized that the sudden loss of her father from a heart attack was quite traumatic for her. She also began to read about post-traumatic stress disorder and realized that the physical abuse she experienced at her father's hands may have actually caused her brain to be more sensitive to the big reaction she had at his death. It was hard for her to imagine the two were linked.

The biggest piece of work in the therapy was learning about relationships. She was able to see how the silence and isolation in her own home did not prepare her to go out into the world to interact with others. She realized she rarely knew what her feelings were; putting words to them was impossible. Some of the basic concepts of relating such as trust, empathy, listening, and respect were all foreign to her. By the end of the third month she could see she was in "Relationships 101," with much to learn and much to experience if she were to develop the rich life she imagined for herself.

REFERENCES

Alexander, G., Goldman, P. (1978). Functional development of the dorsolateral prefrontal cortex: An analysis utilizing reversible cryogenic depression. *Brain Research, 143* (233–249).

American Psychiatric Association (1994). *Diagnostic and statistical manual of mental disorders* (4th ed.) Washington, DC: American Psychiatric Press.

Banks, A. (2002). Polypharmacy in post-traumatic stress disorder. In Ghaemi, N. (Ed). *Polypharmacy in psychiatry.* New York:. Marcel Dekker, Inc.

Banks, A. (2001). PTSD: Post-traumatic stress disorder: Relationships and brain chemistry. *Project Report #8. Stone Center Publications.* Wellesley, MA.

Benes, F., Turtle, M., Khan, Y; et al. (1994). Myelination of a key relay zone in the hippocampal formation occurs in the human brain during childhood, adolescence and adulthood. *Archives of General Psychiatry, 51* (477–484).

Bremner, J., Staib, L., Kaloupek, D., Southwick, S., Soufer, R., Charney, D. (1999). Neural correlates of exposure to traumatic pictures and sound in Vietnam combat veterans with and without post-traumatic stress disorder: A positron emission tomography study. *Biological Psychiatry, 45* (806–816).

Bremner, J., Narayan, M., Staib, L., Southwick, S., McGlashen, T., Charney, D. (1999). Neural correlates of memories of childhood sexual abuse in women with and without post-traumatic stress disorder. *American Journal of Psychiatry, 156* (1787–1795).

Bremner, J., Randall, P., Scott, T., Bronen, R., Seibyl, J., Southwick, S., Delaney, R., McCarthy, G., Charney, D., Innis, R. (1995). MRI-based measurement of hippocampal volume in patients with combat-related post-traumatic stress disorder. *American Journal of Psychiatry, 152* (973–981).

Cashman, L., Molnar, E., Fa, B. (1995). *Comorbidity of DSM-III-R Axis I and Axis II disorders with acute and chronic post-traumatic stress disorder* (abst.) Presented at the 29th annual convention of the Association for the Advancement of Behavior Therapy, Washington, DC.

Chi, J., Dooling, E., Gilles, F. (1977). Left-right asymmetries of the temporal speech area of the human fetus. *Archives of Neurology, 34* (346–48).

Centers for Disease Control (1988). *Vietnam Experiences Study: Psychological and Neuropsychological Evaluation.* Atlanta, GA: Centers for Disease Control.

Cooper, I., Upton, A. (1985). Therapeutic Implications of modulation of metabolism and functional activity of the cerebral cortex by chronic stimulation of cerebellum and thalamus. *Biological Psychiatry, 20* (811–813).

Davis, M. (1992). The role of the amygdala in fear-potentiated startle: Implications for animal models of anxiety. *Trends Pharmacol Science, 13* (35–41).

DeBellis, M., Keshavan, M., Clark, D., Casey, B., Giedd, J., Boring, A., Frustaci, K. A. (1999). Bennett Research Award. Developmental traumatology. Part II: Brain development. *Biological Psychiatry, 45* (1271–1284).

Diorio, D., Viau, V., Meaney, M. (1993). The role of the medial prefrontal cortex (cingulate gyrus) in the regulation of the hypothalamic-pituitary-adrenal response to stress. *Journal of Neuroscience, 13* (3839–3847).

Erikson, E. (1963). *Childhood and Society* (2nd ed.). New York: Norton.

Eriksson, P., Perfilieva, E., Bjork-Eriksson, T., Alborn, A. M., Nordborg, C., Peterson, D. A. (1998). Neurogenesis in the adult human hippocampus. *Natural Medicine, 4* (1313–1317).

Famularo, R., Fenton, T., Kinscherff, R., Augustyn, M. (1996). Psychiatric co-morbidity in childhood post-traumatic stress disorder. *Child Abuse and Neglect, 20* (953–961).

Ferber, R. (1985). *Solve your child's sleep problems.* New York: Simon and Schuster.

Finkelhor, D., Browne, A. (1984). The traumatic impact of child sexual abuse: A conceptualization. *American Journal of Orthopsychiatry, 55* (530–541).

Fuster, J. (1980). *The prefrontal cortex: Anatomy, physiology and neurophysiology of the frontal lobe.* New York: Raven Press.

Galaburda, A. M. (1984). Anatomical assymetries in the brain. In Geshwind, N., Galabruda, A. M. (Eds.). *Biological foundations of cerebral dominance.* Cambridge, MA: Harvard University Press.

Gilligan, C. (1982). *In a different voice.* Cambridge, MA: Harvard University Press.

Gould, E., Woolley, C. S., McEwen, B. S. (1991). Adrenal steroids regulate postnatal development of the rat dentate gyrus: I. effects of glucocorticoids on cell death. *Journal of Comp. Neurology, 313* (479–485).

Gurvits, T. V., Shenton, M. E., Hokama, H., Ohta, H., Lasko, N. B., Gilbertson, M., Orr, S. P., Kikinis, R., Jolesz, F. A., McCarley, R. W., Pitman, R. K. (1996). Magnetic resonance imaging study of hippocampal volume in chronic, combat-related post-traumatic stress disorder. *Biological Psychiatry, 40* (1091–1099).

Guyton, A. (1986). *Textbook of medical physiology* (7th ed.) Philadelphia: W. B. Saunders Company.

Heath, R. G. (1977). Modulation of emotion with brain pacemaker. Treatment for intractable psychiatric illness. *Journal of Nervous and Mental Disorders, 165* (300–317).

Jacobson, M. (1973). Genesis of neuronal specificity. In Rockstein, M. (Ed.). *Development and aging in the nervous system* (105). New York: Academic Press.

Jordan, J., Kaplan, A., Miller, J. B., Stiver, I., Surrey, J. (1991). *Women's growth in connection.* New York: The Guilford Press.

Jordan, J. (1991). The movement of mutuality and power. *Work in Progress No. 53.* Wellesley, MA: Stone Center Working Paper Series.

Jordan, J. (1987). *Clarity in connection: empathic knowing, desire and sexuality.* Wellesley, MA: Stone Center Working Paper Series.

Jordan, J. & Dooley, C. (2000). Relational practice in action: A group manual. *Project Report #6.* Wellesley, MA: Stone Center Publications.

Kaplan, H. & Sadock, B. (1991). *Synopsis of psychiatry behavioral sciences clinical psychiatry.* Baltimore: Williams and Wilkins.

Kessler, R. C., Sonnega, A., Bromet, E., Hughes, M., Nelson, C. B. (1995). Post-traumatic stress disorder in the National Comorbidity Survey. *Archives of General Psychiatry, 52* (1048–1060).

Kiser, L. J., Heston, J., Millsap, P. A., et al. (1997). Physical and sexual abuse in childhood: relationship with post-traumatic stress disorder. *Journal of American Academy of Child and Adolescent Psychiatry, 36* (1236–1243).

Kraemer, G. W. (1992). A psychobiological theory of attachment. *Behavioral and Brain Science, 15* (493–541).

Kulka, R. A., Schlenger, W. E., Faiurbank, J. A., Jordan, K., Weiss, D. Cranston, A. (1990). *Trauma and the Vietnam war generation.* New York: Brunner/Mazel.

Lazarus, R. S. (1985). The psychology of stress and coping. *Issues in Mental Health Nursing, 7* (399–418).

Ledoux, J. E. (1990). Information flow from sensation to emotion: Plasticity in the neural computation of stimulus value. In Gabriel, M. & Moore, J. (Eds.). *Learning and Computational Neuroscience: Foundations of Adaptive Networks.* Cambridge, MA: MIT Press.

Le Doux, J. E, Iwata, J., Cicchetti, P., Reis, D. J. (1988). Different projections of the central amygdaloid nucleus mediate autonomic and behavioral correlates of conditioned fear. *Journal of Neuroscience, 8* (2517–2529).

Liberzon, I., Taylor, S. F, Amdur, R., Jung, T. D., Chamberlain, K. R., Minoshima, S., Koeppe, R. A., Fig, L. M. (1999). Brain activation in PTSD in response to trauma-related stimuli. *Biological Psychiatry, 45* (817–826).

Mahler, M. S., Pine, F., Bergman, A. (1975). *The psychological birth of the human infant.* New York: Basic Books.

Maiti, A., Snider, R. S. (1975). Cerebellar control of basal forebrain seizures: Amygdala and hippocampus. *Epilepsia, 16* (521–533).

McFarlane, A. (1988). The longitudinal course of post-traumatic stress morbidity. *Journal of Mental Disorders, 176* (30–39).

Miller, J. B., Stiver, I. (1997). *The healing connection.* Boston: Beacon Press.

Miller, Jean B. (1976). *Toward a new psychology of women.* Boston: Beacon Press.

Molfese, D. L., Freeman Jr., R. B., Palermo, D. S. (1975). Ontogeny of brain lateralization for speech and nonspeech stimuli. *Brain Language, 2* (356–368).

Morgan, M. A., LeDoux, J. E. (1995). Differential contribution of dorsal and ventral medial prefrontal cortex to the acquisition and extinction of conditioned fear in rats. *Behavioral Neuroscience* (681–688).

Pinchus, J. H., Tucker, G. J. (1978). *IN: Behavioral neurology.* New York: Oxford.

Pitman, R. K. (1989). Post-traumatic stress disorder, hormones and memory. *Biological Psychiatry, 26* (645–652).

Polan, H. J., Hofer, M. A. (1999). Psychobiological origins of infant attachment and separation responses. In Cassidy, J. & Shaver, P. H. (Eds). *Handbook of attachment: Theory, Research and Clinical Applications* (162–180). New York: The Guilford Press.

Pollack, W. (1998). *Real boys: Rescuing our sons from the myths of boyhood.* New York: Random House.

Rakic, S., Zecevic, N. (2000). Programmed cell death in the developing human telencephalon. *European Journal of Neuroscience, 12* (2721–2734).

Rauch, S. L., Van der Kolk, B. A., Fisler, R E., Alpert, N. M., Orr, S. P., Savage, C. R., Fischman, A. J., Jenike, M. A., Pitman, R. K. (1996). A symptom provocation study of post-traumatic stress disorder using positron emission tomography and script-driven imagery. *Archives of General Psychiatry, 53* (380–387).

Resnick, H. S., Yehuda, R., Pitman, R. K, Foy, D. W. (1995). Effects of prior trauma on acute hormonal response to rape. *American Journal of Psychiatry, 152* (1675–1677).

Roozendaal, B., Koolhaas, J. M., Bohus, B. (1992). Central amygdaloid involvement in neuroendocrine correlates of conditioned stress responses. *Journal of Neuroendocrinology, 4* (46–52).

Schiffer, F., Teicher, M. H., Papanicolaou, A. C. (1995). Evoked potential evidence for right brain activity during the recall of traumatic memories. *Journal of Neuropsychiatry Clinical Neuroscience, 7* (169–175).

Schore, A. (2003). *Affect dysregulation and disorders of the self.* New York: W. W. Norton & Company.

Sears, W., & Sears, M. (1995). *The discipline book.* Boston: Little Brown and Company.

Sesack, S. R., Deutch, A. Y., Roth, R. H., Bunney, B. S. (1989). Topographical organization of the efferent projections of the medial prefrontal cortex in the rat: An anterograde tract-tracing study with Phaseolus vulgaris leucagglutin. *Journal of Comp. Neurology, 290* (213–242).

Shin, L. M., Kosslyn, S. M., McNally, R. J., Alpert, N. M., Thompson, W. L., Raven, S. L., Macklin, M. L., Pitman, R. K. (1997). Visual imagery and perception in post-traumatic stress disorder. A positron emission tomographic investigation. *Archives of General Psychiatry, 54* (233–241).

Shin, L. M., McNally, R. J., Kosslyn, S. M., Thompson, W. L., Rauch, S L., Alpert, N. M., Metzger, L. J., Lasko, N. B., Orr, S. P.; Pitman, R. K. (1999). Regional cerebral blood flow during script-driven imagery in childhood sexual abuse-related PTSD: A PET investigation. *American Journal of Psychiatry, 156* (575–584).

Sidman, R. L., Rakic, P. (1973). Neuronal migration, with special reference to developing human brain: A review. *Brain Research, 62* (1–35).

Simonds, D. M., Scheibel, A. B. (1989). The postnatal development of the motor speech area: A preliminary study. *Brain Language, 37* (42–58).

Snider, R. S., Maiti, A. (1975). Septal afterdischarges and their modification by the cerebellum. *Exp. Neurology, 49* (529–539).

Solomon, S. D., Davidson, J. R. (1997). Trauma, prevalence, impairment, service use and cost. *Journal of Clinical Psychiatry, 58* (5–11).

Southwick, S. M., Krystal, J. H., Morgan, C. A., Nicolaou, A. L., Nagy, L. M., Johnson, D. R, Heninger, G. R., Charney, D. S. (1997). Noradrenergic and seroternergic function in post-traumatic stress disorder. *Archives of General Psychiatry, 54* (749–758).

Stein, M. D., Koverola, C., Hanna, C., Torchia, M. G., McClarty, B. (1997). Hippocampal volume in women victimized by childhood sexual abuse. *Psychological Medicine, 27* (951–959).

Stern, D. (1985). *The interpersonal world of the infant.* New York: Basic Books, Inc.

Teicher, M. H., Anderson, S., Polcari, A., Anderson, C., Navalta, C. (2002). Developmental neurobiology of childhood stress and trauma. *Psychiatric Clinics of North America, 25* (397–426).

Teicher, M. H., Ito, Y., Glod, C. A., Anderson, S. L., Dumont, N., Ackerman, E. (1997). Preliminary evidence for abnormal cortical development in physically and sexually abused children using EEG coherence and MRI. *Annals of the New York Academy of Science, 821* (160–175).

Teicher, M. H., Anderson, S. L., Dumont, N. L., et al. (2000). Childhood neglect attenuates development of the corpus collosum. *Social Neuroscience Abtract, 26* (549).

U.S. Department of Health and Human Services, Administration on Children, Youth and Families (2001). *Child Maltreatment 1999: Reports from the states to the National Child Abuse and Neglect Data System.* Washington, DC: U.S. Government Printing Office.

Van der Kolk, B. (1988). The trauma spectrum: The interaction of biological and social events in the genesis of the trauma response. *Journal of Traumatic Stress, 1* (273–290).

Wada, J. A., Clarke, R., Hamm, A. (1975). Cerebral hemispheric asymmetry in humans: Cortical speech zones in 100 adults and 100 infant brains. *Archives of Neurology, 32* (239–246).

Ward, J. (2000). *The skin we're in.* New York: The Free Press.

Watson, J. B., Mednick, S. A., Huttunen, M., Wang, X. (1999). Prenatal teratogens and the development of adult mental illness. *Developmental Psychopathology, summer, 11(3),* 457–66.

Weinberger, D. R. (1987). Implications of normal brain development for the pathogenesis of schizophrenia. *Archives of General Psychiatry, 44* (660–669).

Weiss, E. L., Longhurst, J. G., Mazure, C. M. (1999). Childhood sexual abuse as a risk factor for depression in women: Psychosocial and neurobiological correlates. *American Journal of Psychiatry, 156* (816–828).

Widom, C. S. (1999). Post-traumatic stress disorder in abused and neglected children grown up. *American Journal of Psychiatry, 156* (1223–1229).

Yehuda, R. (2000). Biology of post-traumatic stress disorder. *Journal of Clinical Psychiatry, 61* (14–21).

Yehuda, R., Mcfarlane, A. C., Shalev, A. Y. (1998). Predicting the development of post-traumatic stress disorder from the acute response to a traumatic event. *Biological Psychiatry, 44* (1305–1313).

10

◉

Sexuality Across the Life Span

By Stella Kerl, Ph.D., and Ana Juárez, Ph.D.

REFLECTION QUESTIONS

1. *In your opinion what is healthy human sexuality?*
2. *How did you come to learn about sex and who taught you about healthy sexuality?*
3. *Does this definition apply to both women and men? To women of different races? To other cultures?*
4. *Explain two problems that pervade dominant understandings of sexuality in psychology.*
5. *How is Latina sexuality portrayed in much of the dominant literature?*
6. *Which kinds of stereotypes have negatively affected African American sexuality?*
7. *What might researchers do to create less biased models of sexuality?*
8. *Are you aware of assumptions and stereotypes you may have about sexuality in women of color? If so, how do you think you these stereotypes were created?*
9. *What assumptions do you have about male sexuality?*

INTRODUCTION

I (Stella) am a licensed psychologist, a Latina, and an associate professor of counseling at Texas State University-San Marcos. My experiences as a psychologist and a therapist have brought me into contact with many conceptualizations and conversations about women's sexuality in general and Latina sexuality in particular. My anger over the misconceptions, blatant stereotyping, and devaluing of Latino culture related to sexuality (while glorifying sexual freedom and liberation present in the United States) is one of the reasons I began to write in the area. I hope to help therapists and students think more contextually about sexuality, and avoid ethnocentric and essentialized understandings of human sexuality that may translate into biased ways of working with clients.

I (Ana) am an anthropologist and an associate professor of anthropology at Texas State University-San Marcos. I rely on several years of comparative ethnographic fieldwork focused on race, gender, and sexuality among Mexicans and Mayas in Quintana Roo, Mexico, to inform my current work. I have observed cultures with significantly different conceptions of sexuality than those in popular American thought, and find it particularly relevant that Mayas believe males and females have fundamentally similar and balanced sexual natures. They do not construct distinct identities based on same-sex partner choice, do not particularly eroticize women's breasts as is so common in U.S. culture, and they base gender primarily on the kind of work people do, rather than on the kinds of bodies people have. Mexicans, in contrast, believed that men's sex drive is stronger than women's and constantly gossiped, discussed, and joked about sexual desire, activities, and liaisons. My Mexicana/Tejana background and fieldwork experiences made me realize the complexity and diversity of sexuality even within "Mexican" culture and society, the ethnocentrism and essentialism of much of the research, and that analyses of sexuality must always be contextually situated. These are the perspectives that frame our work.

Many of the ideas for this chapter were derived from conversations between the two of us. We began our writing about sexuality when I (Stella) entered Ana's office one day and Ana teasingly asked, "Who do you think is more sexual: White women or Latinas?"

I thought for a moment. "Hmm . . . I guess I'd have to say Latinas."

"What makes you say that?" questioned Ana.

"Well, I think of the food and the dancing, and . . . I don't know. It just seems like we're more sexual."

"I thought we were pretty sexual, too," replied Ana. "But I've been doing all this reading lately that says we are not sexual and are completely repressed!"

So we started exploring the area, first critiquing current conceptualizations of Latina sexuality (Juárez & Kerl, 2003), and then looking at broader understandings of sexuality.

ESSENTIALISM AND ETHNOCENTRISM IN DOMINANT UNDERSTANDINGS OF SEXUALITY IN PSYCHOLOGY

What is sexuality? People often think of sexuality in different ways. The dictionary defines sexuality as "the condition of being characterized and distinguished by sex, concern with or interest in sexual activity" (*American Heritage Dictionary of the English Language,* 2000). Sex, of course, includes activities involving genital organs, intercourse and the sexual urge, desire or instinct as it manifests itself in behavior.

As the recent Clinton/Lewinsky scandal demonstrated, though, not everyone thinks of sex in the same way. Some people limit "sex" to intercourse, while others use it more inclusively and broadly to include kissing, touching, and even sexual feelings. These differences in meaning become a problem when looking at the issue of sexuality since the same kind of problem occurs when people talk or write about sexuality. In other words, many people may feel they understand what "healthy sexuality" looks like, but we believe that the meaning of "healthy" and of "sexuality" depends on context, that is, society, cultures, and individual contexts. People who do not consider context believe that sexuality exists in humans in a particular way.

We have called this *essentialized understandings of sexuality.* The notion of essentialism generally refers to the underlying assumption that "individuals or groups have an immutable and discoverable 'essence'—a basic, unvariable, and presocial nature" (Moya, 1997, p. 381). In our work we have critiqued the notion many people have that women's sexual disposition and experience is natural, and that women share an innate, universal essence of sexuality, which is fundamentally the same for them all.

We question the assumption that there is a "natural" (essential) sexuality that is basically the same for all women (and men). Much of the research in psychology seems to take this view, and research is often designed to help uncover or figure out how sexuality works in people. This view looks at sexuality as a biological drive or instinct, and tends to focus on reproductive issues, sexually transmitted diseases, and physical sexual functioning. Several researchers looking at how writings about sexuality have evolved over the years have made this point (Bullough, 1990; Duggan, 1990; Giami, 2002; Milligan, 1993).

Duggan (1990) observed that little was written about sexuality before the 1960s and the early 1970s. However, when literature in the area began to emerge, it "generally retained the assumption that sex is a biological drive . . . [Researchers] usually portrayed the Victorians as repressors of the sexual instinct, and their successors, the Freudian modernists, as liberators of a healthy sex drive" (p. 96). Duggan reiterated that the researchers "did not question the prevailing assumptions that sex was biologically based, heterosexually organized, and rooted in 'natural' gender roles . . . they identified the central point of conflict over sexuality as that between advocates of repression and proponents of the expression of a natural, healthy sexuality" (p. 96).

Scholars such as Foucault (1980) and D'Emilio and Freedman (1988) recognized that sexuality became less connected to reproduction during the Industrial Era, and this trend was intensified by the 1960s. Giami (2002) shows how the 1960s further changed concepts about healthy sexuality. Before, "sexuality was primarily conceived as oriented toward procreation . . . in the context of marriage, expressed both as a biological function and a moral value for society" (p. 2). He argued that the introduction of oral contraception in the 1960s socially legitimized sexual activity for reasons other than procreation in and outside of marriage. With its role in the emergence of the concept of sexual health in 1975, Giami (2002) stated that the World Health Organization helped create the perception of sexuality as a public health issue instead of looking at it from a moral perspective.

Kinsey, Pomeroy, and Martin (1948) and Masters and Johnson (1966) are well known as pioneers in the field of human sexuality. At the time, public morality severely restricted open discussion of sexuality as a human characteristic. Specific sexual practices, especially sexual behaviors that did not lead to procreation, were not part of public dialogue. Kinsey's research changed that. He and his colleagues reported findings on the frequency of various sexual practices and today the Kinsey Institute for Research in Sex, Gender, and Reproduction remains the most frequently cited source on sexual behavior (Reinisch, 1990). Masters and Johnson (1966) first studied the physiology of human sexual response, and then began to look at the treatment of sexual problems (Masters & Johnson, 1970). Their work defined sexuality as a healthy human trait and normalized the pursuit of pleasure during sex as a socially acceptable goal. Many researchers followed this path, and today human sexuality and sex therapy have become common fields of study, with numerous books and journals assisting to explore and promote the goal of healthy sexuality.

We believe that having a seemingly objective view of a common "healthy sexuality" can be problematic. Looking at sexuality as a biological or instinctual drive leaves out the issue of how power and context can affect meaning, and the myth of objectivity can lead to the stereotyping of people who experience sexuality differently.

For this reason, Tolman and Diamond (2001) strongly suggested that researchers should examine both biological factors and sociopolitical/cultural factors when looking at sexuality. They stated that "neither a purely biological nor a purely sociocultural approach can encompass the complexity of sexual desire" (p. 34) and that sexuality is embedded in "relationships nested within societies nested within cultures and historical epochs . . . and always embedded in particular biological contexts" (p. 34). This approach was attempted when the National Commission on Adolescent Sexual Health released a statement, which went beyond biological issues to include relationships and values (Pamar & Haffner, 1998). They stated that a definition of sexual health for adolescents included the ability to develop and maintain meaningful interpersonal relationships, to appreciate one's own body; to interact with both genders in respectful and appropriate ways, and to express affection, love, and intimacy in ways consistent with one's own values.

However, researchers who examined sexuality in a broader manner (including values and relationships) have also interpreted meaning and behavior in ways that portray sexuality in nondominant groups as being deficient. Dealing almost exclusively with heterosexuality, these interpretations are characterized by essentialized understandings of both culture and human sexuality. There seems to be an assumption that modern White sexuality has progressively become more liberated and is the healthy, right way to be sexual.

Along with the proliferation of information about sexuality, the women's liberation movement of the late 1960s and 1970s contributed to the view that "liberated" women should have more modern ideas about sex rather than the more conservative values of chastity and purity that were prevalent at the time. However, current researchers have documented that the expectation of chastity continues to coexist with the newer model of "healthy sexuality" in a way that poses a dilemma for many women (Tolman, 2002). The resulting confusion may often put women in a double bind: be sexual, but be chaste.

Much of the literature about Latina sexuality (Alonso 1995; Alonso and Koreck 1989; Castillo, 1991; Espín, 1999; Limón, 1997) suggested that Latino culture itself was repressive to women. Others writers and researchers (Gil & Vasquez, 1996; Gómez & Marín, 1996; Salgado de Snyder et al., 2000) have implied that, in order to become sexually healthy, Latinas should follow the more sexually liberated sexual perspectives of American culture. These values include acknowledging and embracing one's sexuality, taking pleasure in sex, and being able to voice one's own sexual likes, dislikes, and desires. In other words, a person who has a healthy sexuality practices these behaviors regularly.

Our initial work (Juarez & Kerl, 2003) concluded that essentialized understandings of sexuality lead to ethnocentric bias when it came to looking at Latina sexuality. More specifically, we argued that dominant representations narrowly portray Latina sexuality as inherently negative and/or dichotomize Latinas as either being traditional and sexually repressed, or being acculturated and sexually liberated.

We believe that the dominant view of sexuality (with the goal of a "healthy sexuality") is based on White, Western understandings of sexuality and that the movement from a passive, repressed sexual expression to a progressively liberal way of expressing sexuality is assumed to be the right way to be sexual. When people from other ethnic groups or cultures do not fit the "healthy" model, these dominant understandings have at times suggested that the other culture should become more like the dominant culture.

In addition to having essentialist assumptions of sexuality, we believe that the dominant models also demonstrate *ethnocentrism*. Ethnocentrism refers to the assumption that one's own culture or ethnic group is better, or more natural, and other cultures are judged in terms of one's own cultural perspective. For example, we pointed out that in current scholarly work about Latina sexuality, it was clear that ethnocentrism was evident in the assumption that *less* acculturated Latinas were *more* repressed—and that acculturation toward the "natural" (essentialized) White American way was seen as the answer (Juárez & Kerl, 2003).

One of our favorite examples of this type of bias is evident in a popular self-help text used by therapists, *The Maria Paradox: How Latinas Can Merge Old World Traditions With New World Self-Esteem* (Gil & Vazquez, 1996). I (Stella) first encountered this book when I began to talk to my therapist friends and colleagues about the article I was writing with Ana.

"Oh, you are writing about Latina sexuality?" I was asked. "You probably know about this book, then." At least three of my therapist friends had mentioned the book to me, which made me begin to notice it on the shelves of the site supervisors as well (therapists in the community who were supervising my students). A colleague at the student counseling center mentioned that they had just had an in-service based on the book.

"It's been so helpful!" she enthused.

So I read it, and couldn't believe what I was reading! Were my colleagues really using this book as a reference for working with Latina clients? It seemed to me that the book blamed Latino culture for many of the "problems" that Latinas faced, including sexual problems. It urged them to adopt "new world" values that were consistent with a more "healthy sexuality." One chapter about sex had the title, "Do not forget that sex is for making babies—not for pleasure: Old world marriage vs. real live passion" (p. 126). The authors stated that "the strong element of repression running through Hispanic tradition helps women ignore their own erotic impulses before and after marriage" (Gil & Vazquez, 1996, p. 130).

This chapter also stated that the denial of pleasure for Latinas was an old world value related to traditional Latina/o culture. It argued that *marianismo* was a cultural value that mandated what a good Latina should be: submissive, subservient, self-sacrificing, self-renouncing, sexually pure, and erotically repressed. This, combined with the emphasis on fertility, and "the inability to feel and express eroticism, becomes characteristic of sexual development of many Latinas—far different from what North American girls are taught . . . for the Hispanic woman to enjoy sexual pleasure, even in marriage, is wrong because it indicates a lack of virtue" (Gil & Vazquez, 1996, p. 135). They recommend that readers, "try to determine as honestly as you can whether you're still holding on to *marianista* sexual beliefs" (Gil & Vazquez, 1996, p. 134), thus suggesting that more liberal views are characteristic of more acculturated women.

In another place in the same chapter, Gil and Vasquez (1996) tell a story about a woman who was repressed due to her "old world" values. However, she had started dating an American man who helped her make her own choices rather than make choices for her like ordering for her at dinner like her domineering late husband (who was Latino). The authors state that she felt understood by his supportive comments, such as "I know that's the way you were taught to think and feel about yourself, but here in the United States you have options regarding your likes and dislikes" (p. 150). Her American boyfriend seems to be a North American savior rescuing her from her life of culturally determined Central or South American values of repression and submission that kept her from exercising her own choices.

As we stated previously, several authors have pointed out the American values toward women's sexuality are not as ideal as the American boyfriend professed (Tolman, 2002). There are ongoing contradictions between chastity and liberated sexuality for American (White) women, yet women's sexuality in that literature seems to be portrayed in a more moderate manner than literature that looks more specifically at the sexuality of women of color. Our current belief is that the "healthy sexuality" model appears to be mediated by race: it seems that White sexuality is portrayed as moderate, not extreme. When situated in a racially mediated context, it is neither too hot, nor too cold, but just right.

Historically, the darker women are, the more promiscuous they are presumed to be. For example, Castañeda (1990) shows that when Euro-Americans gained control of California in the latter part of the nineteenth century, darker mestiza women were considered promiscuous, but whiter Spanish women were considered chaste and marriageable. Thus, even within a model of a liberated White sexuality, the emphasis on chastity may convey an avoidance of extremes and promote moderation and control or reserve (sex should occur in a love relationship; don't be too sexual or too prudish, etc.).

Images of African American sexuality have also been historically portrayed in extremist ways. When Grimes (1999) interviewed sexuality researcher Wyatt (1999) about her work, Wyatt stated that "Black women were typically portrayed [either] as asexual beings, like mammy, or oversexed temptresses with uncontrollable lust" (Grimes, 1999, p. 66). She went on to say that:

> Sex research that came after Kinsey focused almost exclusively on comparing Black and White female sexuality. It did not explore the many dimensions of Black female sexuality and usually supported the mythology that Black females are more promiscuous than White females. It was research that emphasized the difference between Blacks and Whites, usually with Blacks being the "atypical" ones, and Whites being the "norm." (p. 27)

Two early researchers explained racial differences between African Americans and Whites in different ways. Kinsey, Pomeroy, and Martin (1948) stated that Black men and women are more sexually permissive than White men and women and that these differences were due to social class. Coleman (1966) focused on the differences between men and women in sexual permissiveness: He believed that Black men and women were more similar in sexual permissiveness than White men and women because Black women have more power in the family and are less subject to male domination (and thus it was more difficult to sustain a double standard).

Weinberg and Williams (1988) tested these two theories and found that Black men and women were more likely than White men and women to engage in premarital sex earlier and more frequently, and that Black men had more partners. They also found that Black men and women were more likely than White men and women to engage in extramarital sex, and that Black men had extramarital sex with more partners. They also found that Black men and women were more open about sex than White men or women and had fewer emotional problems associated with sex. They determined that social class was not a factor in differences between Blacks and Whites.

This study demonstrates ethnocentric bias in several ways. The authors state that "we may expect . . . that Black women are more sexually permissive (in behavior as well as norms) than are White women" (Weinburg & Williams, 1988, p. 198). The use of the word permissive, which is used throughout the article, is an evaluative word. It implies that the standard, or "normal" model is one that is better, not as indulgent or lacking in discipline. Additionally, the article reinforces stereotypes by suggesting that sexual stereotypes originated from true events:

> The black woman, like the black man, is subject to a sexual stereotype that portrays her as sexually "loose" . . . the disorganization of the black family beginning with slavery and continuing under later economic and social deprivation made the black woman particularly vulnerable to sexual exploitation by both black and white men. Such conditions did not support the cult of chastity that had structured white male-female relationships. (p. 200)

While the passage does not directly blame African American women for being "loose," it frames sexual behavior in a way that appears to apologize or excuse it (i.e., they would be more chaste if only the conditions were better!). That is, if events like slavery hadn't damaged the culture, Black men and women would behave more like White people. This again is an ethnocentric and essentialized perspective that looks at the White, Western standard as being the (better) norm.

Additionally, Black female slaves were often sexually exploited by their White male owners (Staples, 1972; Wyatt, 1997). While this event certainly contributed to the "promiscuous" stereotype given to Black women, it also may have contributed to the related belief that White women are pure and virginal. White male slave owners took their "desires" to Black women (whom they sometimes raped); thus White women were able to be seen as chaste. Greene (1994) wrote:

> Although all women were considered the originators of sexual sin, white women were elevated to a pedestal of sexual purity and virtue, whereas African-American women came to be depicted as the embodiment of sexual promiscuity and evil. . . . By depicting their black female victims as morally loose, white males rationalized what had become their routine and accepted practice of rape. (p. 340)

Noted psychologist and sexuality researcher Gail Wyatt (in Grimes, 1999) argued that looking at Black and White differences was less informative than looking at the diversity of sexuality within the African American population. Her research emphasized within-group differences among Black women. She agreed that all women of color experience sexual stereotyping, but pointed out that stereotypes have been especially damaging regarding African American women and their sexuality (Wyatt, 1997). She noted that while White women are often depersonalized by being labeled as less intelligent or incompetent, depersonalization of Black women focuses first on their sexuality (Wyatt, 1999). Stereotypes about Black women perpetuate myths about sexual irresponsibility.

Lawrence (1992) wrote that "stereotypes about race and sex . . . are deeply embedded in the American psyche" (p. 65). Stephens and Phillips (2003) wrote that the racialized and sexualized stereotypes of African American women, including Jezebel, Mammy, Matriarch, and Welfare Mother images have evolved into similar, yet more sexually explicit images that include the Freak, Gold Digger, Diva, and Dyke. Mecca and Rubin (1999), in their study of sexual harassment in African American students, found that participants delineated a new category of sexual harassment previously untapped by largely Caucasian studies: comments or sexual attention based solely on racial stereotypes or racially based physical features.

Perhaps due to the strength of the impact of sexual stereotyping for African American women, it is difficult to find research about African American sexuality from the lived experience of the participants. Rose (2003) recently studied the sexual testimonies of 20 Black women. She acknowledged that it was difficult to find research about Black women's sexual experiences because their association with sexual deviance and excess has kept African American women from talking about sexual desires and experiences—her participants reported fear that their words would be misused. Her study of Black women's sexual lives show them to be aware of distorted myths about Black sexuality that fuel racist, demeaning stories about Black men and women.

George and Martínez (2002) found that stereotypes about Black sexuality affected rape victim blaming. The researchers first measured the racism scores of 332 predominantly White and Asian college students, then had them evaluate a rape vignette with varying victim race, perpetrator race, and rape type. They found that racial makeup of the victim and perpetrator determined victim blaming, in that interracial rapes were less uniformly judged as "definitely rape" and were judged as having more culpable, less credible victims, and less culpable perpetrators. Additionally, racism scores in male participants positively predicted victim blaming in all rapes. Racism scores in female participants moderated victim blaming in interracial acquaintance rapes. The researchers concluded that the racial stereotypes about Black sexuality could have strong influences on discriminatory adjudication outcomes. The researchers did not report differences between participants of White or Asian ethnicities.

While it is clear that much of the sexuality research involving African Americans focuses on the powerful effects of negative stereotypes, research about the sexuality of ethnic groups in general is fairly sparse. Wiederman and Maynard (1996) looked at ethnicity in published sexuality research from the years 1971 to 1995 in two major sexuality journals. They found that ethnicity was included as a variable in only 7.3 percent of the articles. Among those articles in which ethnicity was included as a variable, Whites and Blacks were equally likely to have been included and were more likely to be represented compared to Latinos, Asians, or other ethnic groups. Searches in popular psychology databases also demonstrate the insufficiency of work in this area, and when it is present, it is vulnerable to the essentialism and ethnocentrism previously discussed.

POWER RELATIONS AND SEXUALITY

If essentialized views of sexuality lead to ethnocentric bias, what is an alternative? Researchers objecting to the essentialist paradigm often turn to social contructionist perspectives, arguing that reality and meaning are products of the particular context of the phenomenon being studied: the context of political power, of dominant ideologies, of cultural and historical periods, and of social forces that shape how we see "truth." In this perspective, sexuality is created through the social context in which it exists, and there is no "natural" or universally "healthy" sexuality apart from the context within which it was created.

Foucault (1980) argued that sexuality (the experience of sexuality as well as research and writing about it) is constructed by relations of power that produce discourses and create knowledge that maintains power. For example, the fact that the sexuality of people of color is either invisible or seen as negative contributes to the assumption that ethnic groups must acculturate in order to be healthy, which serves the view of the dominant group. Silverman (2003) wrote that Foucault believed social institutions as well as language, rituals, and practices of ordinary communicative experiences were all "vehicles for the subtle domination of knowledge . . . knowledge serves power and the dominant hierarchy" (p. 242). Her work suggested that power relations have shaped female sexuality and influenced our scholarly and popular knowledge and understanding of female development. Miller (1976), in her groundbreaking work on the psychology of women, also developed this perspective. She wrote that "a dominant group . . . has the greatest influence in determining a culture's overall outlook—its philosophy, morality, social theory, and even its science" (p. 8).

Many feminists have looked at sexuality from this perspective, arguing that gender hierarchies create experiences of sexuality that oppress and repress women. As we stated earlier, researchers blamed Latino culture for passivity and repression in Latina sexuality and suggested that acculturation would help them develop a healthy sexuality. However, passivity and lack of sexual voice has also been characterized as a problem for women in U.S. culture.

Fine (1988) and Tolman (1991) have stated that in U.S. culture, women in general have a "missing discourse of desire." This discourse is defined by failing to acknowledge that women have sexual feelings, and that discussion about sex centers around male desire and women's response to male desire instead of women's own desires. A recent study of advice columns in teenage magazines (Garner, Sterk, & Adams, 1998) showed that advice related to sex and sexual relationships had changed very little from 1974 to 1994. The discourse within the advice columns encouraged young women to be sex objects and teachers of interpersonal communication rather than friends or lovers.

Although some researchers recognize the sexist discourse, American culture is not blamed, marianismo is not mentioned, nor is it ever suggested that they need to acculturate to another culture. Acculturation, though often seen as necessary and desirable, does not always mean improved mental health. Some researchers have found that acculturation can be associated with negative outcomes, including a higher risk for eating disorders (Cachelin et al., 2000; Chamorro & Flores-Ortiz, 2000; Dulce, Hunter, & Lozzi, 1999) and delinquent behavior (Cross, 2003; Wong, 1999; Sommers, Fagan, & Baskin, 1993). We believe that acculturation is not the way toward a "healthy sexuality" in women of color; moreover, we question the current construct of "healthy sexuality."

In fact, some mainstream texts on American sexuality paint women's (presumably White women's) sexuality as passive and repressed, not the expected "healthy model" of sexuality. Low and Sherrard (1999) surveyed gender role stereotypes in textbooks on human sexuality and marriage and family. They compared the content of photographs in the textbooks from the 1970s to the 1990s. Even though photographs from the 1990s contained some feminist messages, photographs with more "traditional" messages regarding sexuality and marriage and family still dominated.

Similarly, a popular book, *Women's Experience of Sex* (Kitzinger, 1983), a classic primer dedicated to helping women define and liberate their own sexuality, also described very passive and repressed sexual practices that are not characteristic of the "healthy sexuality" model. However, rather than validating differences in sexuality, the author urges women to overcome passivity and move toward a "freer" sexuality (defined using the "healthy sexuality" model). This book is directed toward women in general, and probably White women in particular (of over 60 pictures, only 4 included people of color, and no mention is made of race or ethnicity). Kitzinger asserts that, early on, "a woman learns to do what she is asked/told, be pleasing to a man, care for others before herself, be compliant, say yes, let others make choices for her, act helpless" (p. 13). Referring to U.S. culture, she wrote:

> We live in a society in which men get serviced and women provide the service. Women are the nurturers. It is assumed that men go out to work and provide economic support in return for housekeeping, child-rearing and sexual availability and they are active and dominant, while women are relatively passive and submissive, having been led to expect men to know best how to satisfy them. Many people do not live like this, of course, but everything we believe and do is affected by that basic template from our culture which shapes and limits our choices. (p. 10)

Tolman and Diamond (2001) point out that these kinds of examples are seen by feminist researchers working from a social constructionist perspective as being tied to gender hierarchies. She wrote that "from this perspective, experiences of sexual desire are inextricably linked to the historically and culturally specific belief systems in which we are embedded. . . . To the extent to which we perceive our desires as fundamentally 'natural' and context-independent, this is only because the sociocultural forces that shape our subjective experiences of sexuality are largely invisible to us" (p. 39). In other words, our sexualities seem "natural" because we have little opportunity to question them. We cannot see that they are interpreted to us through our sociocultural framework, so they just seem "normal." However, she also suggests that a purely social constructionist approach to studying sexuality would miss important contributions that biological approaches offer, such as addressing how sexual desire is experienced within the bodies of individuals.

ESSENTIALISM AND MALE SEXUALITY

Recent theorists have also been critical of essentialized views of male sexuality, in particular, heterosexual male sexuality. Brooks (1997) states:

> Essentialists see men's sexuality as composed of a set of basic and innate preferences that characterize all men across the ages—the idea that men's sexuality is hard-wired. Whatever root cause is cited—evolution, genetic make-up, brain structure, hormones, or God's will—the essentialist position hold strongly to the idea that women and men are fundamentally very different. Not always stated, but clearly implied, is the idea the culture should not discount these differences and should make suitable accommodations. Significant change is viewed as either unlikely or many centuries away. (p. 37)

Sexual behaviors that are seen as "normal" for men are those that fit into stereo-typically traditional views of masculinity. There is much quantitative research indicating that men think about sex more than women, that men masturbate more than women, that men buy and use more pornography, and that they engage in more sex without commitment and with more partners. Essentialist views include looking at biological differences, such as the significance of testosterone, to explain these differences between male and female sexuality (Barash, 1979). Other views point to evolutionary explanations for essential aspects of male sexuality (evolutionary psychologists/sociobiologists such as Buss, 1995). Many researchers do not look for explanations of sexuality, but simply take it for granted that sexual impulses/drives/instincts that fit stereotypically masculine roles are innate (i.e., Zibergeld, 1992).

An example of this view is the very popular book, *Men Are From Mars, Women Are from Venus* (Gray, 1992), and the same author's follow-up work, *Venus and Mars in the Bedroom* (Gray, 1995). Gray's thesis seems to be that men and women need to understand and accept each other's innate differences in order to get along. From this perspective, a man who cannot express his feelings and a woman who is not interested in professional success are seen as natural, as is the view that men are sexual "hunters" and need to channel their (innate) powerful sex drives into more romantic behaviors in order to connect with women.

However, rather than looking at these sexual drives as normal and innate, Levant (1997) argued that they are products of traditional gender ideology, which:

> serves that purposes of a patriarchal society by encouraging the socialization of males to develop those characteristics that are functional for attaining and maintaining power, such as toughness, competitiveness, self-reliance, and emotional stoicism. . . . In a patriarchal society, men are given certain preroga-tives vis-à-vis women, are allowed certain 'rewards' for their hard work (such as objectified sex) and even more, are judged on their manliness by their sexual conquests. (p. 14)

He called this dominant view of male sexuality "non-relational sexuality," writing that "non-relational sexuality can be defined as the tendency to experi-ence sex as lust without any requirements for relational intimacy, or even for more than a minimal connection with the object of one's desires" (p. 10). Brooks (1995, 1997) has a similar view and also bases his theory on gender differences in power and resulting socialization:

> The Centerfold Syndrome is a pervasive distortion in the way men are taught to think about women and sexuality. It is an outgrowth of the social construc-tion of male sexuality and the dysfunctional ways that men initially relate with women and intimacy and later encounter women in the sexual arena. Although men are programmed to relate with women in ways consistent with the Centerfold Syndrome, these patterns are not exclusively the result of early training, as they are assiduously reinforced by the many destructive ways that contemporary culture portrays women. (Brooks, 1997, p. 31)

Elements of "the Centerfold Syndrome" include voyeurism or looking at women's bodies, objectification of women's bodies, sex as a validation of masculinity, trophyism, and the fear of true intimacy (Brooks, 1995). Both Levant and Brooks stress that, although a nonrelational sexuality may be normative for men, it is not innate, but socially constructed.

Lewis and Kertzner (2003) make similar arguments regarding male African American sexuality. They write:

[W]e do not believe there is an essential African American sexuality . . . rather, we believe that similarities in African American males' sexual lives are more likely due to similar social constructions and/or social experiences faced by African Americans in particular social and historical milieus. (p. 383)

They state that the research literature examining the sexual behavior of African American men makes incorrect assumptions of homogeneity of sexual behavior in African American men, presumes a static nature of sexuality in general, ignores contexts in which behavior occurs, and focuses on overt sexual behaviors as opposed to their meanings. These criticisms may be directed at researchers who have found greater problems regarding male sexuality in ethnic groups, including findings reporting that Black students were more accepting of rape myths than were Whites (Giacopassi & Dull, 1986) and that Hispanic college students held more accepting attitudes toward forcible date rape than did White students (Fischer, 1987). Cowan and Campbell (1995) reported that "Hispanics viewed male sexuality as a force more overwhelming and uncontrollable than did Whites. These findings suggest stronger beliefs in the idea that rape is motivated by sex among Hispanics than among Whites" (p. 151).

Research into understanding sexuality in male adolescents also often looks at overt behavior and attitudes rather than contextualizing those reports. Cowan and Campbell (1995) looked at beliefs about the causes of rape in male and female adolescents and found that, while girls rated male pathology the highest of the five causes of rape, boys rated female precipitation the highest. Epps and Haworth (1993) found that male adolescents convicted of sex offenses reported similar attitudes as male adolescents convicted of other types of offenses and stated that their findings supported the view that negative and stereotypical attitudes toward women were commonplace and not specific to sex offenders. Hall, Howard, and Boezio (1986) found a stronger relationship between a sexist attitude and tolerance of rape than the relationship between tolerance of rape and an antisocial personality.

Thus, it appears that the literature about male sexuality suffers from some of the same problems as does that involving women. In particular, the tendency to observe sexual behaviors and attitudes, decontextualize them, and see them as innate and normal appears to be common. Although it seems as though this "normal" male sexuality, (i.e., nonrelational sexuality) appears to be linked to undesirable consequences, "there is scant published psychological literature that focuses squarely on the problem of men's nonrelational orientation to sexuality and views it critically" (Levant and Brooks, 1997, p. 3).

TOWARD COMPLEX, DIVERSE, RELATIONAL, AND CONTEXTUALLY SITUATED SEXUALITIES

What are some ways then that would allow us to study sexual experiences, yet still integrate questions of context? While we found several integrated models that look at Latina sexuality in ways that did not suggest ethnocentrism or essentialism, we believe that these perspectives might be used to look more specifically at other cultures and ethnicities as well.

Unfortunately, most of these newer, more complex analyses are available primarily through alternative presses specifically designed to represent marginalized people, such as Third World Press and Aunt Lute Press, not in mainstream academic journals and presses. We suspect that perspectives focusing on African American, Asian, or Native American sexualities in ways that do not reflect the dominant voice might be similarly "hidden." We will discuss some of these newer approaches and we hope that reading and incorporating the insights of alternative analyses will help move us beyond stereotypical, ethnocentric, and essentialized understandings of Latina sexualities in particular and other sexualities as well.

In the mid-1990s, some Latina writers—many of them contributing to lesbian theory—increasingly questioned the need to emulate modern White models of sexuality. Instead they began to affirm the possibilities for healthy liberated expressions that draw on and incorporate their own cultural logic, and move away from the negative model found in much of the literature. Anzaldúa (1998), a leading theorist of gender, sexuality, and ethnicity, reformulated dominant ideas by questioning the naturalness of "lesbian" as both a term and an identity:

> For me the term lesbian *es problemon*. As a working-class Chicana, mestiza—a composite being, *amalgama de culturas y de lenguas*—a woman who loves women, "lesbian" is a cerebral word, white and middle-class, representing an English-only dominant culture, derived from the Greek word *lesbos*. I think of lesbians as predominantly white and middle-class women and a segment of women of color who acquired the term through osmosis, much the same as Chicanas and Latinas assimilated the word "Hispanic." When a "lesbian" names me the same as her, she subsumes me under her category. I am of her group but not as an equal, not as a whole person—my color erased, my class ignored. *Soy una puta mala*, a phrase coined by Ariban, a *tejana tortillera*. "Lesbian" doesn't name anything in my homeland. Unlike the word "queer," "lesbian" came late into some of our lives. Call me *de las otras*. Call me *loquita, jotita, marimacha, pajuelona, lambiscona, culera*. . . . (p. 263)

Anzaldúa went on to discuss and contextualize cultural differences but does not imply that modern White practices are healthier:

> Yes, we may all love members of the same sex, but we are not the same. Our ethnic communities deal differently with us. I must constantly assert my differentness, must say, This is what I think of loving women. I must stress, the dif-

ference is my relationship to my culture; white culture may allow its lesbians to leave—mine doesn't. This is one way I avoid getting sucked into the vortex of homogenization, of getting pulled into the shelter of the queer umbrella. (p. 264)

Anzaldúa presented an affirmative aspect of Latina culture and sexuality, implying that White families and communities may push out or abandon lesbians, while Latinas/os might ostracize them but allow them to stay. By analyzing the context of social dynamics, she challenged the notion that it is automatically freer and easier to be lesbian in White communities, and instead pointed out the high price that is often paid for that freedom.

Chávez-Leyva (1998) responds to the assumption that cultural silences about sexual practices indicate a greater level of sexual repression and oppression, specifically looking at the silence surrounding sexual practices such as lesbianism. Silence, she suggests, may have different meanings in different cultures and contexts. Describing the common practice of family and community members silently "knowing" and "imagining" that someone is lesbian, she quotes one of her informants' stories about her relationship with her mother: "Basically, you know, she doesn't like my way of life so we don't talk about it. She respects me, she loves me, she spoils me. But it's something we just don't discuss. I think I don't do it out of respect for her, and she doesn't do it out of respect for me" (p. 432). Chávez-Leyva (1998) reinterprets the meaning of silence in that context. Rather than seeing it as oppressive, silence became to her a way to put love first, a way not only for the daughter to defer to her mother but also a way for the mother to show respect for her daughter (p. 432).

Instead of essentializing silence as always negative and oppressive, Chávez-Leyva (1998) begins to culturally and historically situate its meaning:

Silence has its own contours, its own texture. We cannot dismiss the silences of earlier generations as simply a reaction to fear. Rather than dismiss it, we must explore it, must attempt to understand it. We must learn to understand the ways it has limited us and the ways it has protected us.

This is not, however, a call to continue the silence, nor to justify it. Naming ourselves, occupying our spaces fully, creating our own language, is essential to our continued survival . . . This is a challenge to explore the contradictions of silence within Latina lesbian history, to understand the multiple meanings of silence, to uncover the language of silence. (p. 432)

Arguelles (1998) has used her memories of gender-crossing, women-loving-women in Cuba to question basic assumptions about healthy and liberated sexuality. One of these assumptions was that sexual intimacy is based in romantic love, rational considerations, or initial sexual chemistry. Relationships based on "trans-biographical experience or shared spirituality," which involve ways of relating that run contrary to dominant notions of intimacy, allowed her to:

question the current convention of assessing the health of a relationship by measuring the autonomy of each of the partners. Woven into the poetry of Teresa's songs and played out on her veranda were images of the blending of two strong women, the union of thoughts and feelings, the absence of boundaries. Fusion in relationships became a viable and acceptable possibility for me, as had spirit-oriented and trans-biographical unions. (p. 208)

Jordan, Kaplan, Miller, Stiver, and Surrey (1991) also questioned the pervasive dominant value of autonomy in defining healthy relationships. Their work has been influential in reframing the dominant psychological model of a "healthy self" as being one that is autonomous, self-sufficient, and individualistic. Instead, they developed a theory of women's development of self (Relational-Cultural theory) that relies on mutual empathy to build strong interdependence. RCT might be used as a model with regard to mutuality that could facilitate more relational aspects of sexuality, particularly for men.

Jordan (1997) pointed out that Western psychological theories have traditionally emphasized the "ascendancy of individual desire as the legitimate basis for definition of self and interpersonal relationships" (p. 57). These views can lead to the "possibility of creating violent relationships based on competition of need and the necessity for establishing hierarchies of dominance, entitlement, and power" (p. 57–58), which has led to limited understandings of sexuality. She wrote that "an alternative way of conceptualizing desire suggests that our wishes or wants are always contextually embedded and arise for that interactive context." This leads to alternative ways of interpreting desire and sexuality, ways that lead to "the experience of joining toward and joining in something that thereby becomes greater than the separate selves" (p. 60).

RCT also has implications for the broadening of understandings of sexuality in general. Jordan (1991) wrote:

> In sexuality the significance as well as the failure of mutuality can be experienced keenly. To see sexuality as a process of discharge, a pressure of impulse toward gratification, is to take an extraordinarily limited view. If this were the primary motive for sexuality, surely masturbation would be the preferred modality. But in fact it is the intersubjective, mutual quality of sexual involvement that gives it its intensity, depth, richness, and human meaning. (pp. 89–90)

The more complex, contextually situated approaches presented here begin to move us beyond ethnocentric and essentialized models that view sexuality from dominant perspectives. These approaches begin to examine ways of expressing sexualities without equating liberation with acculturation, but neither do they shrink from documenting sexual and gender inequality. They recognize that human sexuality does not have to take one form or another for it to be healthy or "undistorted." We believe that humans have no "undistorted" sexuality, that sexuality is fluid and changing and as diverse as the contexts in which it is embedded.

CONCLUSION

How might we apply these ideas to our work as therapists and scholars? First, is it important to be able to recognize essentialism and ethnocentrism as it occurs both in popular and research literature. We need to ask, "Who benefits by looking at sexuality in this way? Does this view reinforce bias or stereotypes?" Second, we need to look at the context in all models of sexuality. Context might include culture, adaptation, history, environment, values, power imbalances, etc. Third, we need to be open-minded when hearing of alternative models of sexuality, moving beyond our own values and experiences and working to understand and appreciate the diversity in all our sexualities.

Note: This chapter is based on Juárez, Ana María and Kerl, Stella Beatríz (2003), What is the Right (White) Way to Be Sexual? Reconceptualizing Latina Sexuality, *Aztlan: A Journal of Chicano Studies, 28,* 1: 7–37.

REFERENCES

Alonso, A. (1995). *Thread of blood: Colonialism, revolution, and gender on Mexico's northern frontier.* Tucson: University of Arizona.

Alonso, A., and Koreck, M. T. (1989). Silences: "Hispanics," AIDS, and sexual practices. *Differences, 1*(1), 101–124.

American Heritage Dictionary of the English Language (2000). Boston: Houghton Mifflin.

Anzaldúa, G. (1998). To(o) queer the writer—Loca, escritora y chicana. In *Living Chicana Theory.* Trujillo, C. (Ed.). 263–76. Berkeley: Third Woman Press.

Arguelles, L. (1998). Crazy wisdom: Memories of a Cuban queer. In *The Latino studies reader: Culture, economy, and society.* Darder, A. and Torres, R. D. (Eds.). 206–210. Malden, MA: Blackwell Publishers.

Barash, D. (1979). *The whisperings within.* New York: Penguin Books.

Brooks, G. (1995). *The centerfold syndrome: How men can overcome objectification and achieve intimacy with women.* San Francisco: Jossey-Bass.

Brooks, G. (1997). The centerfold syndrome. In Levant, R. and Brooks, G. (Eds.). *Men and sex.* New York: John Wiley & Sons.

Bullough, V. (1990). History and the understanding of human sexuality. *Annual Review of Sex Research, 1,* 75–92.

Buss, D. (1995). Psychological sex differences: Origins through sexual selection. *American Psychologist, 50,* 164–168.

Cachelin, F., Veisel, C., Barzegarnazari, E., & Striegel-Moore, R. (2000). Disordered eating, acculturation, and treatment-seeking in a community sample of Hispanic, Asian, Black, and White women. *Psychology of Women Quarterly, 24*(3), 244–233.

Castaneda, A. (1990). *The political economy of nineteenth century stereotypes of Californianas in between borders: Essays on Mexicana/Chicana history.* Del Castillo, A. (Ed). Encino, CA: Floricanto Press.

Castillo, A. (1991). La Macha: Toward a beautiful whole self. In *Chican lesbians: The girls our mothers warned us about.* Trujillo, C. (Ed.). 24–48. Berkeley: Third Woman Press.

Chamorro, R. & Flores-Ortiz, Y. (2000). Acculturation and disordered eating patterns among Mexican American women. *International Journal of Eating Disorders, 28* (1), 125–129.

Chávez-Leyva, Y. (1998). Listening to the silences in Latina/Chicana lesbian history. In *Living Chicana Theory.* Trujillo, C. (Ed.). 429–434. Berkeley: Third Woman Press.

Coleman, J. (1966). Letter to the editor: Female status and premarital sexual codes. *American Journal of Sociology, 72* (2), 217.

Cowan, G. & Campbell, R. (1995). Rape causal attitudes among adolescents. *Journal of Sex Research, 32* (2), 145–153.

Cross, W. (2003). Tracing the historical origins of youth delinquency & violence: Myths & realities about black culture. *Journal of Social Issues. Special Issue: Youth perspectives on violence and injustice, 59* (1), 67–82.

D'Emilio, J. & Freedman, E. (1988). *Intimate matters: A history of sexuality in America.* New York: Harper and Row.

Duggan, L. (1990). From instincts to politics: Writing the history of sexuality in the U.S. *Journal of Sex Research, 27* (1), 95–109.

Dulce, J., Hunter, G., Lozzi, B. (1999). Do Cuban American women suffer from eating disorders? Effects of media exposure and acculturation. *Hispanic Journal of Behavioral Sciences, 21* (2), 212–218.

Epps, K., & Haworth, R. (1993). Attitudes toward women and rape among male adolescents convicted of sexual versus nonsexual crimes. *Journal of Psychology, 127* (5), 501–506.

Espín, O. (1999). *Women crossing Boundaries: A psychology of immigration and transformations of sexuality.* New York: Routledge.

Fine, M. (1988). Sexuality, schooling, and adolescent females: The missing discourse of desire. *Harvard Educational Review, 58* (1), 29–53.

Fischer, G. (1987). Hispanic and majority student attitudes toward forcible date rape as a function of differences in attitudes toward women. *Sex Roles, 17,* 93101.

Foucault, M. (1980). *The history of sexuality.* New York: Vintage.

Garner, A., Sterk, H., & Adams, S. (1998). Narrative analysis of sexual etiquette in teenage magazines. *Journal of Communication, 48,* 59–60.

George, W. & Martínez, L. (2002). Victim blaming in rape: Effects of victim and perpetrator race, type of rape, and participant racism. *Psychology of Women Quarterly, 26* (2), 110–120.

Giacopassi, D. & Dull, R. (1986). Gender and racial differences in the acceptance of rape myths within a college population. *Sex Roles, 15,* 63–75.

Giami, A. (2002). Sexual health: The emergence, development, and diversity of a concept. *Annual Review of Sex Research, 13* (1), 1–35.

Gil, R. M. & Vazquez. C. (1996). *The Maria paradox: How Latinas can merge old world traditions with new world self-esteem.* New York: The Berkley Publishing Group.

Gómez, C. & Marín, B. (1996). Gender, culture, and power: Barriers to HIV-prevention strategies for women. *The Journal of Sex Research, 33* (4), 355–362.

Gray, J. (1992). *Men are from Mars, Women are from Venus: A practical guide for improving communication and getting what you want in your relationships.* New York: HarperCollins.

Gray, J. (1995). *Venus and Mars in the Bedroom: A guide to lasting romance and passion.* New York: HarperCollins.

Greene, B. (1994). Diversity and difference: The issue of race in feminist therapy. In *Women in context: Toward a feminist reconstruction of psychotherapy.* Mirkin, M., (Ed.). New York: The Guilford Press.

Grimes, T. R. (1999). In search of the truth about history, sexuality, and Black women: An interview with Gail E. Wyatt. *Teaching of Psychology, 6* (1) 66–70.

Hall, E., Howard, J., & Boezio, S. (1986). Tolerance of rape: A sexist or antisocial attitude? *Psychology of Women Quarterly, 10,* 101–118.

Jordan, J. (1991). Empathy and self-boundaries. In Jordan, J., Kaplan, A., Miller, J. B., Stiver, I., and Surrey, J. (Eds.). *Women's Growth in Connection: Writings from the Stone Center,* 67–80. New York: The Guilford Press.

Jordan, J. (1997). Clarity in connection: Empathic knowing, desire, and sexuality. In *Women's growth in diversity: More writings from the Stone Center.* Jordan, J. (Ed.). New York: The Guilford Press.

Jordan, J., Kaplan, A., Miller, J. B., Stiver, I., and Surrey, J. (1991). *Women's Growth in Connection: Writings from the Stone Center.* New York: The Guilford Press.

Juárez A. & Kerl, S. (2003). What is the right (white) way to be sexual? Reconceptualizing Latina sexuality. *Azlán: A Journal of Chicano Studies, 28* (1), 7–37.

Kinsey, A., Pomeroy, W., & Martin, C. (1948). Sexual behavior in the human male. Philadelphia: Saunders.

Kitzinger, S. (1983). *Women's experience of sex: The facts and feelings of female sexuality at every stage of life.* New York: Penguin Books.

Lawrence, C. (1992). Cringing at myths of black sexuality. *Black Scholar, 22* (1/2), 65–66.

Levant, R. (1997). Nonrelational sexuality in men. In Levant, R. and Brooks, G. (Eds.).

Men and sex. New York: John Wiley & Sons.

Levant, R. & Brooks, G. (1997). Introduction. In Levant, R., and Brooks, G. (Eds.). *Men and sex.* New York: John Wiley & Sons.

Lewis, L., & Kertzner, R. (2003). Toward improved interpretations and theory building of African American male sexualities, *Journal of Sex Research, 40* (4), 383–395.

Limón, J. (1997). Selena: Sexuality, performance, and the problem of hegemony. In *Reflexiones 1997: New directions in Mexican American studies.* Foley, N. (Ed.). 1–27. Austin: Center for Mexican American Studies Books.

Low, J., & Sherrard, P. (1999). Portrayal of women in sexuality and marriage and family textbooks: A content analysis of photographs from the 1970s to the 1990s. *Sex Roles, 40* (3-4), 309–18.

Masters, W., & Johnson, V. (1970). *Human sexual inadequacy.* Boston: Little, Brown.

Masters, W. H., and Johnson, V. E. (1966). *Human sexual response.* Boston: Little, Brown.

Mecca, S. & Rubin, L. (1999). Definitional research on African American students and sexual harassment. *Psychology of Women Quarterly, 23* (4), 813–817.

Miller, J. (1986). *Toward a new psychology of women, 2nd ed.* Boston: Beacon Press.

Milligan, D. (1993). *Sex-life: A critical commentary on the history of sexuality.* Boulder, CO: Pluto Press.

Moya, P. (1997). Postmodernism, "Realism," and the politics of identity: Cherríe Moraga and Chicana feminism. In *Feminist genealogies, colonial legacies, Democratic futures.* Alexander, M., and Mohanty, C. (Eds.) 125–50. New York: Routledge.

Pamar, S. & Haffner, D. (1998). Making the connection: Sexuality and reproductive health. *SIECUS Report, 27* (2), 4–8.

Reinisch, J. (1990). *The Kinsey Institute new report on sex: What you must know to be sexually literate.* New York: St. Martin's Press.

Rose, T. (2003). *Longing to tell: Black Women talk about sexuality and intimacy.* New York: Farrar, Straus, and Giroux, Inc.

Stephens, D. & Phillips, L. (2003). Freaks, gold diggers, divas, and dykes: The sociohistorical development of adolescent African American women's sexual scripts. *Sexuality & Culture, 7* (1), 3–50.

Salgado de Snyder, V., Acevedo, A., Diaz-Pérez, M., and Saldivar-Garduño, A. (2000). Understanding the sexuality of Mexican-born women and their risk for HIV/AIDS. *Psychology of Women Quarterly, 24,* 100–108.

Silverman, D. (2003) Theorizing in the shadow of Foucault: Facets of female sexuality. *Psychoanalytic Dialogues, 13* (2), 243–273.

Sommers, I., Fagan, J., & Baskin, D. (1993). Sociocultural influences on the explanation of delinquency for Puerto Rican youths. *Hispanic Journal of Behavioral Sciences, 15* (1), 36–62.

Staples, R. (1972). The sexuality of black women. *Sexual Behavior, 2* (6), 4-11, 14–15.

Tolman, D. (2002). *Dilemmas of desire: Teenage girls talk about sexuality.* Boston: Harvard University Press.

Tolman, D. (1991). Adolescent girls, women and sexuality: Discerning dilemmas of desire. *Women and Therapy, 11* (3-4), 55–69.

Tolman, D. & Diamond, L. (2001). Desegregating sexuality research: Cultural and biological perspectives on gender and desire. *Annual Review of Sex Research, 12* (1), 33–74.

Weinburg, M. & Williams, C. (1988). Black sexuality: A test of two theories. *The Journal of Sex Research, 25* (2), 197–218.

Wiederman, M. & Maynard, C. (1996). Ethnicity in 25 years of published sexuality research: 1971–1995. *Journal of Sex Research, 33* (4), 339–342.

Wong, S. (1999). Acculturation, peer relations, and delinquent behavior of Chinese-Canadian youth. *Adolescence, 34* (133), 108–119.

Wyatt, G. (1997). *Stolen women: Reclaiming our sexuality, taking back our lives.* New York: Wiley.

Wyatt, G. (1999). Beyond invisibility of African American males: The effects on women and families. *Counseling Psychologist. Special Issue: Invisibility syndrome in African American men, 27* (6), 802–809.

Zibergeld, B. (1992). *The new male sexuality: The truth about men, sex, and pleasure.* New York: Bantam Books.

11

◉

Gifted Development:
It's Not Easy Being Green

*By Linda Kreger Silverman, Ph.D.,
and Sharon A. Conarton, BSN*

INTRODUCTION

Few therapists are prepared for the unique challenges that their gifted clients face. Like the general public, therapists believe the myth that giftedness provides an unmitigated advantage. However, in many ways, those who are advanced have as much difficulty fitting into society as those who are delayed. Giftedness and retardation are opposite sides of the same coin. It is obvious that one needs special training to educate or counsel individuals at the lower extreme of the spectrum, but it seems less obvious that special training is also needed to work with individuals at the upper extreme.

Many people think that giftedness is simply high achievement in school. This misconception is so pervasive that gifted children and adults are continuously pressured to achieve—an impossible yoke for some to bear. Giftedness is the capacity for high abstract reasoning and greater awareness. It bestows on its bearer heightened sensitivity, intensity, complexity, and idealism. High ability may or may not translate into high achievement in school or in life. There are brilliant volunteers, cooks, parents, teachers, therapists, and the like, who devote themselves to helping others rather than to achieving lasting recognition.

A basic life theme for the gifted is, "From each according to ability, to each according to need." Imprinted on the psyche from infancy, this commandment is both internal and external, and its implications are vast. Those who are blessed with more ability usually feel compelled to give more than anyone else. "The Omnipotent Fantasy" begins in early childhood, and becomes a prescription for overresponsibility, overconscientiousness, overextension, and overwork. Perhaps the greatest challenge for the therapist of the gifted is to help them become aware of their own needs and remind them to put on their own oxygen masks before helping others put on theirs.

CHARACTERISTICS OF GIFTEDNESS

Gifted infants enter the world with a distinct set of characteristics that leads to different life experiences. Like retardation, giftedness occurs on a continuum from mild to profound. The further the individual veers from the norm, the greater the impact of these attributes on development and adjustment. As therapists are usually highly intelligent, all those who enter the mental health profession will relate to the majority of these characteristics. And as individuals who voluntarily enter therapy tend to be smarter than average, most clients will exhibit at least some of these traits. The gifted exhibit more of them and to a higher degree. Increase in intensity of expression of any one of these characteristics escalates internal and external conflict. An exceptionally gifted client endowed with most of these traits at an excruciating level will present a very different clinical challenge than the typical client.

While there are many lists of the cognitive qualities of gifted children, the following personality characteristics appear to have the greatest influence on the development and therapeutic issues of gifted individuals throughout the life span:

1. Asynchrony
2. Perceptiveness
3. Complexity
4. Perfectionism
5. Idealism
6. Overexcitability
7. Intensity
8. Sensitivity
9. Need for meaning
10. Moral concern
11. Divergent thinking
12. Questioning authority
13. Argumentativeness
14. Responsibility for others
15. A strong aesthetic sense
16. A tendency toward introversion
17. An extraordinary sense of humor

Asynchrony

Asynchrony is uneven development and feeling out of sync with others (Silverman, 1993b). From early childhood, gifted children advance mentally much faster than they progress physically. They have heightened sensitivity, intensity, and awareness, combined with age-appropriate emotional needs, social skills, physical competence, and life experience. The higher the child's IQ, the more asynchronous is his or her development. Being out of sync within themselves leads them to feel out of place in school, with age-mates and with societal expectations for children their age. These feelings are carried into adulthood, causing the gifted adult to feel like an outsider in any social sphere. It is this lack of belonging that often drives the gifted adult to seek therapy.

Perceptiveness

Greater awareness, *perceptiveness* or insightfulness, combined with honesty, often gets gifted individuals in trouble. They see through hypocrisy; they sense hidden agendas; they see the essence of a situation (Roeper, 1991). But their truth telling is unwelcome. And they are unwilling to play the game of pretending things aren't happening when they see quite clearly what is really going on. They alienate people until they learn that they can't always say what they see. They need guidance in determining when and how to share their perceptions.

Deirdre Lovecky (1990) states:

An ability to view several aspects of a situation simultaneously, to understand several layers of self within another, and to get quickly to the core of an issue are characteristic of gifted adults with the trait of perceptiveness. . . . The gifted adult must come to understand that the question is not whose world view is more accurate but how to use disparate views in ways that enhance connectedness to others and further understanding of the self. Neither the self nor the other is defective or stupid. . . . The dilemma for the perceptive gifted adult is how and when to use the gift with self and others. (Lovecky, 1990, pp. 76–79)

Therapists enjoy the insightfulness of their gifted clients, and the clients enjoy captivating them with their insights. However, the unsuspecting therapist may spend many ecstatic weeks involved in "The Great Insight Chase" before realizing that the client's ready insights are not leading to any real changes in behavior (Silverman, 1993a). It is important for the therapist to continually remind the client, "That was a wonderful insight you had last week. What did you *do* with it that would make a difference in your life?"

Complexity

The *complexity* of gifted minds is mirrored in the complexity of their emotions. Highly intelligent people see so many variables in a situation, so many connections between seemingly unrelated events, and so many potential outcomes that they may not be able to sort through all of the information to find an appropriate path. Decision making is simpler when one has less information. While there has been much psychological investigation of the pitfalls of black-and-white thinking, little has been written about the dilemma of living with an infinite number of shades of gray. If the individual is petrified of making a mistake and believes that all but one of those shades of gray will be a dreadful error, life becomes a perilous walk on a tightrope with no safety net below. The therapist must provide the safety net, while attempting to unknot the multitudinous variables, so that the client can safely choose a path.

Perfectionism

Perhaps the least understood facet of giftedness is *perfectionism*. This trait is maligned in psychological literature; yet, *all* gifted individuals are perfectionistic to some degree. Perfectionism has positive as well as negative implications (Silverman, 1999). It is linked to the passion to achieve excellence in any endeavor, the love of beauty, and the drive for self-perfection, which compels the person to seek therapy. In gifted children, perfectionism is highly correlated with conscientiousness, rather than with neuroticism (Parker, 1997). Perfectionism should not be viewed as a destructive force that must be rooted out during the therapeutic process. Instead, it should be seen as a powerful energy source that can be redirected to serve the individual's highest aspirations. When the therapeutic connection is at its best, paralyzing perfectionism can be transformed into motivation for success. The therapist must become the cheerleader when the client reaches the point of frustration and anxiety where "I can't do this" takes over. "Yes, you can!" at that critical moment can change a person's life.

Idealism

Idealism is a cousin to perfectionism. Individuals with high ideals commit themselves to making this a better world. They are also on a continuous course of self-improvement, which may prompt their entering into a therapeutic relationship. They are able to see what should be instead of only what is, in themselves and in the world (Dabrowski, 1972). If the individual feels capable of making positive changes, then the idealism becomes a potent force for personal development and innovation. However, idealism is often partnered with frustration and disappointment. This begins in childhood, since gifted children do not have the power and resources to materialize their idealistic visions. Therapists must walk a fine line here. In helping gifted clients to discern reality from idealization, they must be careful not impose their own reality and sense of limitation on their clients.

Overexcitabilities

Overexcitability is a term from Dabrowski's theory (Dabrowski, 1967, 1972; Dabrowski & Piechowski, 1977) denoting the heightened neurological response patterns of gifted and creative individuals. There are five forms of overexcitability (OE): psychomotor, sensual, imaginational, intellectual, and emotional. They occur in different combinations and to varying degrees. Individuals endowed with one or more overexcitabilities react much stronger than the norm to various types of stimuli. Psychomotor OE is an abundance of physical energy, which can be displayed as hyperactivity or workaholism. Sensual OE is heightened sensitivity of the senses and desire for sensory pleasures. Imaginational OE may be expressed as vivid memories, terrifying nightmares, artistic talent, or visionary endeavors. Intellectual OE may appear as the love of debate, fascination with theories, intellectualization of emotional experience, or superb problem-solving capabilities. Emotional OE is the intensification of emotional experience, deep capacity for empathy, strong emotional ties and attachments, or concern with death.

Overexcitabilities actually propel emotional growth, as they enhance developmental potential to reach high levels of moral and emotional development (Piechowski, 1979). Children who experience intense physiological reactions to the bombardment of a variety of stimuli must continuously make choices in order to function. This provides daily practice in setting priorities and gaining inner direction, the same skills needed later in life to construct a strong sense of values. As with perfectionism, the OEs need to be accepted and marshaled in gifted clients; they are an essential part of their innate equipment—not something to be cured.

Intensity and Sensitivity

Intensity and *sensitivity* are functions of the overexcitabilities of the gifted. The OEs are sometimes referred to as "intensities." Gifted individuals are wired to experience all of life more powerfully. There is no such thing as moderation. Michael Piechowski (1992) writes,

> One of the basic characteristics of the gifted is their intensity and expanded field of their subjective experience. The intensity, in particular, must be understood as a qualitatively distinct characteristic. It is not a matter of degree but of a different quality of experiencing: vivid, penetrating, encompassing, complex, commanding—a way of being quiveringly alive. (p. 2)

Heightened sensitivity is the essence of Emotional OE. The gifted see multiple meanings in situations, are easily wounded and have greater capacity for empathy. Emotional sensitivity is the most frequent characteristic parents use to describe their gifted children (Silverman, 1983). "She wears her heart on her sleeve." "He has no skin." The major therapeutic issue for the exceptionally sensitive individual is his or her difficulty establishing boundaries. Books such as Elaine Aron's *The Highly Sensitive Person* (1996) can be very helpful for these clients.

Need for Meaning

A most compelling attribute that brings gifted adolescents and adults into therapy is their insatiable *need for meaning*. This is the heart of the work with gifted clientele. The therapist will often ask, "What does this mean to you?" While other clients present adjustment issues or the desire to be happier, gifted individuals are willing to cope with loneliness, being the perpetual outsider, and even lack of joy, if they can perceive that their experience is meaningful.

> In counseling gifted children and adolescents, one becomes part of their quest for meaning, a guide, or companion as it were, on the journey into the unknown self. Like the Knights of the Round Table searching for the Holy Grail, [the] gifted . . . explore the self within and the world without in hopes of reaching understanding. (Lovecky, 1993, p. 29)

As children, the gifted often say that they want to make a difference with their lives. They are willing to make great sacrifices and endure hardship and poverty in the cause of a meaningful existence. Existential therapy or books, such as *Man's*

Search for Meaning, by Viktor Frankl (1963), would be particularly useful for clients who have a powerful need for meaning.

Moral Concern

Most gifted children appear to have deep moral sensitivity and *moral concern*. This may not always translate into moral action, because of the asynchrony between their intellectual awareness and their emotional maturity. Gifted boys, in particular, are at risk for losing their moral sensitivity along the way if they become the target of teasing or bullying (Lovecky, 1994). Some gifted children who have been abused may squelch this part of themselves and taunt other sensitive children. Others are more resilient, with the inner resources and external support to enable them to transcend the abuse. Therapists can assist morally sensitive clients in perceiving the injustices they have suffered or witnessed as grist for the mill of their own development. Those who are endowed with writing talent can share their experiences with the world, enhancing awareness and potentially bringing about change. *The Color Purple*, by Alice Walker (1982), is a good example of the power of this type of witnessing.

Divergent Thinking

It is fortunate that moral concern generally accompanies high intelligence, as a gifted mind without ethics is a positive menace to society (Drews, 1972). And there is a small segment of the gifted population that does engage in antisocial behavior. *Divergent thinking* is the basis of creativity, but it can also cause delinquent behavior. The creative nonconformist can be a challenge for the therapist. Nonconforming, noncompliant teens are often brought to therapists by parents, schools, social services, or law enforcement agencies. If oppositional teens perceive the therapist as a henchman of the Establishment, they will present a sullen exterior and no growth will take place. The therapist who underachieved, had brushes with the law, or had a period of defiance at some point in his or her life will be more successful in developing rapport. Finding out what interests the client and learning enough about it to be conversant helps the process.

Divergent thinking, seeing the world differently, marks an individual as an outsider throughout the life span. Even in the company of other divergent thinkers, this person is alone in his or her unique perception of the world. Lovecky describes the dilemma of divergent thinkers:

> Divergent thinkers also have to deal with being different. Although they do not accept the status quo, conform well, or fit in with peers and are often subjected to teasing, they do not know why they are different, or why they upset other people. Often they feel entirely alone, with no one to understand them, even in their own families. A number of such youngsters become severely depressed in adolescence because both self-esteem and a sense of connection to others is affected. (Lovecky, 1993, p. 33)

Therefore, a major therapeutic issue for creative clients is coping with loneliness. Bibliotherapy with existential literature can be ameliorative. *The Half-Empty*

Heart, by Alan Downs (2003), is also recommended reading. A second issue is postpartum depression (Gowan, 1980), which typically occurs after the birth of a creative product. Creative individuals find Dabrowski's theory (1967, 1972) relevant and uplifting.

Questioning Authority; Argumentativeness

The tendency to question authority is obviously a companion to divergent thinking in youths who get into trouble with the law. However, this trait is so common in the gifted that it does not necessarily lead to antisocial behavior. Questioning is natural for an inquisitive mind. All gifted children argue; for them it is a form of mental exercise. Those who are extremely polite or introverted may not voice their disagreements, but when asked, they will admit to arguing with others in their minds. In some gifted families, argumentation is the basic form of communication. *Arguing* and *questioning authority* are lifelong pursuits of the gifted.

Leta Hollingworth (1939), the first counselor of the gifted, found that the higher a child's IQ, the more argumentative he or she was likely to be. And if the individuals who have authority over the child were perceived as illogical, irrational, erroneous, or unjust, the child was likely to develop a negative attitude toward authority. Hollingworth recommended teaching gifted children the fine art of argumentation: how to argue with oneself; the etiquette and art of polite disagreement in a private setting; and arguing in public. Her emotional education program also included helping them learn to balance candor with tact and teaching them how to handle the apparent foolishness of others with patience and love. Therapists may have to engage in similar educational processes with argumentative gifted clients. They can help these clients temper their sharp intellectual fencing skills with an understanding of other people's feelings and reactions to their penchant for debating.

Responsibility for Others

Those with higher ability are often called upon to take *responsibility for others*. It is a natural role as they are usually empathic, wanting to help others, conscientious, and good problem solvers. However, there are pitfalls in being the responsible one. The person with a take-charge personality may be resented by others and perceived as controlling. Highly responsible people may have difficulty saying "no" to all the demands made of them. They are easily overcommitted and overextended because they see the need and think they are the only ones who can fill it. They may know very little about what they need to take care of themselves. They may feel that they are extraordinary; therefore, they must give more. Even though they are gifted, they need to know that they are ordinary human beings with human needs. People who give a great deal to others need a great deal of support from others as well. They may not be aware that they need this support and, even if they are, they are usually reluctant to ask for help. The therapist needs to teach them how to ask for assistance and when to delegate, help them to set priorities and enable them to discern when they are really needed from when they are habitually taking charge and depriving others of the opportunity to contribute.

Strong Aesthetic Sense

In 1939, Leta Hollingworth wrote that she had never met a gifted person who did not have a love of beauty. *Strong aesthetic sense* comes with the territory and can become a driving force of the personality. The desire to create beauty can express itself in gardening, taste in clothing, the care with which one decorates one's home, delight in music, art, and sunsets, orderliness, and even a love of mathematics.

Beauty is both expensive and time consuming. Gifted individuals with limited funds may become depressed by inelegant surroundings. Family conflicts erupt when perfectionistic mothers with a strong aesthetic need strive to maintain too high a level of order in their homes. For example, one mother insisted that her teenage son hang all of his clothes in the same direction in his closet. It is necessary for the therapist to honor the gifted client's need for beauty and, at the same, assist them in picking their battles.

Introversion

Introversion increases with IQ. In the general population, at least 70 percent are extraverts and 30 percent are introverts. In the moderately gifted population, approximately 40 percent are extraverts and 60 percent are introverts. And in the highly gifted population, 25 percent are extraverts and 75 percent are introverts (Silverman, 1998b). Therefore, it is essential for therapists who work with the gifted to be familiar with the constellation of traits of introverts.

Extraverts get their energy from interaction and introverts from inner reflection. An extravert will spend the entire session talking, not allowing the therapist to get a word in edgewise, and then leave thanking the therapist profusely for being an enormous help. Basically, the extravert wants a sounding board. But the introvert has difficulty talking about problems. Introverts mull over a problem like a broken record and are reluctant to seek therapy until they have run out of ideas on how it can be solved. They have engaged in so much inner dialogue about it that they are weary. Revealing as little information as possible, they want the therapist to cut to the chase and offer a solution they haven't thought about (Silverman, 1993a). The therapist needs to give the introverted client time to think instead of expecting immediate answers to questions. It's wise to say to them, "Think about what we've discussed today and get back to me next week with your thoughts about all this." "The Inner World of Introverts," a chapter in *Upside-Down Brilliance: The Visual-Spatial Learner* (Silverman, 2002), contains a blueprint for living with and teaching introverted children.

Extraordinary Sense of Humor

As gifted individuals are blessed with an *extraordinary sense of humor*, it helps enormously for the therapist to have a good sense of humor as well. Leta Hollingworth (1940) found humor to be the saving grace of the gifted, because it enables them to cope with the foolishness they see all around them. A successful therapist of the gifted can help them to see the humor in their life situations. Humor appears to be most effective after the therapist and the client have developed a solid ground of trust (Berg & DeMartini, 1979).

COUNSELING THE GIFTED

No clients present themselves as gifted. Most gifted adult clients have never known that they were. They may have been "smart" in school or say, "If I was interested, I could get good grades." They present themselves as "together," and able to handle their lives. They may identify their presenting symptom as "I want to know more about myself," or "My wife thinks I'm depressed." They're often elusive and it is several sessions of therapy before the real issues emerge.

The first tip-off that clients are gifted is that they never feel as though they belong. Even if they appear successful and have been high achievers, their lives are lived as outsiders. They may be leaders, but they are not of the group they lead. Because they learned very young that they had more ability than others around them, they were quick to meet others' needs (Miller, 1981). Being able to perceive others' needs and meeting them, others have developed an expectation that the gifted one will take charge.

It's hard for gifted adults to come into therapy. It's not because they think that they're smarter than the therapist; they just feel that the therapist won't be smart enough.

> Psychotherapy with the gifted is not easy. When working with a more average client, I am surprised every so often at how much easier things are without constant questioning of my premises or my ability to understand accurately. Because the basis of accurate empathy is accurate understanding, failures of empathy occur many more times with gifted clients than with more average ones. (Lovecky, 1990)

The gifted are so used to defending with rationalization and intellectualization that they repress their feeling function and are wary of therapists who make them feel ("I can go home and cry. I want the problem solved."). This attitude in itself is a defense against having their belief system challenged. There is an intransigent tenacity in maintaining their system of dealing with a world they are not really part of and managing their internal chaos. They must be right. If they aren't, there is impending doom. Then they've always been wrong.

These are people who do not receive. Often they are so other-oriented, with such a strong energy projection, that the therapist must compete with these clients in order to be there for them. It is very seductive to have a client who comes into the session focused on the therapist, wanting to meet the therapist's needs. It's necessary to convince them that they have needs and that you, as the therapist, are there for them. These clients will have the same relationship with you that they have with everyone else in their lives. Observing how the client relates to the therapist is a good monitor to measure how much the client is learning to receive.

The gifted are idealists. They often do not see the world as it is, but the way they want it to be. And then they strive to make others live up to this idealized reality. Norma was a middle child in a large family. Her parents were abusive and unstable. She took on the task of trying to remake her family the way it ought to be. She wanted it to be fair and balanced. Naturally, this didn't work. She became a scapegoat. It didn't work out in the world either, where she tried to change the professional agencies in which she worked. A creative and diligent professional, Norma was always considered a valuable member of the team until she became incensed by the unfairness and lack of ethics practiced by the administration. She became desperate in her quest to make the organization the way it ought to be. Her truth telling was unwelcome. She would become a scapegoat and was soon eliminated. She never could figure out what happened: "I was right. Everyone could see what was wrong. I get punished for saying that the Emperor has no clothes. Why won't anyone do anything about it?"

It is inconceivable to gifted people that everyone can't see what they see so clearly. And they also have difficulty believing that the therapist can see something they don't. "Yes, Norma, you're right. But it is not your responsibility to correct this situation. It would be the best of all possible worlds if everything was fair, but that's not the way the world works." This message needs to be conveyed over and over in many different ways with every event that comes up during the session. The gifted are so adamant. Their intensity and determination may make you doubt your own beliefs at times.

Gifted individuals often carry their childhood mission on into adulthood. Instead of just trying to change their families, they are compelled to change the world. Henri spent his childhood being responsible for a disadvantaged brother, keeping him safe and protected. As a brilliant gifted adult, he has been highly influential in helping improve the environment. However, he is still unsatisfied. "The world needs me. I haven't done enough." And, for Henri, nothing will ever be enough, much to the detriment of his stress level and balance in life. No matter how much he has done, the bar is always raised higher than he can achieve.

Another primary issue for the gifted is learning that they have limitations. They do not think they have them. They do. We all do. The gifted are reluctant to believe that. "If I set my mind to it, I can do *anything.*" They experience profound disappointment when they encounter an obstacle they cannot overcome or run out of time or energy or when they have to count on others to help them achieve their vision and those people do not follow through.

Honesty is yet another issue in therapy with the gifted. The perceptive client may well pick up some of the therapist's own distress, and they may probe to learn more. This may challenge the therapist's values of self-disclosure and boundaries.

> Interpersonal relationships with these people can feel uncomfortable to others, including the therapist. . . . Therapy can be helpful by freely allowing the exploration of insights and intuition, as well as in learning the tolerance points of the therapist who can be honest about vulnerability to the gifted person's insightful revelations. (Lovecky, 1990, p. 78)

It's a fine line to walk between being honest and honoring your own values. The need to know often intrudes on other people's values and belief systems. Disconcerting questions from gifted clients present opportunities to demonstrate that they have the right to say "no" when other people are intruding on their boundaries.

Often the characteristics of gifted leaders are attributes that others find negative. Tom's friends found his largess beneficial and enjoyed his charisma, but were critical of him. He shared, "Why shouldn't I provide for my friends? I have more money than they do. I know more people of influence. And I really enjoy helping people when I can. I can make things happen for them." Other people saw him as controlling, arrogant, and manipulative, always competing to be the alpha male, until they got to know him better. These characteristics were evident in his behavior as a child. Before he was 4, he insisted that his parents remove the training wheels on his two-wheeler, against their better judgment. He had to prove to his father that he could do it and he did. This became a life theme: throughout his adult life, he is still trying to prove himself. As a gifted athlete, driven to be the best, he *had* to win. He was hyperfocused on his sport, neglecting relationships and never able to learn about interdependence. He had to do everything by himself. Dependence was unknown to him.

In his therapy, the first task was to familiarize Tom with the literature on giftedness in adults. Although he was not a reader—he had only read two or three books in the previous decade—he read seven books and articles in the first two weeks. Among them were *The Gifted Adult*, by Mary-Elaine Jacobsen (1999); *Gifted Grown-ups*, by Mary Lou Streznewski (1999); and various articles from *Advanced Development Journal*. As part of his therapy, Tom was encouraged to keep a journal of his thought processes, the feelings that came up, and his dreams. He kept track every day of his new insights as he read. During sessions, the therapist helped Tom to plan how he would modulate his energy in relationship to his friends and his behavior with his partner.

Josh came into therapy in his late 30s to deal with his procrastination. Having just married a woman with a ready-made family, repairing an old house, starting his doctorate and working full-time, he found he was procrastinating on his academic work. He would have to stay up all night or take time away from the family. He was also having difficulty trying to get everything right for his wife. Josh was living out "The Hero Myth" (Conarton, 1999). He must achieve, be the best he can be, always climbing the mountain. Little did he realize he would have to learn about accepting his limitations and his unrealistic expectations for achievement and perfection. Until then, he would continue to procrastinate. Putting off his writing until the anxiety was so high, he would go into crisis mode to meet the deadline. Then he had a good excuse for not producing a perfect product. He didn't have time.

Like many gifted men, Josh had strong feminine characteristics and wanted his relationship to be the best he could make it. He had to learn that his way of doing things and the values that worked for him were not the same as for his partner. He could never "get it right" for himself and also for her; he had to content himself with doing what worked for him, in spite of her reaction.

These are a lot of frustrations for a man who likes to think of himself as together and able to handle anything. All the frustrations and drive for achievement and perfection generate energy. For many gifted adults, this creates intense chaos and disorder. Sometimes so intense it may engender self-destructive behavior. Perfectionism can be used as a positive energy force, but it may also become negative when the overexcitabilities become overwhelming. Managing this chaos and accelerated energy varies for each client.

Introverts are more prone to turning the energy in on themselves or becoming withdrawn or ill. Extraverts prefer to act out their distress. The client needs help to find ways to modulate this energy by a stress-reduction regime, including self-soothing, meditation, or breathing exercises.

Physical activity or making an emotional connection with someone may work for others. Unfortunately, when the chaos and anxiety cannot be controlled in other ways, the client may resort to self-medicating addictions as a means to disconnect in relationships. The client experiencing the energy of intense overexcitabilities, possibly accompanied by AD/HD, may say, "I only feel like myself when I'm drunk or stoned." Even with addiction therapy, clients may experience a cycle of indulging, abstinence, depression, anxiety, and again indulging. The cycle will continue until they learn the warning signs that trigger them into it and use appropriate self-therapy. Instead of going to a place of isolation, clients may also benefit from learning to communicate what is going on with them as a means to reduce any relational stress they may have to repair later on.

GENDER DIFFERENCES THROUGHOUT THE LIFE SPAN

Exceptional ability does not begin when a child is identified as gifted, nor does it end when a student exits a gifted program or graduates from school. The phenomenon, like retardation, involves inherent differences in development from birth through maturity. The same characteristics observed in young gifted children continue to be exhibited in them when they are 20, 40, or 60. The cognitive traits of high abstract reasoning, extensive vocabulary, rapid learning, curiosity, and awareness, as well as fascination with books, unusual sleep patterns, high energy, marked need for stimulation, and intense reactions are all part of the experience of being a gifted adult. The only gifted characteristic that does not survive midlife is excellent memory (Silverman, 1998a).

While the general characteristics of the gifted are similar in both men and women, there are differences as they move through the life span. The concepts of separation and connection are manifested early and appear to account for some of the differences in the behavior of adult males and females. Males begin the separation process in early infancy, whereas females are more relationally oriented (Chodorow, 1978).

A boy child is different from a girl child. He is "the other." And from early on, the mother knows he is the other. The girl child is not perceived as "the other"; she is perceived as being the same as the mother. Manifesting more of the archetypal feminine characteristics than the boy, the girl is more certainly identified with the mother; both mother and daughter operate with more open emotional connection and boundary flexibility. The sense of being the same as mother necessitates her taking on the feelings of mother. If mother is angry, she is angry; if mother is guilty, she is guilty. Without the basic preliminary separation, the intrapsychic bond between mother and daughter becomes stronger and stronger. (Conarton & Silverman, 1988, p. 45)

This pattern in the early separation of boys and connectedness of girls prepares them for adolescent relationships. It sets up the prototype of the male being separate and still receiving the care and nurturing of the mother and, later, his partner, while the female is oriented to being there for the other and doing the connecting to the other. The process is true for all, but, with the gifted, we find more exceptions. These boys and men characteristically have more developed feminine energy (Dellas, 1969; MacKinnon, 1962) and demonstrate more sensitivity and empathy, not fitting the contemporary macho culture. Gifted girls and women, with intensified masculine development (Casserly, 1979), may reject their feminine characteristics and become male-identified women competing in the marketplace.

The majority of relationships continue to live out the pattern of the other-oriented female caring for the male partner as well as the children and the extended family. The male manifests his energy as being separate and living the myth of the hero. For some relationships this works—at least for a time. There is enough energy from being "other-oriented" for a woman to maintain this lifestyle until the late 20s or the beginning of midlife. Then she starts being unhappy, depressed, disillusioned. Like all women at the beginning of midlife, she may find it difficult to put energy into her own needs while managing the needs of others. She may even start to feel resentful and wonder: "What about me?" If she does not have a partner, she may well feel the same about the corporation or profession she has identified with. For marriages, this midlife change brings about profound issues. This period of time is often the first time couples seek therapy.

There are therapeutic differences in treating gifted men and women. There are also several defenses that must be addressed in both. They are used to using rationalization and intellectualization as strategies of disconnection. They are highly resistant to experiencing and being vulnerable with the therapist, as well as others close to them. The therapist must be comfortable and skilled in developing and maintaining an intimate therapeutic relationship while modeling a balanced giving and receiving dynamic. Humor helps. To be able to laugh at how uncomfortable they become at needing and receiving takes away the humiliation of not being the giver and being in control. For the controlled, together-appearing giver, having to ask for help is like holding up a flashing neon sign saying, "Needy! Needy! Needy!"

Another factor leading the gifted to distance themselves, mainly women, is "The Impostor Phenomenon" (Clance, 1985).

Despite external evidence to the contrary, many bright and capable women continue to doubt their competence, downplay or dismiss their abilities, and subscribe to the disabling belief that they are impostors or fakes or frauds. This debilitating syndrome blocks women's ability to realize their full potential. (Bell, 1990, p. 55)

Even when they have continued success, they often attribute this to a mistake or to good fortune, rather than high ability. Their beliefs are so strong that they protect themselves from being found out. Paradoxically, the gifted are also forced to be impostors. Individuals who never fit in have to disguise their real natures in order to get through school, keep a job, fit into social groups, or get along with relatives. An introvert is afraid of being exposed or being publicly humiliated by mistakes. Extraverted gifted women are constantly told they are "too much." They may try to tone down, shut their mouths, and not give their opinions.

A gifted woman strives to develop a façade that makes her appear to fit in. The Imposter Phenomenon contributes to a woman developing an "as if" personality. She tries to relate and act as if she is like other people, and feels the proper feeling prescribed by the collective. Feeling like an impostor and never belonging, she represses her authentic self. That, coupled with asynchronous development earlier in life, lays the foundation for depression, a state that many women describe as "a lack of joy."

In 1926, Leta Hollingworth, one of the earliest researchers of gifted development, described the dilemma faced by gifted women as "the woman question." This dilemma has stood the test of time and mirrors much of what modern women face:

> The intelligent girl begins very early to perceive that she is, so to speak, of the wrong sex. From a thousand tiny cues, she learns that she is not expected to entertain the same ambitions as her brother. Her problem is to adjust to a sense of sex inferiority without losing self-respect and self-determination, on the one hand, and without becoming morbidly aggressive, on the other. This is never an easy adjustment to achieve, and even superior intelligence does not always suffice to accomplish it. The special problem of gifted girls is that they have strong preferences for activities that are hard to follow on account of their sex, which is inescapable. Stated briefly, "the woman question" is how to reproduce the species and at the same time to work, and realize work's full reward, in accordance with individual ability. This is a question primarily of the gifted, for the discontent with and resentment against women's work have originated chiefly among women exceptionally well endowed with intellect. (pp. 348–349)

The general public reflects the attitude, "If you're so smart, why don't you solve your own problems?" Most men believe that they should be able to solve their problems alone and that therapy is only for those who are weak. However, gifted men are seeking higher truth, more self-knowledge and awareness. It takes unusual courage for a man to independently seek therapy. Those who have that courage are either in severe depression or are highly intelligent or both.

At midlife it is usually women who make the initial movement toward therapy, with men following, oftentimes reluctantly. The therapy with men must be slow and reassuring. They are used to having goals, being in control and having all the answers. They are extremely uncomfortable not knowing about connection and relational skills. They must develop personal interactions for their relationships that are contrary to their interactions in the patriarchal world of work. Men take pride in opening this new vista if the therapist does not move too fast and lose the client. Group therapy is an effective tool. Learning relational skills with men is less threatening than with the intimate woman in their lives.

The therapeutic approach with women differs greatly. Most gifted women have been relational and connecting all their lives. They must learn not to do all the emotional work in the relationship with their partner and to negotiate their own needs with the needs of others. They must accept that they can't always be all things to all people. It is a difficult task for women who are used to being out there for the other first to learn to put their authentic feelings and needs into relationships. Because the gifted woman is highly perceptive, she is able to pick up her partner's vulnerable feelings, and finds it inconceivable that men, for example, really do not have comparable relational skills or know what women mean when they say they want "more" from the relationship.

Gifted men and women ascribe to the belief of equality between the sexes. Often, this gets translated into men and women being the same. Both men and women need to be taught more about the differences, especially in midlife, when men are pulling back from the work world and women are eagerly embracing it.

All the characteristics of the gifted, both positive and negative, enhance the individual's life at late middle age and early old age. Just like everyone else, they have had to deal with tragedies, chronic illnesses, and repeated losses. With the gifted, their energy to persevere and not accept limitations may well facilitate coping with handicaps. Their voracious interest in reading gives them access to nutrition and medical information. Their drive to be the best has hopefully brought them success in their work lives. It may not have brought them money, because of their altruistic natures and interest in social projects. They may now need financial advice. Their therapy at this stage generally focuses on discovering what they really want for the rest of their lives. It may be a new occupation or developing a latent talent. If they have been financially successful, their time can be spent mentoring younger associates or seeing the rest of the world or doing the things they never had time for. Throughout any of these lifetime endeavors will be the constant of finding meaning for their lives.

SPECIAL POPULATIONS

The general guidelines provided in this chapter will apply to most clients, but there are always exceptions. And the gifted are more diverse than any other population. At least four groups of gifted individuals have specific needs in addition to those discussed above: visual-spatial learners; gifted individuals with learning disabilities; gifted individuals with attention deficit/hyperactivity disorder (AD/HD); and gifted individuals of color. In an overview chapter, these topics can only be touched on briefly.

Counseling Visual-Spatial Learners

Visual-spatial learners (VSLs) are individuals who have stronger right hemispheres and think in images (Silverman, 1989; 2002). It takes significantly longer for VSLs to translate their pictures into words. Think of the time involved for a computer to download graphics and other complex images in comparison to text. This is similar to the process VSLs go through in converting the pictures in their mind into words. They tend to panic under time pressures, such as timed tests.

As they are often poor students, VSLs rarely recognize their own giftedness, although they may receive recognition for artistic or musical talent, mechanical aptitude, imaginative ideas, excellent problem-solving abilities, empathy, or intuitive knowledge. Some VSLs excel at both right- and left-hemispheric abilities, but those who have weak left-hemispheric skills are often wounded by school experiences. Good students master reading, spelling, calculation, rapid handwriting, and memorization—the left-hemispheric curriculum—with ease. These require excellent auditory-sequential skills. The ability to visualize is equally valuable and is gaining increasing importance in the technological 21st century. And studies have shown that at least one third of the population is strongly visual-spatial (Silverman, 2002). However, schools still maintain the 5,000-year-old emphasis on left-hemispheric, linear-sequential skills. This makes visual-spatial learners feel inadequate and stupid.

Wounded VSLs may come into therapy feeling anxious, depressed, or unworthy. The healing process begins with education about their learning style. Sections of *Upside-Down Brilliance: The Visual-Spatial Learner* (Silverman, 2002) can enhance the self-esteem of these creative thinkers. The therapist should encourage these clients to use their powerful visualization abilities in the therapeutic process. To date, there is only one article on counseling visual-spatial learners: "It Takes One to Know One: Counseling Needs of Visual-Spatial Learners," by Michael Davis (2003):

> Healing, then, is a two-fold process. One aspect is to help individuals understand and to validate the pain that they experienced when they received messages that denied their competence and sense of being OK. Having a safe space and caring support to move through these feelings is healing. The second aspect will be corrective. It will provide the needed reflection and validation of competence that individuals would have benefited from receiving in the first place. And since the negative messages were most probably delivered multiple times, over years, these will be conversations worth having a number of times, remembering numerous incidents. (p. 34)

Counseling Gifted Individuals with Learning Disabilities

Giftedness masks disabilities and disabilities mask giftedness. A person who is both gifted and learning disabled is likely to escape detection. It is intensely frustrating to feel smarter than everyone else in some ways and dumber than everyone else in other ways. When a gifted/learning disabled individual initiates therapy, the therapist's first job is to dispel the myth that learning disabled equals "dumb." As with visual-spatial learners (and many of these "twice-exceptional" learners are VSLs), these clients need to be educated about their strengths and their weaknesses and to see them separately, rather than allowing them to cancel each other out. Their picture of high intelligence needs to be shifted from an achievement orientation to abstract reasoning and the characteristics outlined in this chapter. It is helpful for them to read about successful twice-exceptional learners. One resource is *Intellectual Giftedness in Disabled Persons*, by Joann Whitmore and C. June Maker (1985), which describes the learning disabilities of eminent individuals.

Counseling Gifted Individuals with AD/HD

It is extremely difficult to separate attention deficit/hyperactivity disorder from giftedness, as most gifted adults exhibit at least some AD/HD characteristics. Differential diagnosis is extremely important in children and should be conducted by individuals who have expertise in giftedness as well as expertise with AD/HD. The best resource on this topic is Deirdre Lovecky's *Gifted Children with Attention Deficits: Different Minds* (in press).

Gifted individuals with AD/HD may be overlooked, since they are able to concentrate for long periods of time in areas of interest. Problems only occur when they are *not* mentally engaged. On the other hand, the high energy level of gifted children with overexcitabilities—particularly Psychomotor OE—can be misdiagnosed as AD/HD. Symptoms in girls differ from those in boys. Most people recognize the hyperactivity and impulsivity of boys, but miss the quieter, dreamy distractibility of girls with the inattentive form of AD/HD. An excellent resource for women is *Women with Attention Deficit Disorder* by Sari Solden (1995).

Coaching is popular today for helping individuals achieve their goals. Some of the coaching techniques are helpful for adults with AD/HD. Behavior therapy has a good track record with these clients. Supervised medication trials, neurofeedback, and elimination diets and nutritional supplements are useful as well.

Gifted Individuals of Color

Gifted individuals come in all shapes and sizes, races and cultural groups. Some consider it unethical to provide mental health services without training in multicultural counseling. Diversity training is important to enable the therapist to relate to culturally diverse clients. Gifted individuals of African American, Native American, Asian American, or Hispanic descent must move in and out of two different cultures. Often these cultures have conflicting values and expectations (Evans, 1993). The therapist needs to be attuned to the inner conflict this creates. These individuals are marginalized in society and must continuously struggle to believe in themselves.

Gifted children of color are often underrepresented in programs for the gifted (VanTassel-Baska, Patton, & Prillaman, 1989). This situation must be corrected. Therapists can act as advocates for a bright child to gain acceptance into a program for gifted students. They can also assist parents in gaining scholarships to private schools for the gifted. Many private schools have scholarships earmarked for children of color. For more information on counseling culturally diverse children, see Kathy Evans's chapter (1993), "Multicultural Counseling," in *Counseling the Gifted & Talented*.

SUCCESSFUL THERAPY WITH THE GIFTED

The gifted derive meaning through service to others. Their desire to make a difference in the world influences them to be more other-oriented, but they are not as likely to take care of their own needs. The role of being the giver is honorable, but it is equally important to work toward mutuality in relationships and learn how to receive and allow others to give. Therapy may help these individuals to recognize the fine line between serving and controlling. They tend to be compassionate toward others and accept the limitations of others, but hold unrealistic expectations for themselves. In order to attain balance and wholeness, gifted clients need to recognize their own needs and learn how to meet them. Without self-acceptance, they do not develop their soul self.

As is true with all clients, the therapist must help his or her gifted ones to find their authentic selves, help them fulfill the teleological need to become whole, find joy in everyday life, and learn the skills of having fulfilling relationships. While they can be more challenging to work with, gifted clients also have greater potential for growth and development, which is very rewarding to the therapist.

REFERENCES

Aron, E. (1996). *The highly sensitive person.* New York: Broadway Books.

Bell, L. A. (1990). The gifted woman as impostor. *Advanced Development, 2,* 55–64.

Berg, D. H., & DeMartini, W. D. (1979). Uses of humor in counseling the gifted. In N. Colangelo & R. T. Zaffrann (Eds.), *New voices in counseling the gifted* (pp. 194–206). Dubuque, IA: Kendall/Hunt.

Casserly, P. L. (1979). Helping able young women take math and science seriously in school. In N. Colangelo & R. T. Zaffrann (Eds.), *New voices in counseling the gifted* (pp. 346–369). Dubuque, IA: Kendall/Hunt.

Chodorow, N. (1978). *The reproduction of mothering: Psychoanalysis and the sociology of gender.* Berkeley: University of California Press.

Clance, P. (1985). *The imposter phenomenon.* Atlanta: Peachtree.

Conarton, S. A. (1999). After the hero. *Advanced Development, 8,* 97–112.

Conarton, S. A., & Silverman, L. K. (1988). Feminine development through the life cycle. In M. A. Dutton-Douglas & L. E. A. Walker (Eds.), *Feminist psychotherapies: In-*tegration of therapeutic and feminist systems (pp. 37–67). Norwood, NJ: Ablex.

Dabrowski, K. (1967). *Personality shaping through positive disintegration.* Boston: Little, Brown.

Dabrowski, K. (1972). *Psychoneurosis is not an illness.* London: Gryf.

Dabrowski, K., & Piechowski, M. M. (1977). *Theory of levels of emotional development* (Vols. 1 & 2). Oceanside, NY: Dabor Science.

Davis, M. (2003). It takes one to know one: Counseling needs of visual-spatial learners. *Gifted Education Communicator, 34*(1), 33–35.

Dellas, M. (1969). Counselor role and function in counseling the creative student. *The School Counselor, 17,* 34–39.

Downs, A. (2003). *The half-empty heart: A supportive guide to breaking free from chronic discontent.* New York: St. Martin's Press.

Drews, E. M. (1972). *Learning together.* Englewood Cliffs, NJ: Prentice-Hall.

Evans, K. (1993). Multicultural counseling. In L. K. Silverman, Ed. (1993). *Counseling the gifted & talented* (pp. 277–290). Denver, CO: Love.

Frankl, V. E. (1963). *Man's search for meaning: An introduction to logotherapy.* New York: Washington Square Press.

Gowan, J. C. (1980). Issues on the guidance of gifted and creative children. In J. C. Gowan, G. D. Demos, & C. J. Kokaska (Eds.), *The guidance of exceptional children* (2nd ed., pp. 58–66).

Hollingworth, L. S. (1926). *Gifted children: Their nature and nurture.* New York: Macmillan.

Hollingworth, L. S. (1939). What we know about the early selection and training of leaders. *Teachers College Record, 40,* 575–592.

Jacobsen, M-E. (1999). *The gifted adult: A revolutionary guide for liberating everyday genius.* New York: Ballantine.

Lovecky, D. V. (1990). Warts and rainbows: Issues in the psychotherapy of the gifted. *Advanced Development, 2,* 65–83.

Lovecky, D. V. (1993). The quest for meaning. Counseling issues with gifted children and adolescents. In L. K. Silverman, Ed. *Counseling the gifted & talented* (pp. 29–50). Denver, CO: Love.

Lovecky, D. V. (1994). The moral child in a violent world. *Understanding Our Gifted, 6*(3), 3.

Lovecky, D. V. (in press). *Gifted children with attention deficits: Different minds.* London: Jessica Kingsley.

MacKinnon, D. W. (1962). The nature and nurture of creative talent. *American Psychologist, 17,* 484–495.

Miller, A. (1981). *The drama of the gifted child.* New York: Basic Books.

Parker, W. D. (1997). An empirical typology of perfectionism in academically talented children. *American Educational Research Journal, 34,* 545–562.

Piechowski, M. M. (1979). Developmental potential. In N. Colangelo & R. T. Zaffrann (Eds.), *New voices in counseling the gifted* (pp. 25–57). Dubuque, IA: Kendall/Hunt.

Piechowski, M. M. (1992). Giftedness for all seasons: Inner peace in time of war. In N. Colangelo, S. G. Assouline, & D. I. Ambroson (Eds.), *Talent development. Proceedings from the 1991 Henry B. & Jocelyn Wallace National Research Symposium on Talent Development* (pp. 180–203). Unionville, NY: Trillium Press.

Roeper, A. (1991). Gifted adults: Their characteristics and emotions. *Advanced Development, 3,* 85–98.

Silverman, L. K. (1983). Personality development: The pursuit of excellence. *Journal for the Education of the Gifted, 6*(1), 5–19.

Silverman, L. K. (1989). The visual-spatial learner. *Preventing School Failure, 34*(1), 15–20.

Silverman, L. K. (1993a). A developmental model for counseling the gifted. In L. K. Silverman, Ed. (1993). *Counseling the gifted & talented* (pp. 51–78). Denver, CO: Love.

Silverman, L. K. (1993b). The gifted individual. In L. K. Silverman, Ed. (1993). *Counseling the gifted & talented* (pp. 3-28). Denver, CO: Love.

Silverman, L. K. (1998a). Developmental stages of giftedness: Infancy through adulthood. In J. VanTassel-Baska (Ed.), *Excellence in educating gifted & talented learners* (3rd ed., pp. 145-166). Denver, CO: Love.

Silverman, L. K. (1998b). Personality and learning styles of gifted children. In J. VanTassel-Baska (Ed.), *Excellence in educating gifted & talented learners* (3rd ed., pp. 29–65). Denver, CO: Love.

Silverman, L. K. (1999). Perfectionism: The crucible of giftedness. *Advanced Development, 8,* 47–61.

Silverman, L. K. (2002). *Upside-down brilliance: The visual-spatial learner.* Denver, CO: DeLeon.

Solden, S. (1995). *Women with attention deficit disorder.* Grass Valley, CA: Underwood.

Streznewski, M. K. (1999). *Gifted grown-ups: The mixed blessings of extraordinary potential.* New York: Wiley.

VanTassel-Baska, J., Patton, J., & Prillaman, D. (1989). Disadvantaged learners at risk for educational attention. *Focus on Exceptional Children, 22,* 1–15.

Walker, A. (1982). *The color purple.* New York: Simon & Schuster.

Whitmore, J. R., & C. J. Maker (Eds.). (1985). *Intellectual giftedness in disabled persons.* Rockville, MD: Pro-Ed.

12

Grief, Loss, and Death

By Thelma Duffey, Ph.D.

REFLECTION QUESTIONS

1. *What does our U.S. society tell us about grief?*
2. *What are some common responses to grief?*
3. *How do we reconcile our responses to grief with society's expectations for the bereaved?*
4. *How do we best attend to a person in grief?*
5. *What consequences could we face if we ignore our losses?*

INTRODUCTION

No one can ever prepare for the feelings of anguish, vulnerability, fear, and profound loneliness that come with the experience of loss. Indeed, there is no experience that can tap into our primitive, childlike terror than anticipating or experiencing the loss of someone or something we love and value. Depending on the loss and the degree of grief surrounding it, our ability to cope with life can feel utterly unmanageable. During these times, we may even begin to question our sanity.

By its very nature, loss disconnects us not only from the source of our attachment but also from the psychic and relational relief that comes from our connection to our beloved. Experiencing loss brings with it a myriad of feelings, physical sensations, cognitions, and behaviors. The goals of this chapter are to present a brief overview of historical and contemporary grief and loss theories, common patterns of coping in bereavement, and relational considerations in bereavement. As such, the grief experience will be examined through a relational lens. The focus will be on examining the many ways in which loss affects our lives and the means by which we may integrate the experience of loss into our lives.

Dana asked that we bring a personal context to our work in this writing. In so doing, I will begin by saying that although my interest in this subject has been long-standing professionally, I have been no stranger to loss myself. When Dana asked me to write this chapter, it seemed like a natural fit. The project has taken on a more formidable tone for me since we first discussed this writing, and the reality and enormity of loss has been brought home to me again in very sudden and personal ways.

In the midst of this work, my trusted colleague, mentor, and dear friend, Lesley, was killed in a car accident as she was going home from work. Without a doubt, I loved Lesley. Just over two years before, my beloved uncle died unexpectedly from surgery we anticipated as routine, and my sister-in-law, Laura, was killed crossing the street while she was away on business. My uncle was a bright, vibrant, witty, interesting, and very loving member of our family. Laura was a wonderfully devoted sister-in-law, wife to my brother, and mother to my niece, then 3 years old.

Two months after Laura's death, my father's wife died suddenly in her sleep. She suffered a massive heart attack. Seeing my brother and father in grief, knowing that my little niece, now 6, will never see her mother again, seeing my own mother's pain from these losses, and managing my own grief reactions from these and other personal losses and disappointments has made for a very difficult three-year period. Given that, the prospect of writing this chapter is far more sobering for me than I had anticipated. My skin, for the first time in a long time, feels thinner to me. Brutally aware of how our lives can literally turn on a dime, this writing has become a personal challenge. It is from that context that I write.

COMMON RESPONSES TO GRIEF

Because the state of grief can bring a person to believe that he or she is alone and disconnected from any semblance of normalcy, it is important to delineate the many feelings, thoughts, and behaviors that are likely to come during our losses. Worden (1991) describes some common feelings of the bereaved. These include feelings of sadness, anger, guilt, reproach, anxiety, loneliness, and fatigue. We feel in shock, helpless, and numb when we are in grief. We pine for what we miss. Our feelings affect us physically, with many of us experiencing a hollowness in the stomach, tightness in the chest and throat, an oversensitivity to noise, a sense of depersonalization, shortness of breath, muscle weakness, lack of energy, and dry mouth (Worden, 1991).

Bereavement also brings with it a variety of thoughts and behaviors. We experience disbelief, confusion, and preoccupation. We have difficulty going to sleep or staying asleep through the night. We have difficulty eating, with our favorite foods tasting flat or even unpleasant. We may become absent-minded and forgetful, withdraw from others, dream of the deceased, and avoid reminders of the deceased. We search, call out, sigh, engage in restless overactivity, and cry. We may visit places that remind us of the deceased, and hold dear to us objects that belonged to the deceased (Worden, 1991).

Much of the early literature on loss speaks to the ways in which we resolve our losses. A distinguishing tenet of this chapter is that our losses are not "resolved." Resolution would suggest a conclusion that loss from a beloved source simply cannot bring (Kagan, 1998). We posit, however, that depending on a number of factors delineated below, losses may be integrated into the experience of our lives, and, as such, provide us with a context from which we eventually sort through and make meaning out of even the most wounding of life's experiences.

Certainly, the process of grief has been the focus of writers, researchers, and theologians throughout time; and all of us have, in a myriad of ways, lived it. Given that, coming to terms with our losses is perhaps the single greatest challenge we face as human beings.

In spite of the fact that loss is universal, our basic personalities, relational supports, spiritual beliefs, life experiences, developmental contexts, circumstances of our loss, and culture in which we live uniquely influence our experience of loss (Kagan, 1998). These losses may be concrete losses such as in the case of death, or they may be abstract losses, such as in the death of a dream. Some of these are also disenfranchised losses that are not culturally sanctioned by our communities. Members of marginalized groups and people involved in unresolved, unrequited, private, or publicly unsanctioned relationships may find themselves, at the point of great loss, without support, disconnected, displaced, and anonymous (Murray, 2002). Whatever the context, all loss, by its very nature, severs concrete connections to the people, places, and circumstances to which we have become attached. As a result, we cannot address the experience of grief without also addressing the experience of connection.

Indeed, a theme throughout this text is the yearning that we each have for relational connection with the significant people in our lives (Jordan, J. 2001; Miller, 1976; Miller & Stiver, 1993; Stiver, 1990). At best, we are fortunate to experience the profound pleasure, nurturance, safety, and sense of well-being that are by-products of participating in a growth-fostering connection. When we authentically experience this connection and the personal and relational benefits that come with it, and when we then lose these, either through death or through some other form of separation, we experience profound feelings of grief.

Conversely, there are times when we do not experience and enjoy the healthy connections we may have yearned for. When we experience loss or separation from our hope for this connection, our grief speaks to the unrequited nature of our experience. In either case, we suffer distinct forms of disconnection, not only from the source of our loss but also from the part of us that yearned for this connection. In addition, we may experience disconnection from others as we attempt to belong in a world that can feel curiously and painfully distant, remote, and disconnected. We may feel disconnected, at times, even from ourselves, in which case it is no wonder that we enter the experience of isolation. It is these disconnections and the journey back into connection that are explored in this chapter.

DIVERSE CONTEXTS OF LOSS

Humphrey and Zimpfer (1996) define loss as "the state of being deprived of or being without something one has had, or a detriment or disadvantage from failure to keep, have or get" (p.1). They define four categories of loss: relationship, self, treasured objects, and developmental. Our losses come to us through death, divorce, separation of some sort, or rejection. Although the catalyst for the loss differs, the very real feelings of wounding are universal. No one can prepare for the phone call, discussion, letter, or the knock on the door that comes when we find out that our lives are terminally altered.

Our losses are also exacerbated by society's lack of real experience with loss. According to Greenspan (2003), we live in an emotion-phobic society. Greenspan adds:

> We've mastered the internal regulation process of emotion-phobia. We've been taught to believe that emotions are not appropriate in any context but an intimate personal relationship. In public, we compose ourselves, compulsively apologizing when we cry, as though displaying authentic sorrow is bad form, a sign of emotional weakness. (p. 18)

As such, we disconnect from our *selves* and our experiences as we attempt to accommodate the sociocultural and even theory-based expectations that can drive the grief experience.

Although it is beyond the scope of this chapter to address the multicultural implications of grief and loss, Parkes, Launagani, and Young (1997) speak to the diverse customs that surround loss and grief across cultures. They note that in some cultures, such as India, Nepal, China, and Greece, the whole community is involved in rituals of death and dying, and the bereaved are exhorted to display grief openly. In contrast, many North American societies, particularly Britain, Scandinavia, and the United States view rituals of death as private or family based. They add, "Even the funeral ceremonies are seen as private events . . . where free expression of emotions, although not expressly discouraged, is by no means encouraged (Parkes et al., 1997, p. 219). At the same time, these authors discuss how diversity exists within cultures, reminding us that if given an opportunity, people will tell you who they are, how they perceive reality, and what they need in times of loss (Parkes et al., 1997).

In addition to our culture, some families establish norms for expressing painful feelings that impede access to our authenticity, connections, and genuine experience. They develop patterns of relating where children feel a need to shield and protect their parents from hurt by discussing painful realities. Warding off further loss, children internalize their feelings and present flat or even cheerful demeanors in the face of adversity and loss. Afraid of discussing their hurts, they develop patterns of hiding their feelings and vulnerabilities. Losing touch with themselves, they are later at an even greater loss to manage life and all that it brings.

In other families, members experience shame when they express sentiments and feelings that make others uncomfortable. They learn to minimize their feelings, avoid painful discourse, which ultimately results in a lack of authenticity in their relationships. Other families punish its members when they express themselves authentically. As such, members suffer assaults when they risk being vulnerable and feel that holding in these feelings and even dissociating from them is, for them, a more valid option.

Having no access to their authentic experience, and fearful of sharing it when it is available, individuals learn to engage in and develop disconnecting patterns of relating that eventually keep them from the supportive experience that connection can bring. Believing they have no choice, they grow accustomed to relating to others from a walled-off, self-protective position. It is no wonder that, later in life, they are unable to grieve with others in a mutually supportive way. Relational-Cultural theory (RCT) posits that, although these responses are natural coping strategies of survival (Jordan, 2001), they ultimately create a context for profound feelings of isolation. One would expect that facing losses in life would only exacerbate this sense of isolation.

Competing theories exist with regard to the long-standing predominant 20th-century Western perceptions of grief, with some observing that the theories themselves promote unrealistic expectations for the bereaved, and thereby perpetuate disconnections in the grief experience (Hogan, Morse, & Tason, 1996).

THEORIES OF GRIEF

Freud first introduced the concept of bereavement in 1917 (Marwit & Klass, 1995). Since then, theorists and researchers have attempted to operationally define and describe the process of grief through a clinical lens. In addition, the researchers examine the grief experience with respect to the educational, religious, and cultural contexts of the bereaved. Four broad-based theories have emerged. These include the psychoanalytic, psychoanalytic-cognitive, behavioral, and psycho-biological (Hogan et al., 1996).

The psychoanalytic experience of grief is diagnostic and examines the means by which a person recovers from loss. Individuals pass through grief stages of predetermined and appropriate duration. If they do not complete these stages within the designated time frame, they carry diagnoses of delayed or distorted grief (Hogan et al., 1996). This model values the ultimate severance of emotional and psychological ties to the deceased. Marwit & Klass provide two examples from Freud in 1917 and Bowlby in 1971 to make this point. From Freud:

Each single one of the memories and expectations in which the libido is bound to the object is brought up and hypercathected, and detachment of the libido is accomplished in respect to it. . . . The fact is that when the work of mourning is completed, the ego becomes free and uninhibited again (as cited in Marwit & Klass, 1994, p. 245).

Bowlby adds:

Just as a child playing with Meccano must destroy his construction before he can use the pieces again (and a sad occasion it sometimes is), so must the individual, each time he is bereaved or relinquishes a major goal, accept the destruction of a part of his personality before he can organize it afresh toward a new object or goal (as cited in Marwit & Klass, 1994, p. 245).

These theories have profoundly influenced mainstreamed grief therapy perspectives and a culture that is hungry to understand and explain this multifaceted and complex phenomenon (Marwit & Klass, 1994). Other theories, however, have emerged and challenged these. The psychoanalytic-cognitive experience is among them. While having a Freudian base, with the unconscious serving as an active agent, this theory serves to reinterpret the process of grief. Separation anxiety is viewed as being a natural result of losing significant bonds and attachments. However, unlike the psychoanalytic goal of having the bereaved separate from the deceased, the relationship between the two parties continues, with the inner experience with the deceased providing comfort for the bereaved. From this perspective, grief work is deemed successful when we reconstruct our cognitive schema to include a new relationship with the deceased (Hogan et al., 1996).

In a review of the literature on the evolution of grief and loss theory, Marwit & Klass (1995) report that Stroebe, Gergen, Gergen, and Stroebe (1993) also challenged the established psychoanalytic grief-work orientation models, suggesting that contemporary theory is a simple rejection of earlier equally compelling romanticist notions of bonding with the deceased. Marwit and Klass (1994) concluded that "the definition of healthy and unhealthy resolution of grief is more grounded in the various theoretical models of grief than in research on the lives of survivors" (p.284). Researchers also suggest that current theory is unconfirmed by cross-cultural studies, noting that other cultures respect attachments to and guidance from the deceased. Valliant (1985) states, "Contrary to folklore and psychiatric myth, separation from and loss of those we love do not cause psychopathology. Rather, failure to internalize those whom we have loved causes psychopathology" (as cited in Marwit & Klass, 1995, p. 285).

In contrast, the behavioral orientation to grief rejects the notion of intrapsychic factors with respect to grief and instead focuses on the psychological and physiological challenges for the bereaved. For the behaviorist, bereavement is a normal experience affected by environmental resources, such as our social support (Hogan et al., 1996).

Perhaps one of the most influential writers on grief, loss, and stage theory is Elizabeth Kubler-Ross. She conceptualized stages of grief to include denial, bargaining, anger, despair, and acceptance (Kubler-Ross, 1969). According to Kubler-Ross, when we are in the stage of denial, we simply can't believe what has happened to us. We think to ourselves that this must be a mistake. Again, according to this model, as we become more connected with our loss, we move into the stage of anger. We may feel angry at our loss, at the doctors who treated our loved ones, at fate, and we may also feel envy toward others who are not suffering our losses. A question, however silent, during this stage would commonly be, "Why me?"

Bargaining is another stage of Kubler-Ross's (1969) grief model. We attempt to postpone the inevitable or believe that somehow we can earn our way out of loss. When we are unsuccessful at bargaining, depression sets in. Memories of past losses come to the forefront as we grieve over these and our current loss. She purports that in working through these stages, we eventually reach a place of acceptance. This is the final and the most difficult stage with which we must come to terms. At this point the struggle to accept the loss is over.

Kubler-Ross set the stage for other grief theorists to explore the stage and task factors in grief therapy. Further affecting the direction for exploring grief and loss in her groundbreaking book, *Necessary Losses,* Viorst (1986) states that most of us experience shock, numbness, denial, and disbelief when we enter the first stage of grief. We may look to others as if we are "managing well," while in fact, we have not begun to understand the extent of the losses we face. We often focus on positive aspects of the loss (she did not suffer, she is in a better place, she is freed from suffering, etc.). Shock also affects our memory, decision making, and reaction time. Denial keeps us from taking in the loss before we can handle it.

Viorst (1986) reports that psychic pain, longing, and yearning characterize the second stage of grief. As such, we may experience anger, idealization of the bereaved, depression, guilt, troubles with concentration, and we may become disorganized or forgetful. We begin to release our illusions as the reality of the loss becomes more real to us. As we become more aware of the permanence of our loss, our grief may intensify. In the third stage, we begin to recover some energy, some hopefulness, and some capacity to enjoy life. We begin to gain perspective on our loss and to integrate the loss into our experiences (Viorst, 1986).

TRAUMATIC LOSS

Undergoing the process of grief is hard enough, but for those suffering a traumatic loss, it becomes even more challenging. Levinson and Prigerson (2000) describe traumatic grief as that which is sudden, unexpected, and/or violent. They propose that experiencing deep grief may be more extensive for those of us whose losses are traumatic. According to Levinson and Prigerson, symptoms may include:

> difficulty accepting the death, searching and yearning for the deceased, constant preoccupation with thoughts of the deceased, anger, bitterness, numbness and detachment, sense of futility about future, sense that a part of oneself has died, and a shattered sense of trust, safety, and control . . . depressed mood, hypochondrias, psychomotor retardation, and possible suicidal ideation, sleep disruption, and anxiety. (p.32)

Traumatic losses are common, impacting the bereaved in profound ways. For example, after over 24 hours of unsuccessfully trying to locate his wife of 14 years, police informed Gregory that they found her body. Police reports indicated possible homicide. Throughout their marriage, Annie had been as predictable as clockwork. Given that, Gregory had become especially concerned when she did not answer her cell phone and when she never arrived at work as expected. Gregory had become frantic but worked to keep a cool head. In the early hours of the morning the day following his wife's disappearance, Gregory sat in his condominium pacing, hoping against all hope that his best, most authentic, and reliable friend and companion would be home soon. That was not to be the case.

Nothing could prepare him for the news that she was gone from this world, and from him, forever. Gregory now faces a terminal disconnection from the person he has grown to love, trust, and depend on. He is losing the support, nurturance, mutuality, energy, and investment that came from his relationship with Annie. He is also losing the dream that they created together.

Grief experts assert that traumatic losses may result in complex grief experiences. They refer to these experiences as delayed, avoided, and chronic. According to these writers, we experience delayed grief when we hold back our emotions, and carry avoided grief when we do not fully express our feelings during the grief experience. We experience chronic grief when our thoughts and feelings about the loss encapsulate our lives (Leick & Davidson-Nielsen, 1991). Kagan (1998), on the other hand, de-pathologizes the concept of complicated mourning. She posits that experiencing traumatic situations, providing the example of the death of a child, elicits thoughts, feelings, and behaviors otherwise considered abnormal.

Gregory explains his feelings in the aftermath of his wife's tragic death:

> I walk through my days numb if I can help it. People have been great, but really, what can anybody do? They mean well, but how can they have a clue? Everyone goes home to their families, to their lives, as it should be. I don't want a pity party. But really, I don't expect it to get better. I'm not sure I really want it to. The truth is, I wouldn't be here if I could help it except that if I did anything to make that happen it could keep me from seeing my wife later. I don't want to risk that. In case there is more to this world than this, I don't want to mess up my chances of getting there. She was incredible. She'll be there.

Neimeyer, Prigerson, and Davies (2002) explain that "the loss of an intimate relationship through death poses profound challenges to our adaptation as living beings. We respond to our losses through a series of hard-wired reactions and yearnings" (p. 240). Bereaved individuals struggle to re-create lost meaning. Traumatic bereavement poses additional challenges to their adjustment. It proves illegitimate, sometimes permanently, their "assumptive world" (Janoff-Bulman & Berg, 1998) of safety, trust, and optimism (Neimeyer et al., 2002). In discussing the death of her husband, Thompson (1996) writes:

> When Mark's death came, it propelled me into a reality I was totally unprepared for! Because death is so often a taboo subject, our society doesn't prepare us for death and we must struggle to somehow cope on a personal/individual level. In reality, however, nothing really can prepare us for the shock and grief of losing a loved one. . . . I felt as if I was moving on into an empty world, surrounded by people yet so alone. . . . You never get over the death of a loved one, however, you do learn to live with what has happened. . . . Mark had immense impact and inspiration in my life. . . . He inspired me in life and in death. (p. 6)

Davis, Nolen-Hoeksema, and Larson (1998) note that when we are able to find meaning in loss we are better able to ultimately adjust to our losses and to a life where they are sure to occur. For many, religious, spiritual, and psychological issues are part of a lifelong quest for exploring what is meaningful in life, the nature of our essential value, issues of redemption, and the means by which we ultimately find peace. For many, working through grief issues is an important avenue for coming to terms with life. Clearly, unique religious, spiritual, existential, and psychological aspects must be considered when working with the bereaved as many struggle with such considerations as life's meaning, past actions, missed opportunities, fears of living with loss, and fears of leaving this world.

CHILD AND ADOLESCENT REACTIONS
TO DEATH

With grief and loss being the most debilitating of experiences faced by us as living beings, imagine the impact of parental or sibling loss on the life of a child. Children, by nature, are dependent on the love, nurturance, support, and availability of their parents. When they lose these very fundamental needs because of the death of their parent, how isolating, lonely, confusing, and sad it can be as they readjust to life. In these circumstances, children lose aspects of their innocence and must face the very real fact that life is temporary and that we can and do lose the people we love and depend on most.

Coming to terms with the loss of a parent is not something we "get over." The writings of some grief scholars and theorists can appear painfully artificial and remote as they describe the steps children take when adjusting to their losses. In reality, children, like all of us, face the very human condition of learning to live with memories of times we long for still, yearnings for connection with a person no longer with us, and the loneliness that unrequited connection with a deceased loved one can engender.

Silverman, Nickman, and Worden (1992) examined the means by which children form and maintain connections with their deceased parents. They interviewed 125 children between the ages of 6 and 17 one year after a parent's death. Results indicate that children construct images of their deceased parents and derive comfort from their connections with them. They form these connections as they create rituals for concretizing memories of the deceased and integrating them into their current relationships with others. They internalize the experience of their parents, legitimizing the deceased's existence in themselves and in their world. Some children reported believing that the deceased parent could see them and communicate with them in their dreams.

M'Lissa is a beautiful, wide-eyed, lovable, and vivacious 6-year old. Her mother died suddenly and tragically when she was 3 years old. M'Lissa's father has been exceptional in his care for her, in spite of the enormous grief he carries. He is dedicated to anticipating her needs and takes great care in providing for her social, spiritual, relational, material, and recreational opportunities. Her grandparents have also been tremendous sources of support in each of these areas. M'Lissa appears to be a happy young girl, uninhibited in her song, dance, and play.

Still, an observant eye would see a distant look that can come across her face as she sits in her car seat while on a drive. It would see a momentary look of grief that washes over her as she sees a face resembling her mother. It would notice her bright smile that seems to communicate wonder, hope, and then relief as she tells a story involving a former experience with both parents. Scanning the safety of her bold discussion in a culture that is death-defying (De Spalder & Strickland, 1987), at 6 years of age, M'Lissa discerns whether the listener can "handle" her words that communicate how she misses her mother. Bearing intellectual and emotional intelligence, she has picked up the social cues and responds to them. Still, M'Lissa is fortunate in that it appears that most people in her world seem to welcome her voice. Others are not as fortunate.

Greenspan (2003) notes:

> Without a listener, the healing process is aborted. Human beings, like plants that bend toward the sunlight, bend toward others in an innate healing tropism. There are times when being truly listened to is more critical than being fed. Listening well to another's pain is . . . capable of healing even the most devastating of afflictions. . . . Children speak their pain automatically when there is a listener, but learn to hide it when there is no ear to hear. (p.14)

Silverman et al. (1992) report that some children in the study could not believe their parent's death was real and sometimes forgot that the parent had died. Further, children who did not dream of their parent had more difficulties coping with the death of their parent. They concluded that the child's effort to sustain a connection to a deceased parent facilitated the child's ability to make sense of the experience of loss and to integrate the loss as a reality (Silverman et al.).

Further, they found that of the 125 children in the study, 74 percent believed heaven exists (Silverman et al., 1992). Altschul (1988) describes the experience of children discussing their views of heaven in a group fantasy session, while fantasizing a reunion with the deceased in heaven. Participating in rituals of funeral, spiritual readings, and believing in a hereafter can ultimately contribute to a sense of belonging, comfort, and hopefulness for the bereaved (Altschul).

Certainly, entering into the experience of grief can challenge our spiritual perspectives as we become angry at the unfairness of life. M'Lissa's older cousin, Tara, describes the mixed feelings she felt when she would hear M'Lissa sing "Jesus Loves Me" on the heels of her mother's death:

> I have faith and do believe that M'Lissa's mother is in heaven. I do believe she was such an incredible person, that if any of us got there, she would. Still, it hurt me to see M'Lissa's innocent little face shine brightly as she sang a song about God and love and her blessings. I felt protective of her, horrified for her loss, and again confused by how it came to be that this little girl, like so many others, would never see her mother again in this lifetime. My anger is, in large part, based on my helplessness and on my utter protest that things have, do, and will continue to happen to terrific, undeserving people. I have no doubt that it's part of my soul's mission to come to terms with this. I feel too much belligerence around this to think I can continue to live this way comfortably.

Mishne (1992) reports that losing someone we love symbolizes a loss of omnipotence and control over our world. This loss affects our self-esteem and optimism about our lives and our futures. Although we ultimately have no control over our destinies, we can and must control our fears so they do not disempower or overwhelm us. Zambelli and De Rosa (1992) add that encouraging children to distinguish between facts and beliefs about death seems to help them gain more control over their fears.

Grief affects older children and adolescents in unique ways. Families often reorganize existing roles during times of transition and change. For example, when a parent dies, adolescents may inherit a role or roles previously held by the deceased family member. The eldest son may be the new "man of the house" following his father's death (Worden, 1991, pg. 126). The surviving parent may be overcome by grief and unable to attend to the needs of the child (Zambelli & De Rosa, 1992). Children and adolescents can take on responsibilities that are beyond their years, experiencing further losses of childhood.

In one example, two adolescent girls were visiting family one holiday when their father died tragically and unexpectedly. The older child, Abby, came home to help with arrangements. Soon thereafter, their mother died of a sudden illness. Although there were other adults involved, Abby, approaching 18, was primarily responsible for taking care of the sale of the house and the details surrounding the arrangements. Her classmates were horrified at her loss but were ill equipped to connect with her around it. As many approached the excitement of going off to college, one quietly asked, "Who is going to take Abby to school? Who will help her set up the room?"

Adolescents are often at a loss with regard to expressing their grief. Common expressions include exhibiting behavior and performance problems at school, missing classes, failing to complete assignments, reduced quality of academic performance, rebellious behavior, withdrawal into depression, ego deficits, expression of inferiority in relation to others with intact families, low self-esteem, and bewilderment and frustration. Other expressions include poor problem-solving abilities, feelings of guilt and anger, difficulty in relationships with parents or other family members, difficulty in relating to peers, and dealing with fears of one's own or other's death (De Minco, 1995).

According to Mishne (1992), the expression of rage is the most common reaction to loss by adolescents. They express rage toward others in the environment or toward the self, culminating in depression. When they express rage toward others, a cycle often occurs. Adolescents, like all of us, need emotional connection with others while they are experiencing grief. Yet, these cycles serve to distance others, serving as barriers to connection.

Another common grief reaction in adolescence is the experience of guilt. De Minco (1995) reports that some adolescents experience guilt by regretting that they did not spend more time with the deceased or feeling contrite for the times they were angry with the deceased. They also experience guilt when they question whether they had expressed sufficient love toward the deceased. In some cases, some adolescents feel responsible for the death itself.

Some adolescents also experience guilt in relation to the surviving family members. They may believe that if they talk to their parents about their own grief, it will upset them. They take care of their parent's feelings by protecting them from their own pain. This can result in a parent's misinterpretation that the adolescent is coping well with the loss (DeMinco, 1995).

One could expect that authority figures, such as teachers, would be available to listen to adolescents in their grief, but depending on their own unresolved grief issues, these authority figures could unwittingly shut down the expression of grief in their students. They may do so by refusing to answer questions, using diverting techniques, or making negative nonverbal responses to their attempts to talk about death and dying. Likewise, positive relational images about grief and loss for the child are created when figures of authority express a willingness to deal with the feelings of the child.

Parents who are not comfortable with their child's feelings of helplessness, anger, or loss may minimize or even directly discount the viability of what the child is expressing. They may shame their children by calling them names such as "selfish" or "self-centered" rather than listening to what the child is trying to communicate and helping the child sort through their feelings. However unconscious, any time a parent disagrees with or cannot handle the feelings of a child, the parent is communicating to the child that there is something wrong with him or her.

Dynamics of culture, spirituality, and gender also influence an adolescent's experience of grief. Morin and Welsh (1996) examined the cultural differences of adolescents' perception of death. They explored the experiences of urban and suburban youth to determine how such factors as family mythologies, cultural background, life experiences, and their environments influenced their notions of death. Results indicated that 25 percent of urban youth associated violence with death and 16.6 percent associated religion with death, while suburban adolescents virtually reported no associations of violence with death (0 percent) and few reported religion with death (5.3 percent) (Morin & Welsh).

Children and adolescents face unique challenges as they make their way in the world without the love, support, and living connection with a parent. Not only family members but also teachers, counselors, other parents, and members of the community all have an opportunity to create for them a different form of safety, stability, and relationally rewarding experience. Children may never "recover" from the death of their parents. Yet, they can and do thrive in spite of it.

A PARENT'S LOSS OF A CHILD

To lose a child is one of the most profoundly devastating tragedies that a family must overcome. I began my work as a counselor by working with children who were chronically and terminally ill. I have never felt such utter humility as I did when invited and included in the experience of coming to terms with their impending death. I cannot count the times that I am reminded of something they said or did as they were dying that later gave perspective to my understanding of life. The dignity, raw honesty, vulnerability, and trust that these children and their families displayed were examples of the authenticity and freedom that can come in even the darkest of moments.

"The death of a child is considered one of the most difficult and traumatic events that a family can experience" (Farrugia, 1996, p.30). Parents who lose their children experience a myriad of feelings such as guilt, lack of control, and intense anxiety. They suffer symptoms not unlike those of clinical depression, such as loss of sleep, lack of appetite, lethargy, withdrawal, and suicidal thoughts. Losing a child can create tension in marriages, particularly if tension was present to some degree in the relationship before the death. Given that individuals differ in the ways that they manage bereavement, couples must communicate their needs if they are to provide mutual support (Kamm & Vandenberg, 2001). These researchers found that couples that maintained open communication experienced more intense grief initially compared with couples that did not communicate but that the grief lessened over time.

Miller and Ober (1999) explain that our culture does not have the words that describe the experience of losing a child. This could complicate the communication and expression between couples and with others.

Although suffering the death of a child creates tremendous stress, according to Farrugia (1996), the death of a child does not necessarily result in the parents' divorce. It may intensify marital patterns, whether they were positive or negative. Schwab (1998) cautioned readers not to oversimplify a complex situation by assuming that divorce after a child's death is a result of the loss. In a review of the literature on childhood death, he found that many marital relationships are strong and can survive the death of a child.

Losing a child creates a unique form of bereavement (Kagan, 1998). The pain from this loss is unspeakable. Comstock and Duffey (2003) note that "pain is a condition highly avoided in our culture . . . people undergoing painful circumstances feel isolated and lost . . . many of us, during these times, feel like fish out of water . . . oftentimes we are" (p. 118).

Jeffrey was a 12-year-old, bright-eyed young man of Asian-German descent. An avid sportsman, Jeffrey enjoyed football, baseball, and soccer. He rode his bike, played on the computer, and was an excellent student. He was the oldest of the three children, the only boy. Complaining of headaches and undergoing a series of tests, Jeffrey's doctors confirmed that he had a brain tumor. His father was in the military and the family relocated to provide him with the best possible care. It was at that time that I met him.

I spent many days with Jeffrey. He was funny, reserved, brilliant, and sweet. He enjoyed playing piano, doing homework, and having a purpose, a job to complete. For fun, we would take trips to his favorite pizza restaurants, talking, laughing, and cutting up. One day I drove him by the university I was attending. We bought candied apples at a nearby stand and sat on the lawn by the trees.

Jeffrey had undergone chemotherapy but was regaining his strength. His faith was unstoppable, genuinely believing that he would survive his illness. Unfortunately, Jeffrey's condition worsened. He went through another round of chemo. His hair fell out and he lost his strength. A shuffle replaced his usual bounce. Jeffrey was still determined that he would survive.

Jeffrey's parents shared his belief. They maintained a positive attitude throughout the entire process. They turned to their faith and prayed for a miracle. Jeffrey attended a church with his family that supported this hope. Still, his condition worsened. One day, not long before his death, he told me that he could not die because he had not done what God wanted him to do on earth. I remember sitting in the car next to him and saying that I had come to know God better for having known him. He smiled. I got a call days later that he was in the hospital . . . that he had asked for me. Although I went immediately, his condition had deteriorated and I was never to have another discussion with him again.

The grief experienced by Jeffrey's family was indeed overwhelming. Jeff's mother could not begin to accept what had happened. She wanted to take his dead body to the altar at church. How could this be true? God could not betray her. Jeff's father sat slumped in the chair of the hospital room, sitting by the empty bed. I stood beside him, helpless but present.

The loss that his family carried was haunting for me. I *knew* their son. He was incredible. To have such a child and to lose him seemed unspeakable. His family's adjustment was naturally difficult. So much of their effort, their hope, had gone into believing he would survive. Picking up the pieces was a formidable task. I had very little opportunity to work with Jeffrey's family because they moved away soon thereafter.

Jeffrey is a part of my experience, a part of me. So are his parents and sisters. During our time together, they shared such courage, vulnerability, authenticity, and humanness. They faced not only an adjustment to life without him but also faced reconciliation with a spiritual belief system and a faith they no longer understood.

I can only hope for good things for Jeffrey's family. Martinson, Davies, and McClowry (1987) provide hope for bereaved parents and siblings. They report that siblings of children who died from cancer seven to nine years prior to the study illustrate greater self-concept scores than children who had not suffered this loss. Data indicate that siblings attribute greater psychological growth to the experience. Transforming loss does not suggest resolving loss. It brings with it the potential to appreciate love when you find it; to enjoy each day that we have.

CONCLUSION

So many times we live life without a full awareness of how fortunate we are to be loved; of how fortunate we are when we connect with, invest in, and simply love others. Without a doubt, many of us are blind to the reality that experiencing heartfelt connection, deepened relationships, and relational comfort are gifts to be appreciated and enjoyed, rather than privileges to be assumed or taken for granted. With that said, because experiencing loss is an inescapable part of life, we run a myriad of risks by which our connections can be lost, leaving our lives and *selves*, immeasurably altered. According to Greenspan (2003):

> Grief is a psychospiritual process. . . . People do not get "back to normal" after a child dies, or after any profound loss. . . . We don't choose grief, it chooses us. But we do have a choice in how we deal with it. We have the choice to let it be, not to rush it, to honor it in the way that we are called to. (pp. 95–96)

In fact, no one is exempt from loss or from the feelings of grief that accompany it. Our country has suffered losses in war, by terrorism, and through natural disasters. As a collective people, we grieve the illusions of security, safety, and freedom that we have grown to expect. Western culture's notion of autonomy, independence, and freedom is a myth. We are vulnerable to innumerable crises on any given day. To embrace our vulnerabilities and to find strength in them, without succumbing to immobility is a challenge to consider.

Too many people suffer losses of a lifetime. To minimize their experience by explaining the circumstances in a manner other than the sober and profound would be a disservice. We have a wake-up call, of sorts, every time we experience a minor loss, every time we learn of another person's loss. Still, our culture cautions us to silence. Will we allow our grief to go underground? Will we respond to our wake-up call? If there is indeed meaning to find through these experiences, what could it be?

When I see my little niece's big green eyes look up at me with a trusting gaze, I feel hope. When I imagine how hard it must be for my brother to live without his beautiful, loving wife, I feel grief. When I imagine how much better this world is because she had been in it, I am grateful. When I look at my own family and the love we share, I am humbled. Fortunate I am, indeed.

Loss is too real to trivialize with hollow words. Authentic meaning for many of us comes in the awareness of the gifts we have through the relationships we form. Appreciating them, nurturing their growth, and loving freely—these are the challenges of loss. Knowing that life can turn on a dime, how do we want to spend our next day? With whom do we want to spend it?

My friend, Lesley, wrote these words just days before she unexpectedly died:

At the intersection of our journeys,
we have the opportunity, the divine privilege
to allow Love to become manifest in the world,
and to be a part of the True Light of Creation.
I choose, as each of us may choose, to live
with an awareness of this remembering, and
a desire and intention to be a conduit for that love (Jones, 2002)

We must each find meaning and purpose not only for our experiences in loss but also for our actions in life. Living with loss is a reality. Finding our way through it is the challenge. In finding faith in loss, we experience grace. And in allowing ourselves to love, we ultimately find our courage. Therein lies our strength.

REFERENCES

Altschul, S. (1988). *Childhood bereavement and its aftermath*. Guilford, CT: International University Press.

Comstock, D., & Duffey, T. (2003) Confronting adversity. In J. A. Kottler, & W. P.

Jones (Eds.), *Doing better: Improving clinical skills and professional competence* (pp. 67–83). Philadelphia: Brunner/Rutledge.

Davis, C. G., Nolen-Hoeksema, S., & Larson, J. (1998). Making sense of loss and benefiting from the experience. *Journal of Personality and Social Psychology, 75,* 561–574.

De Minco, S. (1995). Young adult reactions to death in literature and life. *Adolescence, 30,* 179–185.

De Spalder, A., & Strickland, A. (1987). *The last dance: Encountering death and dying.* Palo Alto, CA: Mayfield.

Farrugia, D. (1996). The experience of the family when a child dies. *The Family Journal: Counseling and Therapy for Couples and Families, 4*(1), 30–36.

Greenspan, M. (2003). *Healing through the dark emotions: The wisdom of grief, fear, and despair.* Boston & London: Shambala.

Hogan, N., Morse, J., & Tason, M. (1996). Toward an experiential theory of bereavement. *Omega, 33*(1), 43–65.

Humphrey, G., & Zimpfer, D. (1996). *Counseling for grief and bereavement* (Counseling in Practice Series). London: Sage.

Janoff-Bulman, R., & Berg, J. (1998). Disillusionment and the creation of values. In J. H. Harvey (Ed.), *Perspectives on loss* (pp. 35–47). NY: Brunner/Mazel.

Jones, L. (2002). *Untitled.* Unpublished manuscript.

Jordan, J. (2001). A relational-cultural model: Healing through mutual empathy. *Bulletin of the Menninger Clinic, 65*(1), 92–103.

Kagan, H. (1998). Gili's book: A journey into bereavement for parents and counselors. New York: Teachers College Press.

Kamm, S., & Vandenberg, B. (2001). Grief communication, grief reactions and marital satisfaction in bereaved parents. *Death Studies, 25,* 569–582.

Kubler-Ross, E. (1969). *On death and dying.* New York: Springer.

Leick, N., & Davidson-Nielsen, M. (1991). *Healing pain: Attachment, loss and grief therapy.* London: Routledge.

Levinson, D. S., & Prigerson, H. (2000). Traumatic grief and the spousal loss model. *Illness, Crisis, and Loss, 8*(1), 32–46.

Martinson, I., Davies, B., & McClowry, S. (1987). The long-term effects of sibling death on self-concept. *Journal of Pediatric Nursing, 2,* 227–235.

Marwit, S., & Klass, D. (1995). Grief and the role of the inner representation of the deceased. *Omega, 30*(4), 283–298.

Miller, J. (1976). *Toward a new psychology of women.* Boston: Beacon Press.

Miller, S., & Ober, D. (1999). *Finding hope when a child dies: What other cultures can teach us.* New York: Simon & Schuster.

Miller, J., & Stiver, I. (1993). A relational approach to understanding women's lives and problems. *Psychiatric Annals, 23*(8), 424–431.

Mishne, J. (1992). The grieving child: Manifest and hidden losses in childhood and adolescence. *Child and Adolescent Social Work Journal, 9*(6), 471–490.

Morin, S. & Welsh, L. (1996). Adolescent perceptions of experiences of death and grieving. *Adolescence, 31*(123), 585–595.

Murray, J. A. (2002). Communicating with the community about grieving: A description and review of the foundations of a broken leg analogy of grieving. *Journal of Loss and Trauma, 7,* 47–69.

Neimeyer, R., Prigerson, H., & Davies, B. (2002) Mourning and Meaning. *American Behavioral Scientist, 46*(22), 235–251.

Parkes, C., Launagni, P., & Young, B. (1997). *Death and Bereavement Across Cultures.* London: Routledge.

Schwab, R. (1998). A child's death and divorce: Dispelling the myth. *Death Studies, 22,* 446–468.

Silverman, P., Nickman, S., & Worden, W. (1992). Detachment revisited: The child's reconstruction of a dead parent. *American Journal of Orthopsychiatry, 62* (4), 426–494.

Stiver, I. (1990). Dysfunctional families and wounded relationships. *Works in Progress, No. 41,* Wellesley, MA: Stone Center Working Paper Series.

Thompson, S. (1996). Living with loss: A bereavement support group. *Groupwork. 9*(1), 5–14.

Viorst, J. (1986). *Necessary losses: The loves, illusions, dependencies, and impossible expectations that all of us have to give up in order to grow.* New York: Simon and Schuster.

Worden, W. J. (1991). *Grief counseling and grief therapy.* New York: Springer.

Zambelli, G., & De Rosa, A. (1992). Bereavement support groups for school aged children: Theory, intervention, and case example. *American Journal of Orthopsychiatry, 62*(4), 484–493.

13

Familial and Relational Transitions Across the Life Span

By JoLynne Reynolds, Ph.D.

REFLECTION QUESTIONS

1. *When you think of "family," describe the pictures or images that come to your mind.*
2. *What would you say is the "ideal" family for a child to grow up in and develop into adulthood?*
3. *In your opinion does the marital status, sexual orientation, gender, ethnicity, or income level make some people more suited for parenthood than others?*
4. *What qualities contribute to the strength and health of families?*
5. *How can communities better support the resilience and health of families?*

INTRODUCTION

The purpose of this chapter is to examine the meaning of "family" in postmodern Western culture, the various ways it is formed, and how it survives. I will review common developmental tasks that all types of families experience. Special attention will be given to various types of nontraditional family structures and family experiences that place families outside of the mainstream of support in their communities. The sharing of their lives and experience will help to break down some of the stigmas and prejudices that all of us, even helping professionals, unknowingly use to define what is good or bad, healthy or unhealthy, and normal or maladaptive, in today's families.

MY STORY

In order to give you a context to think about traditional and nontraditional forms of families, I begin this chapter by sharing a little about my own family, both from my childhood, and now as a parent. My hope is that my story will help you examine your own experiences of family and become more open to the experiences of others around you.

"Family" evokes a mixture of emotions for many people. For me, the author, I feel a surge of love, and paradoxically, a sense of longing. In my mind's eye I see images from my childhood: my parents, sister, and me laughing together, sharing meals, holidays, and vacations. I also think of my current family, which includes my daughter and me, our collection of animals, and our network of extended family and friends. I feel tremendous love and gratitude for having them all in my life.

I am a 46-year-old Caucasian mother acting as head of a single-parent non-traditional family. As a single parent, I struggle daily to balance the roles of motherhood, college professor, and psychotherapist, and attempt to find time to enjoy the friendship and support of others, some days with more success than others. I adopted my daughter, now age 8, seven years ago. The circumstances of her birth and first year are a mixture of miracles and losses. She was born to a young single woman in northern Russia. Her first year was spent in an orphanage in the care of the doctors and nurses who staffed the regional "baby house." Although most of her physical needs were met during her first year, she lacked having the advantage of many experiences that are so important the first year of life.

My daughter has thus had some significant losses. She lost the experience of being raised by a biological parent and extended family in her native homeland. She lost the opportunity to be lavishly loved her first year. However she is also very fortunate in many ways. She miraculously survived her biological mother's unwanted pregnancy, her difficult birth, and the vulnerable first year of life in an impoverished orphanage. She gained an adoptive mother who loves her beyond comprehension, supportive and loving extended family members, and a home and wonderful community, all of whom are deeply vested in her success and happiness. My daily goal as her parent is to support in her the development of a positive, competent sense of self and her abilities.

At times, I have an anxious sense that I cannot give enough time to my daughter to assist her in the many challenges of growing up. I also worry that I fall short of meeting the demands of my career. A recent example that stirred my feelings of inadequacy was when as we were settling down to go to sleep, she quietly said to me: "Mom, I think we have only 'half' a family, because I do not have a father." Wow, I felt my heart sink and then break for her. I immediately gained my composure and attempted to acknowledge her sadness and then reassure and support her sense that our family is "good" and "normal." However, it wrenched my heart to hear her sadness and longing. Later, I re-asked myself her painful questions. Are we really half a family? What would it take for us to be a "whole" family in her eyes? Will she be ever be able to feel that we are complete, having everything we need as we are right now, or will her longing for a father cloud her happiness all through childhood? These are not easy questions for me to answer as a single mother. I know that her journey has just begun. And whatever she concludes about our family and her life will be shaped by me, our family and friends, and all her experiences growing up in the cultural diversity of the United States. She shapes my life and I in turn shape hers. It is a beautiful loving dance we know as our family.

The family my daughter and I share now was not the internal template of "family" developed for my life at a young age. I grew up in what many would consider to be a traditional family. My father was career navy and my mother worked in the home. Although we moved frequently, most outsiders thought I grew up in an ideal situation. As was true for many children, the outside appearance was more attractive than my actual experience in my home. Many of my memories are not pleasant. Our home was filled with my parents' intense anger, frightening family arguments, and sometimes, unjust punishments I suffered, followed by deep feelings of aloneness. I had several roles in the family. At the worst of times I became the scapegoat of my father's rage. Other times I felt a sense of responsibility as the caretaker of my mother's feelings and emotional needs. My favorite survival strategy was to become withdrawn and "invisible." This coping skill was invaluable at the time, helping me to stay out of the path of my parents' unpredictable rages that ran through our home. I graduated a year early from high school and immediately went to college to escape my home life.

My parents, for religious and economic reasons, have continued to maintain their marriage now for almost 50 years. By all appearances it is a "successful marriage." Some would look at of our family structure and label our family as "very resilient" and even "healthy"; based on the longevity of my parents' marriage, the long-term marriage of my sister to her husband, and my professional and personal accomplishments. However, the emotional costs of holding together our "traditional" family have left my sister and me doubtful that the choice for them to stay together while we were growing up was in any sense of the word "good" for us.

Throughout the journey of my life, I have developed a core of resiliency and an optimism I hope to pass on to my daughter. Most of the time, it seems to work well for me, and I feel that life is wonderful and full. However, in spite of all I have, I can still measure all my past and present experiences of family with an ideal family fantasy etched in my mind. My fantasy has always been to be a part of a family made up of a happily married couple with several children, approaching each day with a great love and sense of adventure. Even as I write it now it sounds almost comical to me, but yet so difficult to let go of internally. It most probably has elements of what I felt was missing from my childhood. This image is the "ideal" family I carry deep within me. For me, and for all of us, the fantasies and experiences of what we define as "family" mix together and give us a unique sense of who we are and what we hope to have in life.

THE PURPOSE AND MEANING OF FAMILY

Why are family structures such an important part of our way of living? If we look at families throughout history, we understand that families have been essential to long-term human survival. Families share a collective history and identity, as well as a strong sense of loyalty to one another. A primary purpose of family is maintaining and maximizing the functioning of all family members, regardless of their ages or abilities to contribute equally to the group. In order to accomplish this purpose, family members share various roles in the home. Children can carry out some household responsibilities according to their abilities, but contribute to the group enormously by adding elements of fun, play, and unconditional affection. Adults are responsible for the support and care of the children, aging parents, and other adults or partners in the home. When family members do well in these roles, each person feels safe, connected, and loved. Ultimately, families enhance each member's social, psychological, and biological development (Sprenkle & Piercy, 1992; Epstein, Bishop, Ryan, Miller, & Keitner, 1993).

Individuals and their lives are better understood by viewing them within a relational and cultural context of family and community. An individual's life cycle is not an isolated process, but unfolds in the context of relationships (Jordan et al., 1991; Jordan, 1997, 1999; Carter & McGoldrick, 1999).

Western American culture has not historically emphasized or valued the relational way in which people grow and develop. Instead, our culture has emphasized the development of the individual or "self" apart from a Relational-Cultural context (Jordan, 1997). A Relational-Cultural understanding of people is a perspective that views individual growth tied to relational connection and emotional joining to others. Primary to the model is the multidimensional interaction of the individual's development with the family, culture, and the environment. This model is dynamic, and invites us to explore how multiple levels of interaction impact individuals, family, and society. In other words, family and culture shape the individual, and the individual, in turn, shapes the family and the society.

The "ideal family" portrayed in Western-American media, literature, and political rhetoric, has been described as a union between two same-race, heterosexual, legally married adults and their healthy biological children. This structure has been so accepted as the prototype of "family" that it is referred to as the "traditional" family model. The traditional family model fuels a powerful political agenda for groups who wish to advocate for their own "biased" views of status, privilege, and human rights legislation. For many, it has become a weapon to justify prejudice, oppression, and exclusion from society.

Nontraditional or alternative family groups come in a variety of arrangements including single-parent families, blended or stepparent families, gay and lesbian headed families, biracial/multiethnic families, as well as many other configurations (Carter & McGoldrick, 1999; Minow, 1998). For those who live in family groups considered to be nontraditional, the traditional family prototype is a source of feeling different, not good enough, and shame. Their experiences are often marginalized by society as a whole. Weingarten, Surrey, Coll, and Watkins (1998) define *marginalization* as the social phenomenon of being diminished and devalued in comparison to others. Marginalization also encompasses the experience of having one's ideas, feelings, practices, or actions rendered less valid or useful in relation to the dominant ideal. When families are marginalized, their lives and experiences are represented in ways that do not accurately reflect their views. As a result, they lose their voice and are less visible in society. Ultimately, to be marginalized is to feel that you are unworthy of connection (Hartling, Rosen, Walker, Jordan, 2000).

Families experience marginalization in many forms. Some experience the exclusion and shame of marginalization when their children are excluded from parties and group play. Some experience it when they or their family members are frequent subjects of stares and whispers. Sometimes the family group is treated as if they were invisible or nonexistent in a group setting. Other times these families experience hypervisibility, with strangers frequently commenting or complimenting their children who appear *"different"* from other family members (Lambe, 1999; Wardle, 1999).

If we are to end the marginalization and shame of family groups, we must break down myths, and reconstruct the truth about what is "normal." An accurate view of today's family compositions in the United States is a necessary first step in this process.

The U.S. Census Bureau makes two distinctions between groups of people living together as a family. The first group, labeled "family households," are people living together having members of the household related to the householder by biology, adoption, or marriage. A "family group" is much broader, and includes all family units regardless of whether or not the householder is related to one or more of the members.

Perhaps surprisingly, the typical family group is not a "traditional" family. According to the recent U.S. Census Bureau report summarized by Fields and Casper (2001), only half of today's family groups have children living in the home. Additionally, many of the children living in families live with two parents (69 percent); however, only approximately half of those live with both biological parents under the same roof. The other half of two-parent families include adoptive, blended, stepparent, and grandparent headed families. Single-parent families make up the remaining 31 percent of families with children; 26 percent headed by single mothers, and only 4 percent headed by single fathers.

It is clear that the "normal" or "average" family structure is far different from what we have been led to believe. With such differences in the real living arrangements of children, how is it that we have held on so long to this unrealistic image of an "ideal" family? What are the reasons that families who do not meet this ideal are stigmatized, shamed, and pathologized by social institutions and the media?

Carter and McGoldrick (1999) assert that the stigmas attached to nontraditional family structures result from differences in power and status between groups in Western culture. Family differences of race, class, gender, and sexual orientation are not simply *differences*, but are categories, arranged hierarchically according to power, validation, and maximum opportunity. Another explanation provided by Scanzoni & Marsiglio (1993) helps us understand that the roots of the dichotomy between acceptable versus nonacceptable forms of family groups, can be found in how society relies on structural-functionalism. In functionalist thought, certain structures are legitimized and become institutionalized, and variations of these structures are seen as deviant. Inherent in this belief is that the established pattern of how "the good family" is better for society because it promotes more stability and social order than alternative forms of family.

Classifying families as "not good enough" on the basis of structure creates serious problems for everyone. Primarily, using a structural definition of the "goodness" of a family perpetuates false definitions of family health and dysfunction. This kind of classification system heaps enormous criticism, rejection, and shame on families who do not fit the idealized structure. It is important that we remember that we participate in oppression when we are part of communities that overtly or covertly use race, class, gender, or sexual orientation or any other form of classism to categorize and label families as a basis for inclusion and exclusion in activities and resources.

McGoldrick and Carter (1999) advocate that it is time for us to give up our traditional concept of family and expand our very definition of the term. Instead of holding on to narrow idealized images, we should begin to redefine family from how children view strong attachments in their lives. From a child's point of view he or she forms strong attachments without concern as to marital status, biological or nonbiological connection, or sexual orientation of caretakers. The concept of family to a child is the people he or she is attached to and who cares for him/her. Viewing family from this wise perspective opens our minds to *all* forms of various groups that are committed to caring for one another without bias or preconceived notions. Changing our view means that family groups could be "good" and "healthy" whether they are headed by single parents, gay and lesbian parents, whether they are created by adoption or the blending of families together to make stepfamilies, or whether they have members of multiple races or ethnicities.

DEVELOPMENTAL FAMILY TASKS: A CRITICAL ANALYSIS

When considering the many factors that impact the development of a person, it is important to consider how each person's individual life cycle intersects with the family life cycle. McGoldrick and Carter (1999) explain the systemic and relational nature of this process by stating that the family life cycle shapes the individual's development, which in turn impacts family functioning. How a family defines roles, assigns responsibilities, and makes decisions can impact how they weather normal developmental changes of the family as well as how they cope with more hazardous or stressful events.

All families, regardless of whether they are traditional or nontraditional in structure, deal with a variety of problems or tasks during the course of living. These tasks can be grouped into several categories including basic tasks, developmental tasks, and hazardous tasks (Epstein et al., 1993). Both basic tasks and developmental tasks are viewed as normative family demands or expectable family life cycle changes that occur in all types of families. However, hazardous family tasks occur when families face unexpected or traumatic events. These tasks are also referred to as nonnormative demands, and place the family in highly stressful coping modalities.

Basic Tasks of Families

Basic family tasks by definition include the survival functions of its members: providing food, shelter, money, transportation, and the like. Central to a family's survival is their ability to accomplish basic or survival tasks. Providing family members with food, shelter, money, and transportation are important concerns for families at all socioeconomic levels.

For some families, basic tasks are experienced differently from others. For example, in families with more resources, food becomes an arena for self-expression and connection to family members. It is common for middle or upper middle class families to have special meals together, holiday dinners, picnics, religious celebrations, and family reunions. In poorer families, obtaining food becomes an achievement rather than a celebration that can determine their own and their children's survival (DeVault, 1998).

The experience of poverty, and its effect on families, is important to examine. We know that families who live below the poverty threshold have more environmental stress, negative events, and daily frustrations than other families. The combination of multiple stressors common in poverty pile up, resulting in family members feeling a low sense of personal control and overwhelming helplessness (DeVault). The cumulative effects of inadequate housing, environmental hazards, and illness create a constant state of stress for the poor and place them at high risk for marital conflict and/or child abuse (Patterson, 2002).

Family instability is one common outcome of poverty. Families living with poverty sometimes have frequent shifts in household membership, due to job loss, illness, death, imprisonment, mental illness, or alcohol and drug addiction, all of which can have debilitating consequences for family members (Hines, 1999; Galley & Flanagan 2000; Patterson, 2002). In poor families without many resources, ordinary problems or basic tasks, such as the transportation of a sick child to the doctor, easily become a crisis because the family lacks resources to handle them (Hines, 1999). Living in a community that is insensitive to the needs of those in poverty, families without resources have nowhere to turn for support and their ability to be resilient is undermined. Communities that share collective norms and values of helping disadvantaged families, as well as policies and institutions that serve as resources to them, enhance the ability of families facing poverty to fulfill their functions, learn new capabilities when challenged, and be more resilient over time (Patterson, 2002).

Developmental Tasks Common to All Families

Developmental family tasks arise over time and are usually in the form of stages with transitions. Understanding the predictable phases of family development helps us see in context the changes of individual roles and family challenges. Developmental tasks of families follow pivotal events, often beginning with courtship and marriage, the transition to parenthood, raising young children, raising adolescents, midlife transitions and later life transitions (McGoldrick, 1999; Patterson, 2002). The stages of family development are circular and repetitive in the life cycle. Each of these developmental tasks involves changes that are stressful for family members even if the changes are perceived as positive. Each stage and developmental task requires that various family members change their relational patterns and learn new ones to successfully transition to the next stage with health and flexibility.

Celebrations, traditions, and rituals become important ways a family honors transitions for family members. The cultural background of the family influences the types of traditions, rituals, and ceremonies that prepare them to move into the next developmental stage (McGoldrick, Heiman, & Carter, 1993). Naming ceremonies for children, coming of age celebrations, marriage ceremonies, and death memorials and celebrations are all examples of rituals emphasized by different cultures that mark important developmental transitions for families.

As couples begin partnerships to create a family together, issues surrounding work outside the home and the division of household labor become important for the establishment of mutuality and support in the family. Women often experience workplace and family as two equally stressful competing systems. Even female top executives work in disadvantageous environments lacking the "backstage support" of help with household labor as compared to male top executives who have wives at home or wives who work but take on more of the home responsibilities at the same time. Developing a sense of equity in dual career couples is an important part of strengthening the family. A sense of equity is an individual perception of fairness and limitations in the distribution of resources, such as money, time spent on education or career development, individual, couple, and or family leisure/quality time, and sharing in the overt power of decision making and division of household tasks (Farber, 1996).

Other transitions, though important, are not necessarily marked with ceremony or celebration. An early transition most individuals and couples face during adulthood concerns the decision to become parents. Some couples or individuals by choice or by circumstances do not become parents. However, for those who do become parents, the transition to parenthood is one developmental task that can cause great stress in family functioning. The timing of parenthood can influence the happiness of couples. In one study the effects of parenthood were examined during the early years of marriage (Orbuch, Veroff, & Hunter, 1999). The results of their study revealed cultural differences in the experience of becoming parents. When comparing the stress of White and Black couples, they found that parenthood led to a decrease of marital happiness for White couples but not for Black couples. In the stories couples told regarding the experience of becoming a parent, African American couples were much less likely to think about if and when to have children than White couples. In contrast, when White couples shared their stories, they were much more likely to emphasize whether the birth was planned, and if they wanted to have children when they became pregnant.

As couples build their families, the choices a woman makes in her roles of childbearing, parenting, and career development can be scrutinized and judged more by society than the same choices made by men. According to Weingarten et al. (1998), the more a woman with children deviates from the prototypical image of a "good mother," the more society marginalizes her and her mothering practices. Marginalization and shaming of mothers is often done without regard to how competent her children are or how adaptive the family system functions. In Western-American mainstream culture, the prototypical image of a "good mother" is likely to be White, married, and not working in a job that takes her away "too much" from her parenting responsibilities. Additionally she has only one or two healthy children with no physical or behavioral problems; she conceived her children and is raising them in a heterosexual relationship; she and her spouse are older than 20 years of age and are of the same ethnic and racial background. The very concept that there is such a woman as a "good mother" marginalizes all women because *no* mother can *always* be good by her own or others' standards.

Building a successful long-term partnership is a dream of every couple as they begin their journey together. Many speculate but few people know what contributes to the success of long-term relationships. One study, conducted by Carrere and Gottman (1999), examined longitudinal research of the factors that lead to marital success. Of the couples they followed, they found that the ones who build a system of friendship, with fondness and admiration for each other, and who infuse their relationship with positive affect, especially during times of conflict, are the most successful at making marriage last. Surprisingly, anger is not as corrosive an emotion to long-term relationships as previously thought. However, the use of criticism, contempt, defensiveness, and stonewalling when angry or when in conflict is toxic to the health of the relationship. Unhealthy patterns during conflict, not necessarily the conflict itself, increases the distance and isolation in a relationship that precipitate divorce.

Hazardous Tasks for All Families

Hazardous tasks happen to both traditional and nontraditional families. As families work through hazardous tasks they face significant risks of decline in their ability to carry out healthy functioning (Patterson, 2002). Hazardous family tasks are events that challenge all types of families to cope with unplanned events. These crises disrupt the normal flow of family functioning and shift how members may or may not get their needs met by others. Examples of hazardous tasks include the problems that result from injury or illness, loss of income, job change or job loss, infertility, death, divorce, natural disasters, and other family crises (Carter & McGoldrick, 1999; Epstein et al., 1993).

One example of a developmental task that can become hazardous for families is the process of attempting to become parents. Childbearing is a common developmental role transition for both men and women. However for couples who experience infertility, the transition to parenthood becomes a crisis that threatens the hopes and dreams for themselves as individuals, as well as the future of their life together as a family. The medical community has defined infertility as an inability to conceive after a minimum of one year of unprotected sexual intercourse.

A broader definition includes not being able to carry a pregnancy to term and have a baby (U.S. Dept of Health and Human Services, 2003). The Centers for Disease Control (2003) reports 2.1 million married couples identify themselves as infertile. This statistic may not represent the large number of people who experience infertility in the U.S. In contrast to married couples as a separate group, the CDC reported that 9.3 million women used infertility services during 1995 and the number of women ages 15–44 with an impaired ability to have children was reported to be 6.1 million.

Infertility is a hazardous task that can lead to many physical, financial, psychological, and social stressors in a family (Gibson & Myers 2000). Daniluk (2001) found that acknowledging the finality of infertility can become a critical juncture in a couple's relationship. Although the experience of infertility may strengthen some relationships, many couples begin to encounter severe marital problems, depression and anxiety, and damage to their sexual functioning. If their marital problems are not resolved successfully and the couple cannot reconstruct a mutually satisfying vision of their future together, divorce may follow (Chen, 2002; Daniluk, 2001).

Years of infertility paired with the potential of multiple losses of children due to miscarriages place enormous stress on a couple and family system. Communities often respond to long-term problems of infertile couples with isolation, silence, and disregard for the emotional toll suffered by them (Daniluk, 2001).

Infertility can affect the psychological health of both men and women. Gonzalez (2000) found that women who struggled with infertility described a sense of alienation, powerlessness, negative stigma from society, and feelings of personal failure. The women in this study also suffered from depression, anxiety, anger, and obsessive thinking at a time when they are required to make decisions about invasive and costly fertility procedures. In another study by Webb and Daniluk (1999) men who were diagnosed with male factor infertility immediately went into shock and disbelief followed by a period of denial. During denial many men continue to insist that their wives continue to undergo tests and procedures that are unlikely to solve the problem. Men who began to accept their diagnosis reported intense feelings of grief and loss, powerlessness and lack of control, inadequacy, social isolation, and betrayal in response to being unable to father a child. Individual, group, and couples counseling for infertile couples can ease the sense of isolation and increase the ability to cope successfully.

Although I could devote much more of this chapter to any number of hazardous events in the life of families, I also wish to pay special attention to how families choosing to separate or divorce face significantly higher risks for the decline of functioning of all family members. The decision a couple makes to separate or divorce often comes when they feel there are no other options and resources for their happiness. This decision can be particularly difficult, resulting in long-term negative effects, for families with children.

Divorce is often viewed as a failure of the family to be resilient and survive life's challenges. Rather than a failure, divorce may be better viewed as one possible solution to problems that threaten the survival and health of family members. Separation and divorce viewed this way become better understood as hazardous tasks. If handled well by the adults, it is possible that a separation or divorce could ultimately enhance the long-term health of family members. When handled poorly, divorce can greatly undermine the functioning and future functioning of all family members particularly children.

A decision to stay together as a couple is at times more harmful to children than a parental separation. For example parents may have a highly conflictual marriage but choose not to divorce because of their concerns about the effects of divorce on the adjustment of the children. However, research indicates that children who live with high levels of parental conflict ultimately have more behavior problems than those children whose parents divorce (Emery, 1982; Morrison & Coiro, 1999; Hetherington, 1999).

Even so, separation and divorce are difficult crises and extremely hazardous tasks for families. Going through divorce "well" is a difficult process to negotiate between partners who have vested so much of their lives and future plans together. When children are not protected from the intense hurt, anger, and bitterness of the adults, or are even used as weapons by one or both parents against the other, the result can be long-term emotional damage.

In Wallerstein, Lewis, and Blakeslee's (2000) longitudinal study of children of divorce, they found the long-term negative effects of parental divorce to be cumulative, leading to great difficulties when attempting to establish intimate relationships as adults. Throughout each developmental stage of childhood, the impact of parental divorce was experienced in new and different ways. In many of these children's lives, changes in parenting and in the structure of the family place greater responsibilities on them to care for themselves. Home became a lonely place for many children as parents became busy with work and preoccupied with building their social lives. Both mothers and fathers had less time to spend with their children and were less responsive to their needs and wishes. Many children continually suffered through the instability of the parents' new lovers, live-in partners, and stepparents.

Buchanon (2000) states that there are several factors that moderate how divorce will impact children and adolescents. Economic problems of families can be strong and persuasive elements that account for many of the long-term problems of children who experience divorce. Additional life stresses, custody arrangements, and the parents' remarriages or cohabitation with new partners also affect children profoundly. Buchanon emphasizes it is important to remember that there are also characteristics of the child that make some more vulnerable to poor adjustment after divorce. Age, psychological history, gender, and peer network have different potential effects on children as they move through the stress of this transition. The most consistent predictor of behavior problems in the research on children of divorce is the degree of marital and postmarital conflict they witness (Amato, 1993). In Wallerstein et al.'s (2000) research children of divorce who were able to draw support from school, sports teams, parents, stepparents, grandparents, teachers, or their own inner strengths and resources fared better than children who could not find these supports. Parents who rebuilt happy lives and included the children were able to help them experience better postdivorce adjustment than other children. They also found that children who are living with one committed single parent who was responsive to their needs adjusted better to divorce than the children who experience long-term postdivorce parental conflict and neglect.

RELATIONALLY RESILIENT FAMILIES

As all families work through basic, developmental, and hazardous tasks, they function in health on a continuum of functional/balanced to dysfunctional/unbalanced depending on how members handle changing family life cycle tasks. According to Carter and McGoldrick (1999) family member stress is often at its greatest at developmental transition points from one stage to another as families rebalance, redefine, and realign their relationships. Families are highly challenged when they experience a hazardous task or they attempt to cope with long-term chronic stress. These problems push families to either become more competent and resilient or to decline in competency (Patterson, 2002). If members perceive that the stress of handling the problem is becoming too great, it can hinder family members from resolving the emotional issues of the task and the family group will lose health and balance. If emotional issues and problems continue unresolved, they will be carried on as hindrances to the successful management of future life cycle tasks, and subsequently, the family's ability to meet the needs of its members greatly declines.

We have now seen how all families can experience stress as they go through the normal developmental tasks of living as well as during the unexpected or hazardous tasks they encounter. In order to make it through these transitions successfully, roles and responsibilities of members must be flexible to meet the needs of each person at different ages and stages of development. Often the process itself, or *how* a family uses coping skills to manage stressful tasks is more critical than the task itself.

The families that successfully face adversity and continue to survive as a group are called resilient families. Resilient families have relational strengths and resources they draw on that help them successfully navigate difficult times (McCubbin, 1998; Patterson, 2002; Walsh, 1998). They are able to find successful ways to continue interdependence on one another. Relational resilience is a way of functioning as a family, uniquely defined by the family itself within the context of their values and culture (Walsh, 1998). Central to relational resilience is the family's ability to attribute positive meanings to highly stressful situations (Antonovski, 1998; Patterson, 2002; Walsh, 1998). Patterson and Garwick (1994) indicate that families can reduce stress and increase their coherence when they are able to openly communicate meaning on three levels when faced with adversity: (a) about the specific stressful situations, (b) about their identity as a family, and (c) about their view of the world.

The remainder of this chapter focuses on a variety of family structures and family problems that are typically marginalized by communities. This summary is not an exhaustive list of the types of marginalized families and problems. These were selected as a way to open dialogue about what family means, and the value of family ties to each of us. The family groups selected suffer large amounts of stigma, shame, and isolation imposed on them by society. How these marginalized families cope with their difficulties are testimonies to their resilience. As we hear their experiences and voices, we can learn how resilience takes its form in each family group, uniquely shaped in the context of culture and environment. It is my hope that as you explore their experiences you will appreciate their unique strengths and challenges and become more aware of your own resiliency in life's transitions.

"NORMAL" FAMILIES— "REAL" CHALLENGES

Attached to the stigma of nontraditional family structures is the belief that these families produce children that have more emotional and social problems than children raised in a traditional family structure. It has been long assumed that children in traditional families have the psychological and economic advantages of parents who are heterosexual, married, and biologically linked to them. Through the years there has been some research that has indicated children in nontraditional families do have a higher frequency of emotional/behavioral problems than children raised with both biological parents (Lenhart & Chudzinski, 1994; Thomas & Farrell, 1996; Mott, Kowaleski-Jones, & Menaghan, 1997; Biblarz & Gottainer, 2000). However, this research may not give us a clear picture of *how* and *what* qualities exist in all types of families that enhance children's lives regardless of the family's structure.

When seeking to understand the research on nontraditional and traditional family structures, it is important to examine different types of nontraditional families and the special challenges and needs they encounter in their communities. Examining nontraditional families in this manner breaks down harmful stereotypes and generalizations that marginalize them and their experiences from others.

Families Living with Poverty and Homelessness

Because of the violence and suffering they encounter, poor inner-city and homeless families frequently experience chronic trauma and features of post-traumatic stress disorder (McGoldrick, 1993; Boyd-Franklin, 1993; Koch, Lewis, & Quinones, 1998). Children and family members living in housing projects or shelters often report intense fear walking through darkened halls, deserted buildings and "crack houses." Inner-city children may have to walk through a "needle park" filled with empty crack vials and discarded syringes to get to school.

Children growing up in poverty more often face threats of violence, overcrowded schools, and reduced chances for survival than children in higher socioeconomic levels (Garmezy, 1993). Many of these children witness or are victims of violence, child physical and sexual abuse, drug overdose, and AIDS. The behavior problems, depression, and anxiety of many inner-city children may increase after witnessing traumatic events. Some of the post-traumatic symptoms that can occur because of these events include nightmares, flashbacks of traumatic events, and intense fears of entering areas where traumatic events were experienced.

Parents living in long-term poverty can see very few options for their children. They may feel a growing sense of being trapped, disempowered, and rage. The resulting problems these families face in turn heighten the community's collective stress (McGoldrick, 1993). When economic circumstances become too overwhelming for a family, homelessness may be their only choice. Koch, Lewis, & Quinones (1998) point out that 40 percent of the homeless are mothers and children. Mothers may resort to homelessness after escaping from partners who abuse them or their children. Some become dependent on friends and families for shelter until their goodwill is exhausted. Society views homeless mothers with suspicion, fear, and blame, when in fact a single mother often faces the painful economic choice of whether her children will have food or a roof over their heads. Coping with the effects of poverty and homelessness must become a critical issue that concerns all segments of society, not only the families who suffer its immediate consequences (McLoyd, 1995).

Families Raising Children With Special Needs

Bringing children into a family structure accompanies the dream of conceiving and raising physically and emotionally healthy offspring. However, families who parent children with special needs due to physical, mental, or behavioral challenges, face a different reality from what their fantasies of family may have been.

Parents who raise a child or children with special needs and/or disabilities face enormous challenges helping their child meet the personal, social, and educational goals of daily life. These parents must also somehow balance meeting their own needs and the needs of other children who may be in the family. As if this were not enough, these families often suffer social marginalization and isolation from a lack of support, understanding, and visibility in their communities. The marginalization of such families has been described as an experience of being "ghettoized" socially, fiscally, institutionally, and emotionally from the "mainstream" of society. The isolation is also compounded because many times the parents often prefer to keep their pain quiet, not wanting to fuel the view that children with disabilities are a "burden" to their parents and society (Greenspan, 1998). However, breaking down the isolation and developing a broad network of helpful support can make the difference between families being able to cope or not (Gottlieb, 1998).

Although disabilities of all kinds are a normal part of life, they are designated as abnormal by our culture and society. Our standards of "normalcy" affect the way we view a person's worth to society, therefore devaluing physically challenged and disabled people. Society continues to perpetuate the marginalization of children with special needs and their families by holding on to strongly entrenched cultural myths (Greenspan, 1998). These myths are part of Western-American culture. The first myth advocates that a person's merit in the community is linked to work, achievement, and a narrow definition of intelligence. To base a person's merit on achievements and intelligence alone creates an elitist society that defines the value of life on narrow dimensions. It also greatly devalues the abilities, strengths, courage, and accomplishments of children and adults who live with disabilities.

A second myth that contributes to the marginalization of individuals with disabilities is how Western culture elevates athletes to heroic status. Elevating fitness and athletic excellence over all other forms of abilities discounts the accomplishments of many people who are challenged daily with difficulties in their physical and mental functioning. The third myth that contributes to marginalization is valuing individual autonomy over interdependence. When society denies the normalcy and fundamental interdependence of human beings, a sense of shame and humiliation is formed around relational interdependence and needing "help." However, the truth of the human condition is that interdependence and need for help is part of everyone's experience and even more pronounced in the lives of children with special needs and disabilities. Denying the gift of relational interdependence isolates us one from another and contributes to our exclusion of people who appear or are "different." Ultimately, all three of these myths ignore and deny the special gifts, joys, and talents children with special needs bring into the world.

Miriam Greenspan, a writer and mother of a daughter living with a disability, voices her concerns as a parent: "If there is a nightmare here, it is my fear that the world, when I leave it, will not be sufficient to hold her or to see her truly. The unanswered question is: Will her gifts find their place in a world not yet worthy of her, or of any of our children?" (Greenspan, 1998, p. 59). Her words move us to strengthen community support of families raising children with special needs. Communities must become more inclusive of individuals with disabilities, valuing the contributions special needs children and adults have to offer.

Single-Parent Families

Single-parent families are the largest group of nontraditional families. They have grown to 31 percent of all types of families from only 13 percent in 1970 (Fields & Casper, 2001). The rise in single-parent families has been attributed to the increasing divorce rate of two-parent families.

The U.S. Census Bureau reports that the majority of all single-parent households are led by a mother while father-headed single-parent families make up only 4 percent of the population (U.S. Census Bureau, 2000). The poverty rate for single mothers is three times higher than that of single fathers. Therefore, when we discuss the challenges of single-parent families, the majority of our discussion centers on the awareness that women lead the majority of these families.

A single-parent family faces different hardships from two-parent families. When compared to two-parent families, single-parent families have much fewer resources. These may include less financial resources to devote to the children's upbringing and education, as well as less time and energy to nurture and supervise children. Single-parent families may also have reduced access to the community resources that can supplement and support their efforts to raise children (McLanahan, 1999). Diminished contact with the noncustodial parent can result in a loss of emotional support and supervision from adults. Children in single-parent families generally have a lower economic standard of living than those in two-parent families. These circumstances have a cumulative effect on the way children grow up and how well they are prepared for young adulthood (U.S. Census Bureau, 2000).

Part of the stigma of single-parent families is the body of research that points to a higher incidence of adjustment problems among children of divorce compared to two-parent families (Amato & Keith, 1991; Simons et al., 1999; Thomas & Farrell, 1996; Mott, Kowaleski-Jones, & Menaghan, 1997; Simons et. al., 1999; Wallerstein, Lewis, & Blakesley, 2000). Causes for these adjustment problems have been attributed to many factors. Some causes can be attributed to the ineffective parenting practices of the noncustodial parent, a higher incidence of psychological problems among the divorced parents, father absence, and parental conflict before and after the divorce, and reduced family income (Simons et al., 1999). The U.S. Census Bureau reports that children in two-parent families often have better economic and social environments and may live in neighborhoods that contribute to child well-being. According to the Survey of Income and Program Participation (U.S. Census Bureau, 2000), children in two-parent families are read to more frequently and are more likely than other children to participate in sports, clubs, and lessons.

Since women head the majority of one-parent families, most of the research on single parents has focused on the challenges faced by single mothers and their children. One alarming fact is that single mothers with children are more likely to have family incomes below the poverty level (34 percent compared with 16 percent of two-parent families). In contrast, only 16 percent of single-parent families headed by men have incomes below the poverty level (Fields & Casper, 2001). Whether a single parent is divorced or never married also has implications for family resources. Children living with divorced single mothers typically have an economic advantage over children living with those that never married. Divorced parents are also on the whole more educated, older, and have higher incomes than parents who have never married. White single mothers often have less extended family support than Black and Hispanic single mothers who are more likely to live in a related subfamily and have more resources for child care.

How single mothers are depicted in Western culture, including images of who they are and what they need are intensely politicized issues (Schnitzer, 1998). Political issues arise over who pays to support the needs of children living in single-parent families as well as who's in charge of women's reproductive choices and role definitions.

Schnitzer (1998) states the "heightened concern" of the impact of single mothering influences the way they are viewed by society and even shapes the way the single mother views her competencies. Clinicians also may mistakenly be influenced to treat the problems of single mothers and their children from the premise of lacking a husband/father. This concept of "need for a father" can be used as a way to marginalize single mothers both by society and by mothers themselves. In her clinical work, Schnitzer defined common components of what boys in families headed by single mothers need in therapy. Common themes include a greater need for self-esteem, discipline, gender identity, supervision, and financial support. These clinical themes more clearly break down what "need for a father" may truly mean and further define how a mother's role is seen as deficient in these families. Meeting boys' needs may not be as simple as more involvement by a father. As pointed out by Schnitzer, when clinicians do get the fathers into the family session, they learn that simply the presence of the father doesn't necessarily fix everything. Absent fathers, like single mothers, come in varying forms having different strengths and problems. Many times the fathers' own economic and psychological problems render them incapable of providing immediate solutions to the needs of their sons. Rather than characterizing boys' problems as "father absence" or rendering the competencies of single mothers as lacking, the growth in single mother headed families calls us to transform family life, balancing the rights and responsibilities of both partners and provide a respected place for family support in our national priorities.

ADOPTION

Each of us does not have to look far to see people we know who are touched by the act of adoption. In my own experience, adopting my daughter has been very meaningful and spiritually powerful in my life. Our relationship is the most important one to ever come into my life. Because our bond began from experiences of her loss as well as mine, we share a deep joy and a strong attachment to one another. Every day I am touched by the love I feel for her. The intensity of my joy, gratitude, hope, dedication, and protectiveness for her is as strong or stronger than any mother-daughter bond. The special love that I and other adoptive parents share with our children adds meaning and purpose that enriches families.

Adoption can be a beautiful and meaningful way to bring children into a family. As our society becomes more aware of the vast numbers of discarded and unwanted children in the world, we can be grateful to the families who choose adoption to build their family. Adopting a child is choosing to make a deeply personal level of commitment and devotion to a child who began life under the most unfortunate circumstances (Anderson, Piantanida, & Anderson, 1993). Adoption illustrates how children form strong attachments to caretakers regardless of marital status, biological or nonbiological connection, or the sexual orientation of the adults (Minow, 1998).

The ideology of the American family includes the belief in the paramount importance of genes in matters of human development, identity, and bonding. This ideology has rendered all nongenetic family forms abnormal and pathogenic (Wegar, 1998). In fact most of the research on adoption and families has focused on explanations of problems in adoptive families rather than positive personality characteristics and family strengths. Anderson, Piantanida, and Anderson (1993) point out that adoptive families have many of the same experiences and processes as other types of families. They face the common transitions of parenthood, nurturing their children throughout development, and teaching them values and behaviors that will enable them to make the transition from family to a happy and productive adulthood.

There are unique differences in the process of adoption that are different from biological childbirth. These differences can make normal family transitions more complex particularly in how family members give meaning to events, create a family structure, and respond to change (Anderson, Piantanida, & Anderson, 1993). Adoptive families often differ on their configurations. Individuals may choose adoption because of a long history of infertility; miscarriage(s) or the death of a child; because of personal or humanitarian reasons, or having a deep sense of caring or religious duty; or because they may be helping professionals who believe their parenting expertise will facilitate the task of parenting a special needs child. People who choose to adopt may be childless couples, couples already having biological children, couples past childbearing age, or a single person who desires to be a parent (Anderson, Piantanida, and Anderson, 1993).

Children also come to adoption from a variety of life circumstances. The idealized adopted child is the healthy newborn who is adopted and spared a life of deprivation, insecurity, and suffering. However this scenario is only a small percentage of the types of adopted children. Newborn children are often adopted domestically through private or public adoption service providers. It has become increasingly common for the birth mother to be included in the adoption process from selection of the birth parents to establishing a long-term ongoing relationship with the adopted child's family in open adoption. Some children are more difficult to place with adoptive families. Biracial children or children having physical or emotional problems may place special demands on a family. International adoption is increasing in the United States because of difficulties associated with many domestic adoptions. Many countries having extensive poverty and economic problems have large numbers of adoptable children compared to children eligible for adoption in the U.S. Because of the relative ease in adopting international children, this has become a popular choice for many adoptive parents. International children have varying degrees of physical and emotional health. Many of these children have a range of institutionalization experiences differing in the quality of care from one setting to another. Because of poverty and a lack of resources their birth mother may or may not have received adequate care during the prenatal period. During infancy, international children may have been well taken care of or suffered various degrees of abuse or neglect.

Losses and grieving related to the adoption experience are something adopted children, parents, and birth mothers face at different developmental points. The task of mourning for a couple may include the pain, guilt, and anger associated with infertility. The couple needs to grieve the dream of their biological child that "should have been" as well as face questions from family and friends regarding the reason for choosing adoption. Adoptive parents face unique transitional issues surrounding their entitlement to parent, transitioning and forming a new sense of family, how to tell or inform the adopted child of the circumstances of their birth and adoption, as well as issues surrounding "telling" school professionals and community members of the child's adoption. At adolescence, adopted children may have questions about belonging and identity that adoptive parents will need to be able to support. Transracial or transcultural adopted children may raise special issues of the child's need for and right to a positive racial, ethnic, and cultural heritage (Anderson, Piantanida, & Anderson, 1993). Themes of loss and control may permeate how adoptive families interpret or handle the normal transitions of children and families. All the sensitivities that exist in adoptive families require the kinds of support systems that uphold and develop the strengths that these families uniquely have.

Professionals in the community may not give adoptive families the support they need because of the stigmas associated with adoption. Throughout the family and child's life cycle professionals often magnify and overemphasize grief and loss issues in adoption. Stereotypes and images of the "primal wound" in the child and the inadequate "bonding" of adoptive mothers to their children increase the parent's sense of helplessness and inadequacy (Smith, Surrey, & Watkins, 1998; Verrier, 1993). Rather than viewing adoption as recovery from loss, many helping professionals assume that the adoption process results in pathogenic and enduring grief for the adopted child and adoptive parents who are incapable of meeting the needs of their children. When adoption is viewed from a pathogenic stance, adoptive mothers and fathers face the stigma that they will never be able to impact the grief issues of an adopted child, leaving him or her with enormous feelings of helplessness and inadequacy. The use of a pathogenic focus in therapy can be a block to adoptive families who seek support for a variety of normal parenting and developmental family issues (Smith, Surrey, & Watkins, 1998). A more helpful stance is to view adoption as recovery from grief and loss, building on the strengths and resiliency of the family members.

Researchers have unwittingly contributed to a pathogenic view of adoption when exploring reasons behind the high incidence of adoptive families who seek mental health services. The current literature on adoptive families has failed to take into account the interaction between the community and adoptive families. Our knowledge of the impact of adoption is limited by the narrow focus on pathology in current adoption literature. Additionally much research has explained negative life experiences of adopted children by placing the blame on adoption itself, rather than on the pattern of communities to think of adopted children as "different" (Wegar, 1998). By neglecting or downplaying the impact of social stigmatization researchers have failed to account for all the variables that impact the mental health of adopted children as well as the coping behaviors of adoptive parents.

Adoptive parents live in a culture that stigmatizes infertility and links a woman's mental health to her biological motherhood. Childlessness in American culture often evokes images of couples who hate children, the anti-mother, and the non-nurturing female (Wegar, 1998). Smith, Surrey, & Watkins (1998) identify seven ideological myths in the dominant culture that have shaped views and experiences of adoptive families.

1. The primacy and superiority of sameness; thus the valuing of blood relations over all others and the valuing of racial and cultural sameness.
2. The developmental primacy of environmental nurturance over genetic endowment.
3. The purported importance of "bonding" to the mother–child relationship.

4. The psychological health of the child is dependent solely on the relationship with the "nuclear family" mother.
5. An identity that is simple is superior to one that is complex.
6. Adoptive mothers are defective as mothers, causing psychiatric symptoms in their adoptive children.
7. Adoption is a lifelong grieving process for all members of the adoption triangle.

These harmful ideologies must be deconstructed in order to eliminate sources of marginalization and resistance. Once we release harmful myths and ideologies we can embrace a more experiential and relational definition of what it means to parent a developing child growing up in a family.

GAY AND LESBIAN HEADED FAMILIES

Gay men and lesbians form families for the same reasons all couples do. They seek to validate their identity, establish mutuality, and to exercise choice about their intimate lives. They profoundly challenge the idealized images of how family is and should be structured (Weeks, Heaphy, & Donovan, 2001; Benkov, 1998). They make the choice to form a family in a hostile environment, living under a shadow of homophobia and frightening prejudices.

Gay and lesbian family stories all share a common message of desire to validate the meaningfulness of "connectedness" common to all people (Weeks, Heaphy, & Donovon, 2001). Their growing presence in the world invites us to examine what truly constitutes "family" and our assumptions about what constitutes the fundamental needs of children in order for them to grow into psychologically healthy adults (Benkov, 1998).

Recent legal history reflects the polarities of oppression and support gay and lesbian families face. It was not long ago that Anita Bryant campaigned in Florida to overturn a ban on discrimination of homosexual couples calling gay men and lesbian woman "human garbage" and calling for the courts to "save our children." She was successful in her attempts and overturned a law that would have allowed gay men and lesbian woman to adopt children (Benkov, 1998).

Heterosexual couples have long held the constitutional right for autonomy in establishing family relationships. However, it has only been recently that stronger support for gays and lesbians has been affirmed by their constitutional right to choose how they live their lives. In a recent U.S. Supreme Court ruling, *Lawrence et al. v. Texas,* the Supreme Court recognized the right of homosexuals to seek autonomy in order "to define their own concept of existence, of meaning, of the universe and of the mystery of human life" (*Lawrence v. Texas*, June 25, 2003). The Supreme Court ruling has laid the much-needed groundwork to end discrimination of gay and lesbian couples and to afford them the same freedoms as heterosexual couples to engage in marriage, adoption, and the right to form recognized family groups (American Civil Liberties Union, 2003).

Even as the U.S. begins to make long-awaited steps toward recognizing the rights of homosexual couples to have freedom and autonomy of choice when they seek to form open and legal family groups, they will continue to face extreme discrimination and prejudice from communities on all levels. At the heart of this discrimination is a belief that homosexual parents cannot competently raise children. In fact, the deepest fear expressed by many groups is that homosexual parents may harm children. People holding these beliefs passionately assert that homosexual parents cause children to be impaired in gender development, or are destined to develop homosexual partnerships themselves, or that they are at risk of child sexual molestation because of the sexual lifestyle of the parent (Cameron & Cameron, 1996).

The small amount of research on the psychological adjustment of children growing up in gay and lesbian headed families shows little or no evidence that these children differ in the level of their psychological health when compared with children in heterosexual parent headed families (Chan, Raboy, & Patterson, 1998; Patterson, 1995, Patterson & Redding, 1996). Chan, Raboy, and Patterson (1998) found when comparing children in lesbian headed families with children in heterosexual families the only factors predictive of better adjustment in both groups of children were the parents' reports of greater relationship satisfaction, higher levels of love and lower levels of parental conflict in both groups of parents.

There is common agreement that children growing up with homosexual parents have special needs and concerns that other children do not share, especially during adolescence. Research and interviews of older children, particularly adolescents, indicate that many of them express high anxiety about the reaction of others to their parents' sexual orientation, even though they personally felt tolerance for their parents' lifestyles. Cameron and Cameron's research (1996) indicates that they worry and feel panic that if their parents' sexual orientation were to be disclosed, the results would be disastrous for the entire family. They believed they and their families would face social ostracism, loss of jobs, and even their parents' loss of custody of them. In the longitudinal study by Tasker and Golombok (1997) adolescents with lesbian mothers also expressed anxiety about peers knowing their mothers' sexual orientation but less anxiety if the peer was a close friend. A key long-term outcome of their study was that these adolescents continued to have good mental health into adulthood and were not any more likely be gay or lesbian in their sexual orientation than adolescents living with heterosexual parents.

Benkov (1998) believes that lesbian and gay parents challenge society to examine our assumptions about what forms a healthy family. She states that instead of categorizing people and accepting unfounded prejudices, we should become interested in the quality of relational interactions as a measure of family health. If we as a society are able to move toward a view of family health based upon relational processes rather than structure, it is more likely we will better meet the needs of all children.

MULTICULTURAL/BIRACIAL FAMILIES

Families having parents and children with mixed cultural and ethnic backgrounds are more common than ever before. "Multiracial" or "multiethnic" families are frequently used terms to describe families having individuals of mixed racial, ethnic, or cultural ancestry. These families can also include those that are formed by interracial couples. Families with cross-racial or transracial adoption or foster care arrangements are called blended families (Kinney, 1999; Schwartz, 1999). Race is an invention of culture, a social convention where lines are drawn to separate people according to class and social standing (Reddy, 1994). Mixed ethnic family groups evoke strong emotions in others. They challenge the willingness of people to accept the common humanity of us all.

Multiethnic families are marginalized by society because race is such a powerful political and social construct in the United States. Race and ethnicity is a kind of a social fiction, formed by misconceptions about the genetic basis for arbitrary social groupings. Scientists now reject the traditional racial classifications in part because so few people fit into any one racial category (Reddy 1994). However, the power and status society attaches to ethnicity has not disappeared.

Our history is interwoven with racial tension and rejection of crossing or blending color lines. America has traditionally discouraged interracial coupling and even the adoption of transracial children based upon arguments of belonging and acceptance, especially for the sake of the bi/multicultural and transracial children (Orbe, 1999). We do not have to look far in our history to find laws that forbid interracial marriage. It wasn't until June 12, 1967, that the Supreme Court decided the case of *Loving vs. Virginia* ruled that laws forbidding interracial marriage were unconstitutional. It is not clear how many families in the U.S. are multiethnic or multiracial. Wardle (1999) refers to multiethnic families as an *invisible* population (1999). Even the U.S. Census Bureau does not record numbers of multiracial children, because there is no accepted definition of multiracial. It is unclear how many people are multiethnic due to the way people of mixed ethnic or racial backgrounds are pressured to identify with only one of their ethnicities. For example, many institutional forms or government applications allow for only one selection of racial or ethnic identification. Since "multiethnic" or "multiracial" is rarely included as an option in the forced choice selection, individuals who wish to claim a multiethnic background can choose to designate their ethnicity as "Other," a highly impersonal and objectifying label (Herring, 1995; Orbe, 1999).

Although multiethnic families function in many of the same ways that all families do, they face a never-ending barrage of curiosity, insensitivity, bias, and racist intrusions by people who react to their differences. One of the common challenges these families face is how to support the positive identity development of all its members in a culture that values Euro appearance over that of other ethnicities.

In her recent book on raising multicultural children, Donna Jackson Nakazawa (2003) shares the following experience of her biracial son during preschool:

At three and a half my son—Japanese American and Caucasian experienced his first emotional paper cut as a multiracial child. That summer, Christian attended a multiage camp two mornings a week, where kindergartners and toddlers played together outside during recess. One day, when I picked him up, he said—his voice gulping as he gestured with both hands to his eyes— "one of the big boys asked me why I only have little black *holes* for eyes. Do I only have little black *holes* for eyes, Mommy?" His eyes were filled with a desperate questioning I had never seen in them before. My heart did a free fall in my chest. As a multiracial family, we had experienced our share of awkward comments and intrusive stares, but this was the first time anyone had made my son feel—face to face, child to child—that something about him was odd, different. My mind raced through a thousand responses: What could I—what should I—say to him? And what should I not say? When we got home, I guided Christian to the mirror, where I pointed out his beautiful soft brown eyes, his acorn brown hair. "Besides," I said, "the color and shape of your eyes don't matter. Even if they were black, it wouldn't mean a thing. Daddy's eyes are dark brown and mine are blue and yours are the color of honey. Our eyes are all different colors in our family. Does it make any difference in how we love each other?" Dutifully, he shook his head no. Still, I was grasping at straws: Was this anything close to what he needed to hear? I felt painfully ill equipped. I thought (well I hoped) I had helped him process the incident. A few days later, though, Christian was busily sketching loopy "happy faces" when, just as he finished, he scratched out the eyes he'd just drawn in one furious scribble. For weeks, no matter how quickly I tried to salvage his drawings, he worked his marker back and forth until the eyes were all but obliterated, the paper tore through. Drawing after drawing looked like stick figure fugitives with their eyes blackened out to mask their true identities. Although he didn't understand what happened on that playground—and although the kindergartner's words were almost definitely not intended to be hurtful—it had wounded him all the same. He was venting his hurt and confusion and anger in the only way he knew how. (pp. 1–2)

Christian's story poignantly illustrates the powerful influence on children of a dominant culture that values Euro ethnicity over other ethnicities. Very early, children become aware of how others view them, and how to view others who are different from themselves. Parents cannot shield their children from the way a society reacts to difference. However, parents can provide a safe haven where children can express their frustrations as they encounter racism, helping them understand all aspects of racism, sexism, bias, and discrimination, discussing them openly and often, and working to develop within them the courage and self-esteem that will sustain them (Nakazawa, 2003; Lambe, 1999; Wardle, 1999).

Consistently providing the kind of family base that supports positive identity development is a daunting task for multicultural families. Multiethnic identity concerns can complicate other developmental tasks such as establishing positive peer interactions, deeming sexual orientation, deciding on career options, and separating from parents (Herring, 1995). Learning to communicate about "race," racial identity, racism, and related issues is a process multiethnic families work out in various ways. Some multiracial individuals and families oppose the concept of racial labeling altogether, preferring to classify their members solely as "human." This view not only simplifies a complex identity problem but also can be a family's statement of the desire to reject racism in any form, believing that any designation other than "White" regulates them to a lower status in our society (Kenny, 1999).

Orbe (1999) describes four patterned ways interracial families choose to communicate about race to their children. In many biracial families the parents may choose to raise the child with one racial identity over another. This tradition is common in European Americans and African Americans that have commonly adopted the "one drop" rule that states that one drop of Black blood makes one Black. Parents who follow this tradition raise their biracial Euro–African American children as Black, with little or no attention to their White ancestry. Families who choose this view often are protective of their children and are seeking to prepare them for a society that will define them as Black regardless of their appearance in hopes of assisting them in learning how to survive. Raising Euro–African American children as Black also allows biethnic children to benefit from the strength of the Black community and to expose them to the positive, and inspiring accomplishments of African Americans in this country.

Other biracial families take a "commonsense" approach, addressing issues of race depending upon what makes the most sense in each child's situation, including factors such as the characteristics of the community, and the child's appearance. Advocating a "color blind" society promotes a view that the concept of race is illogical, unscientific, and a man-made tool to conquer others. Families using the "color blind" approach teach that "love has no color" and now is the time to advocate for a society based upon color blindness. Another approach is to seek to affirm the multiethnic experience of the child. Parents who talk to their biracial or multiethnic children affirm their diverse heritage and the uniqueness of each culture they represent.

Affirming the multiethnic experience of each child has the potential for a stronger identity development than encouraging one ethnic identity over another. Poston's (1990) biracial model of identity development embraces a process of working toward honoring all of one's ethnic and cultural ancestry. In this model, the biracial child potentially moves through five stages of identity development, arriving at a final stage where there is an embracing of a secure and fully developed biracial identity. In this final stage individuals recognize and value all of their ethnic and cultural identities.

Children who develop a positive multiethnic identity do not do so in isolation but rather within the context of supportive adult relationships. How multiethnic families choose to interpret and teach social realities to children is complicated and often confusing to parents and children alike (Orbe, 1999). Spencer, Jordan, and Sazama (2002) use Relational-Cultural theory to emphasize the importance of parents and caring adults during a multiethnic child's development. Parents can be most supportive when the more powerful parent encourages the authenticity and full voice of the less powerful developing multiethnic child. The parent assists the child toward the development of the fullest expression of her or his personhood. How this process operates explains much about resilience in the face of the adversity multiethnic children will encounter. The child experiences a relationship with her/his parent that results in a deep awareness of being cared for simply for being who they are. At the core of this process is mutual empathy. The adult demonstrates respect for the child, engages with her/him authentically, actively listens, and has a willingness to allow the child to directly and openly impact her or him, which shapes their relationship with one other (Spencer, Jordon, & Sazama, 2002). These types of relationships not only enhance the healthy identity development of multiethnic children, but provide psychological safety and opportunities for growth and learning, which ultimately empower them for life.

Parents can act intentionally in many other ways to enhance a positive multiethnic identity. Nakazawa (2003) suggests that parents begin to stimulate healthy dialogue about race and ethnicity during the preschool years. A positive dialogue about racial differences will set the foundation for possessing a healthy self-concept that will enable them to discard negative views from others in the community. Parents can use toys, books, and play items that reflect the child's unique racial and ethnic heritage as conversation starters. Nakasawa also states it is essential that parents speak up when they see negative, stereotypical imagery. As parents answer questions about appearance and racial differences they should provide answers tailored to the question the child has asked. The answers should be simple, specific, and truthful, and communicate ideas appropriate to the child's developmental level.

In grade school, multiethnic children begin to interact with more children and will begin to have more questions about racial differences and differences in families (Nakasawa, 2003; Wardle, 1999). Parents need to remind children at this age that there are many different kinds of families and whether or not parents and kids "match" family relationships are permanent. Parents should communicate that love defines a family group, not racial sameness.

CONCLUSION

The experiences of the families explored in this chapter challenge each of us to reformulate our definitions of "family." Fundamentally, I hope that we now share a common view that healthy families can exist in a variety of structures. It is very clear that many family structures have been shamed and marginalized because of their divergence from mainstream culture. Some of the marginalized families discussed in this chapter included families of mixed ethnicities and cultures, single parents, gay and lesbian headed families, and families with adopted children. Regardless of structure, the most important consideration in assessing its health and value includes the sense of group identity and acceptance it can provide for all its members, as well as the mutual respect, commitment, and care for one another that can exist. A true understanding of family health then is not based upon its structure, but on the quality of its relational processes among members.

We have also seen that using a broader lens to assess a family's health is dependent upon the context in which the family resides. Consideration of how a family copes must be examined in the context of the multiple systems that impact its health and functioning, including individual members, the family group itself, the school and community, and society as a whole. For example, a single mother who makes many sacrifices to care for a brother with AIDS may be considered healthy and resilient given what it takes for each to survive in an environment where drug abuse and poverty are rampant (Sprenkle & Piercy, 1992).

For many of the family groups explored in this chapter, resilient behaviors are part of what helps them continue to cope in the face of social isolation, shame, and marginalization by members of their community. Resilient families have relational strengths and resources they draw on that help them navigate through difficult times. They continue to find ways to be interdependent, find meaning from stressful events, and openly communicate about one another and the world in which they live. As we look back over how these families have discussed their lives and their children, there is a powerful theme of love and optimism that connects them as a family group and gives them a sense of cohesion and strength.

As professionals and future professionals in positions to be of assistance to families, it is critical that we examine our own biases and beliefs about family groups that are prejudicial and oppressive to them. Learning to focus on their processes that connect them and the coping skills they use provides us with open minds to hear their voices, emphasize their strengths, and bear witness to their resiliency. We must also find ways to actively voice opposition to the marginalization of diverse family structures and become advocates for families who do not have visibility within our communities and nation.

REFERENCES

American Civil Liberties Union (2003). Lesbian and gay rights. Retrieved on August 21, 2003, from http://www.aclu.org/LesbianGayRightsHome/html

Amato, P. R. (1993). Children's adjustment to divorce: Theories, hypotheses, and empirical support. *Journal of Marriage and the Family, 49,* 327–337.

Amato, P. R., & Keith, B. (1991). Parental divorce and the well being of children: A meta-analysis. *Psychological Bulletin, 100,* 26–46.

Anderson, S., Piantanida, M., & Anderson, C. (1993). Normal processes in adoptive families. In F. Walsh (Ed.), *Normal family processes* (pp. 254–281). New York: The Guilford Press.

Antonovsky, A. (1998). The sense of coherence: A historical and future perspective. In H. I. McCubbin, E. A. Thompson, A. I. Thompson, and J. E. Fromer (Eds.), *Stress, coping and health in families: Sense of coherence and resiliency* (pp 3–20). Thousand Oaks, CA: Sage Publications.

Benkov, L. (1998). Yes, I am a swan: Reflections on families headed by lesbians and gay men. In C. G. Coll, J. L. Surrey, and K. Weingarten (Eds.), *Mothering against the odds.* (pp. 113–133). New York: The Guilford Press.

Biblarz, T. J., & Gottainer, G. (2000). Family structure and children's success: A comparison of widowed and divorced single-mother families. *Journal of Marriage and Family, 62,* 533–549.

Boyd-Franklin, N. (1993). Race, class and poverty. In F. Walsh (Ed.), *Normal family processes 2nd edition* (pp. 361–376). New York: The Guilford Press.

Buchanon, C. M. (2000). The impact of divorce on adjustment during adolescence. In R. D. Taylor and M. C. Wang (Eds.), *Resilience across contexts: Family, work, culture, and community,* pp. 179–216. Mahwah, NJ: Lawrence Erlbaum Associates.

Calhoun, E., & Friel, L. V. (2001). Adolescent sexuality: disentangling the effects of family structure and family context. *Journal of Marriage & Family, 63,* 669–682.

Cameron, P., & Cameron, K. (1996). Homosexual parents. *Adolescence 31,* pp. 757–756.

Carrere, S. & Gottman, J. M. (1999). Predicting the future of marriages. In E. M. Hetherington (Ed.), *Coping with divorce, single parenting, and remarriage: A risk and resiliency perspective* (pp. 2–22). Mahwah, NJ: Lawrence Erlbaum Associates.

Carter, B. & McGoldrick, M. (1999). Overview: The expanded family lifecycle, individual, family and social perspectives. In B. Carter and M. McGoldrick (Eds.), *The expanded family life cycle: Individual family and social perspectives* (pp. 1–26). Needham Heights, MA: Allyn & Bacon.

Centers for Disease Control. (2003). 2001 Assisted reproductive technology success rates: national summary and fertility clinic reports. Retrieved April 28, 2004, from http://www.cdc.gov/reproductive-health/ART01

Chan, R. W., Raboy, B., & Patterson, C. J. (1998). Psychosocial adjustment among children conceived via donor insemination by lesbian and heterosexual mothers. *Child Development 69,* pp. 443–447.

Chen, P. (2002). Frequently asked questions about infertility. The National Women's Health Information Center: U.S. Department of Health and Human Services. Available at http:www.4women.gov/faq/infertility.htm. Retrieved from the world wide web May 21, 2003.

Daniluk, J. C. (2001). Reconstructing their lives: A longitudinal, qualitative analysis of the transition to biological childlessness for infertile couples. *Journal of Counseling & Development 79,* pp. 439–450.

DeVault, M. L. (1998). Affluence and poverty in feeding the family. In K. V. Hanson & A. I. Garey (Eds.), *Families in the U.S.: Kinship and domestic politics* (pp. 171–187). Philadelphia: Temple University Press.

Emery, R. E. (1982). Interparental conflict and the children of discord and divorce. *Psychological Bulletin, 90,* 310–330.

Epstein, N. B., Bishop, D., Ryan, C., Miller, I., & Keitner, G. (1993). The McMaster model: View of healthy functioning. In F. Walsh (Ed.). *Normal family processes* (pp. 138–160). New York: The Guilford Press.

Farber, R. S. (1996). An integrated perspective on *women's* career development within a family. *The American Journal of the Family, 24,* pp. 329–342.

Fields, J. & L. M. Casper (2001). America's families and living arrangements: Population characteristics. Revised June 2001, p. 20–537, Washington DC: U.S. Census Bureau.

Galley, L. S., & Flanagan, C. A. (2000). The well being of children in a changing economy: Time for a new social contract in America. In R. D. Taylor & M. C. Wang (Eds.), *Resilience across contexts: Family, work, culture, and community* (pp. 3–33). Mahwah, NJ: Lawrence Erlbaum Associates.

Garmezy, N. (1993). Children in poverty: Resilience despite risk. *Psychiatry, 56,* 127–136.

Gibson, D. M., & Myers, J. E. (2000). Gender and infertility: A relational approach to counseling. *Journal of Counseling and Development 78,* pp. 400–411.

Gonzalez, L. O. (2000). Infertility as a transformational process: A framework for psychotherapeutic support of infertile women. *Issues in Mental Health Nursing, 21,* 619–633.

Gottlieb, A. (1998). Single mothers of children with disabilities. In H. I. McCubbin, E. A. Thompson, A. I. Thompson, and J. E. Fromer (Eds.), *Stress, coping, and health in families: Sense of coherence and resiliency.* Thousand Oaks, CA: Sage Publications.

Greenspan, M. (1998). "Exceptional" mothering in a "normal" world. In C. G. Coll, J. L. Surrey, and K. Weingarten (Eds.), *Mothering against the odds: Diverse voices of contemporary mothers* (pp. 37–60). New York: The Guilford Press.

Hines, P. M. (1999). The family life cycle of African American families living in poverty. In B. Carter & M. McGoldrick (Eds.), *The expanded family life cycle: Individual, family, and social perspectives* (pp. 327–345). Needham Heights, MA: Allyn & Bacon.

Hartling, L. M., Rosen, W., Walker, M., & Jordan, J. V. (2000). Shame and humiliation: From isolation to relational transformation. *Work in Progress, No. 88.* Wellesley, MA: Stone Center Working Paper Series.

Herring, R. D. (1995). Developing biracial ethnic identity: A review of the increasing dilemma. *Journal of Multicultural Counseling, 23,* pp. 29–39.

Hetherington, E. M. (1999). Should we stay together for the sake of the children? In E. M. Hetherington, (Ed.), *Coping with divorce, single parenting, and remarriage: A risk and resiliency perspective.* (pp. 93–116) Mahwah, NJ: Erlbaum Associates.

Jordan, J. V. (Ed.) (1997). *Women's growth in diversity: More writings from the Stone Center.* New York: The Guilford Press.

Jordan, J. V. (1999). Toward connection and competence. *Work in progress, No. 83.* Wellesley, MA: Stone Center Working Paper Series.

Jordan, J. V., Kaplan, A. G., Miller, J. B., Stiver, L. P., & Surrey, J. L. (1991). *Women's growth in connection: Writings from the Stone Center.* New York: The Guilford Press.

Kenny, K. (1999). Multiracial families. In J. Lewis & L. Bradley (Eds.), *Advocacy in Counseling: Counselors, clients, & community* (ERIC Document Reproduction Service No. ED 435 910).

Koch, R., Lewis, M. T., & Quinones, W. (1998). Homeless: Mothering at rock bottom. In C. G. Coll, J. L. Surrey, & K. Weingarten (Eds.), *Mothering against the odds: Diverse voices of contemporary mothers* (pp. 61–84). New York: The Guilford Press.

Lansford, J. E., Ceballo, R., Abbey, A., & Stewart, J. (2001). Does family structure matter? A comparison of adoptive, two-parent biological, single-mother, stepfather, and stepmother households. *Journal of Marriage and Family, 63,* pp. 840–852.

Lambe, M. (1999). Pride or prejudice: When is a compliment a compliment? One family struggles to find the answer. *Mothering,* pp. 74–77.

Lawrence et al. v. Texas, No: 02-102 (U.S. filed June 26, 2003).

Lenhart, T. L., & Chudzinski, J. (1994). Children with emotional/behavioral problems and their family structures. Columbus, OH: *Clearinghouse of Handicapped and Gifted Education* (ERIC Document Reproduction Service No. ED 377 634).

McCubbin H. I. (1998). Families at their best. In H. I. McCubbin, E. A. Thompson, A. I. Thompson, and J. E Fromer, (Eds.), *Stress, coping and health in families: Sense of coherence and resiliency* (pp. xiii–xv). Thousand Oaks, CA: Sage Publications.

McGoldrick, M. (1993). Race, class, and poverty. In F. Walsh (Ed.), *Normal family processes, 2nd ed.,* (pp. 361–376). New York: The Guilford Press.

McGoldrick, M. (1999). Becoming a couple. In B. Carter and M. McGoldrick (Eds.), *The expanded family life cycle: Individual, family and social perspectives* (pp. 231–248). Needham Heights, MA: Allyn & Bacon.

McGoldrick, M., & Carter, B. (1999). Self in context: The individual life cycle in systemic perspective. In B. Carter & M. McGoldrick (Eds.), *The expanded family life cycle: Individual, family, and social perspectives* (pp. 27–46) Needham Heights, MA: Allyn & Bacon.

McGoldrick, M., Heiman, M., & Carter, B. (1993). The changing family life cycle: A perspective on normalcy. In F. Walsh (Ed.), *Normal family processes 2nd ed.* (pp. 405–443) New York: The Guilford Press.

McLanahan, S. (1999). Father absence and the welfare of children. In E. M. Hetherington (Ed.), *Coping with divorce, single parenting, and remarriage: A risk and resiliency perspective* (pp. 117–145). Mahwah, NJ: Lawrence Erlbaum Associates.

McLoyd, V. C. (1995). Poverty, parenting, and policy: Meeting the support needs of poor parents. In H. E. Fizgerald, B. M. Lester, & B. S. Zuckerman (Eds.), *Children of poverty: Research, health, and policy issues. Reference books on family issues* (pp. 269–298). New York: Garland Publishing.

Minow, M. (1998). Redefining families: Who's in and who's out? In K. V. Hanson & A. I. Gary (Eds.), *Families in the U.S.: Kinship and domestic politics* (7–19). Philadelphia: University Press.

Morrison, D. R., & Corio, M. J. (1999). Parental conflict and marital disruption: Do children benefit when high-conflict marriages are dissolved? *Journal of Marriage and the Family, 61,* pp. 627–637.

Mott, F. L., Kowaleski-Jones, L. & Menaghan, E. G. (1997). Paternal absence and child behavior: Does a child's gender make a difference? *Journal of Marriage and the Family, 59,* 103–118.

Nakazawa, D. J. (2003). *Does anyone else look like me?* Cambridge, MA: Perseus Books.

Orbe, M. P. (1999). Communicating about "race" in interracial families. In T. J. Socha & R. C. Diggs (Eds.), *Communication, race, and family: Exploring communication in Black, White, and biracial families* (pp. 167–180). Mahwah, NJ: Lawrence Erlbaum Associates Inc.

Orbuch, T. L., Veroff, J., Hunter, A. G. (1999). Black couples, White couples: The early years of marriage. In E. M. Hetherington (Ed.), *Coping with divorce, single parenting, and remarriage: A risk and resiliency perspective* (pp. 23–43). Mahwah, NJ: Lawrence Erlbaum Associates Inc.

Patterson, C. (1995). Lesbian mothers, gay fathers, and their children. In A. R. D'Augelli & C. J. Patterson (Eds.), *Lesbian, gay, and bisexual identities over the lifespan: Psychological perspective* (pp. 262–290). New York: Oxford University Press.

Patterson, C. J., & Redding, R. E., (1996). Lesbian and gay families with children: Implications of social science research for policy. *Journal of Social Issues, 52,* 29–50.

Patterson, J. M. (2002). Integrating family resilience with stress theory. *Journal of Marriage and Family, 64,* pp. 349–361.

Patterson, J. M. & Garrwick, A. (1994). Levels of family in family stress theory. *Family Process, 33,* 287–304.

Poston, W. S. C. (1990). The biracial identity development model: A needed addition. *Journal of Counseling & Development, 69,* pp. 152–155.

Reddy, M. T. (1994). *Crossing the color line: Race, parenting, and culture.* New Brunswick, NJ: Rutgers University Press.

Scanzoni, J., & Marsiglio W. (1993). Rethinking family as a social form. *Journal of Family Issues 14,* pp. 105–132.

Schnitzer, P. K. (1998). He needs his father: The clinical discourse and politics of single mothering. In C. G. Coll, J. Surrey, & K. Weingarten (Eds.), *Mothering against the odds: Diverse voices of contemporary mothers* (pp. 151–172). New York: the Guilford Press.

Simons, R. L., Lin, K., Gordon, L. C., Conger, R. D., & Lorenze, F. O. (1999). Explaining the higher incidence of adjustment problems among children of divorce compared with those in two-parent families. *Journal of Marriage and the Family, 61,* 1020–1033.

Smith, B., Surrey, J. L., & Watkins, M. (1998). "Real" mothers: Adoptive mothers resisting marginalization and recreating motherhood. In C. G. Coll, J. L. Surrey & K. Weingarten (Eds*.), Mothering against the odds: Diverse voices of contemporary mothers.* (pp. 199–214). New York: The Guilford Press.

Spencer, R., Jordan, J., & Sazama, J. (2002). Empowering children for life: A preliminary report. Project Report for the Robert S. and Grace W. Stone Primary Prevention Initiatives, No. 9, Wellesley, MA: Stone Center Project Report.

Sprenkle, D. H. & Piercy, F. P. (1992). A family therapy informed view of the current state of the family in the United States. *Family Relations, 41,* 404–408.

Tasker, F. L., & Golombok, S. (1997). *Growing up in a lesbian family.* New York: The Guilford Press.

Thomas, G., & Farrell, M. P. (1996). The effects of single-mother families and nonresident fathers on delinquency and substance abuse in black and white adolescents. *Journal of Marriage & Family, 58,* 884–895.

U.S. Census Bureau (2000). Population of the United States: The living arrangements of children, 2000 the current population survey (CPS). Retrieved on July 3, 2003, from http://landview.census.gov/population/pop-profile/2000/chap06.pdf

U.S. Department of Health and Human Services (2003). Frequently asked questions about infertility. The National Women's Health Information Center. Retrieved on June 21, 2003, from http://www.4woman.gov/

Verrier, N. N. (1993). *The primal wound.* Baltimore, MD: Gateway Press.

Wallerstein, J. S., Lewis, J. M., & Blakeslee, S. (2000). *The unexpected legacy of divorce.* New York: The Guilford Press.

Walsh, E. (1998). *Strengthening family resilience.* New York: The Guilford Press.

Wardle, F. (1999). The colors of love. *Mothering, 68*–73.

Webb, R. E., & Daniluk, J. C. (1999). The end of the line: Infertile men's experiences of being unable to produce a child. *Men and Masculinities, 2,* pp. 6–25.

Wegar, K. (1998). Adoption & Kinship. In R. V. Hanson and A. I. Garey (Eds.), *Families in the U.S.: Kinship and domestic politics* (pp. 41–51). Philadelphia, PA: University Press.

Weingarten, K., Surrey, J. L., Coll, C. G., and Watkins, M. (1998). Introduction. In C. G. Coll, J. L. Surrey, and K. Weingarten (Eds.), *Mothering against the odds: Diverse voices of contemporary mothers,* (pp. 1–14). New York: The Guilford Press.

Weeks, J., Heaphy, B., & Donovon, C. (2001). *Same sex intimacies: Families of choice and other life experiments.* New York: Routledge.

14

The Relational Impact of Addiction Across the Life Span

By Thelma Duffey, Ph.D.

REFLECTION QUESTIONS

1. *What factors would lead a person to consider that he or she has an addiction?*
2. *How do we know when our relationships are being impacted by addictions?*
3. *Who is at risk for experiencing addiction?*
4. *How does addiction keep us from connecting authentically with others?*
5. *Within an addiction context, what does it mean to access our vulnerability in order to discover our relational freedom?*

INTRODUCTION

Little did I know that in the midst of writing this chapter I would be struck with the effects of substance abuse, resulting in the loss of an incredibly beautiful person, my long-time mentor, colleague, and trusted friend, Lesley. Let me begin by saying that Lesley was one of the most incredible, bright-eyed, brilliant, jovial, and good-natured people anyone could ever hope to know. She was one of those special people who could light up a room every time she walked in it and who would generate warmth and care, and very practical professional and personal support in what appeared to be an effortless way. She was a mentor par excellence.

Three weeks into this project, Lesley was driving home from work late at night after completing a paper we were to deliver together in Hawaii. Police reports indicate that a young man returning from a party moved into her lane with only his running lights on. He was intoxicated and going 20 miles over the speed limit. No brakes were ever applied. The young man, an honor student and a senior in college, was also killed.

I had been at the coast visiting my father when I received the call. My dear friend, Dana, and her family came to accompany me home. When I returned to the university I went into a stress-based coping mode, where I could organize and "do" something, a feeble attempt to find some sense of control in what was an uncontrollable situation. I was asked by Lesley's family to represent the university at her service. Very intently, I told them and anyone who would listen how I loved her and what a gift she had been for so many of us—what a gift she had been for me.

The truth is that with Lesley, gifts abounded. As we will see in this chapter, it is a wonderful gift to be capable of connecting authentically with others and sustaining such connections throughout the long haul. Repeated accounts describe how, regardless of what life would bring her, Lesley made room to communicate "I love knowing you are in the world. I am so glad you are my friend. You matter to me." Indeed, Lesley was not only capable of connecting; she was abundantly capable of loving.

That week, and for weeks to come, students, past and current, wandered the halls of our program bringing notes and memories of kindnesses that Lesley had so graciously and generously left behind. Faculty who knew and loved her stood in horror, helpless and hurt that such a thing could be true. New students who did not know her quietly and somberly watched as a collective community of their peers and faculty grieved deeply for the loss of their gifted leader and friend. It has been almost one year since the accident and I am reminded daily of how much I miss her, and of the legacy and the void she has left behind.

Many have asked, "How can this be?" Dana, in briefing me for this chapter, queried me to consider and pose to you, "What is the lure? How do such deplorable things continue to happen?" I am pointedly reminded of my loss of Lesley; her loss of life, the losses of her son and husband, and her siblings' loss of their gregarious, fun-loving, and loyal baby sister. I was asked to rewrite this section . . . to make our loss of Lesley more real . . . to describe the sadness of this loss. I have tried to do just that. But sometimes when the loss feels so profound, we have few words available. And when we do begin to speak, it is difficult to stop.

THE EVOLVING FACE OF ADDICTION: A LIFE SPAN ISSUE

In our culture, many people who work hard and live busy, hectic, and stressful lives take an opportunity to let their hair down and enjoy themselves through a variety of means. People relax, unwind, and generally enjoy themselves in a multitude of recreational contexts, such as traveling to new and different places, attending concerts, listening to music, dancing, and dining out with friends or romantic partners. Alcohol is commonly consumed and enjoyed. For many, these are some of life's pleasures that add spice to life and bring levity to an otherwise demanding life.

These experiences can create an enjoyable context for us to lighten our loads, to get to know one another better, and to ultimately develop our relational capacities. Certainly, they can be the very experiences that deepen our bonds and create memories for us to, in years to come, enjoy and recall. Unfortunately, any of these activities, when in the throes of an addictive context, can take on a life of its own. As such, we lose our capacity for choice and our freedom becomes lost to an addictive source.

Indeed, addiction is characterized by an inability to give up a substance, person, or habit, attend to one's emotional needs, reciprocate with loved ones, or even to enjoy a genuine loving experience (Nakken, 1996; Peabody, 1994; Wegscheider-Cruse, 1985). Addictions can result in a lack of capacity to deepen a relationship, to create substantive, mutual love. They involve the virtual lack of freedom to manage ordinary conflicts and tensions in relationships. No longer simple sources of pleasure and entertainment, they represent needs that can lead to feelings of frustration, emptiness, and isolation. And, of course, addictions can also kill.

Researchers and theorists have attempted to describe the addictive experience, with varying definitions evolving through the years. Peele (1985) writes that the term "addiction" was first used in the 19th century to describe the physical dependence that results from *chronic use* of a substance. Evidence of addiction was seen through symptoms of tolerance and withdrawal. Later, the definition of addiction was broadened to describe powerful elements of human behavior and to provide context not only for drug and alcohol abuse but also for a myriad of other compulsive behaviors, such as gambling, sex, work, and eating.

Contemporary theoreticians now examine addiction issues from a more global perspective, including biological, cultural, behavioral, interpersonal, environmental, spiritual, and intrapsychic factors. Indeed, other factors such as genetics, society, maturity, relationship issues, the family, social networks, and the degree of one's own investment in psychological development and growth all impact the development and maintenance of addiction (Bratter & Forrest, 1985; Covington, 1994; Nakken, 1996).

Orford (2001) describes addiction as a process that although initially pleasurable, eventually becomes destructive to self and others. Objects of such addictions include alcohol, drugs, sex, gambling, work, eating, or relationships, among others. In spite of harmful consequences, addicts maintain destructive behaviors, caught up in the illusion that they can ultimately control them. This illusion is reflective of a core condition we will see throughout the addiction literature: *denial.*

Denial, which is sustained through illusions, is central to the maintenance of addictions, tricking addicts into believing they have control of their choices when they, in fact, do not. Comstock adds, "The addict subculture reinforces these illusions. In some respects, addiction becomes socially and economically supported and reinforced. Socially, there will always be somewhere to go where you'll find others to drink and do drugs with. In these contexts, they are called 'regulars,' not 'alcoholics'" (personal correspondence, 2003). Along the same vein, what greater reward could an addiction to work bring than society's love affair with success?

Covington (1994) writes that cultural expectations of women facilitate this denial. "As women, we are expected to direct our attention toward caring for others, not toward self-care, self-knowledge, or our own inner experience" (p.18).

Nakken (1996) suggests that addiction is a failed attempt to control the natural cycles of happiness and loss, boredom and contentment that are basic to the human experience. He adds that individuals attempt to cope with life experiences by abusing substances of one form or another, noting that this form of coping does not support an investment in building relationships or connecting with others in a mutually satisfying way. Instead, it establishes a pattern for developing and sustaining an addictive lifestyle that, because of its disconnecting nature, is egocentric and devoid of meaning. Nakken purports that we have an opportunity to make meaning out of life through our mutually rewarding connections with others. This experience of meaning-making is problematic for addicts because addiction inhibits relational capacities and sustains an addictive process far beyond the dependency on any particular substance or behavior.

Meaning-making is discussed in Relational-Cultural theory (RCT). According to RCT, we make meaning when we know that we matter to the important people in our lives. Given that, we can only imagine the sense of personal insignificance felt by children when their experiences tell them that they do not matter as much as alcohol or work, for example, to the people who are most significant in their lives.

As we can see, theorists have expanded the definitions and perspectives of addiction from its original focus on alcohol and have provided a context for discussing the relational, spiritual, sociocultural, economic, and gendered influences of addiction (Covington & Surrey, 2000). Understanding the relational consequences of addiction provides context for much of the therapeutic work encountered in counseling offices throughout the country.

WHO IS AT RISK?

According to Nakken (1996) any person undergoing periods of loss is at risk for succumbing to addiction. Still, our ability to cope with pain by authentically connecting with and participating in relationships that matter to us significantly decreases this risk. With this in mind, we can see that anyone who is not able to connect with others authentically and participate in a mutually enhancing manner is at greater risk for addiction.

Because our emotional development and relational capacities are stunted at the age that we begin abusing substances in spite of our chronological age, we may not have the capacity to authentically relate to and form connections with others, given our unaddressed substance abusing histories. These relational limitations put us at even greater risk for sustaining our addictive rituals.

Addiction and its consequences are experienced in all factions of the U.S. population. For example, Matano, Wanat, Westrup, Koopman, & Whitsell (2002) examined the substance-related reporting among the well-educated workforce. They found male participants to be more likely to report binge drinking than their female counterparts. They also found Asian Americans less likely to report any drinking, while African Americans were more likely to report binge drinking. The self-reports of binge drinking among younger employees were significantly greater than those of older employees. This research suggested that substance use, particularly in the form of binge drinking, was reportedly problematic among members of the well-educated workforce.

Privilege in our culture also puts us at risk. Because of the privilege enjoyed by the well-educated and economically advantaged members of society, disconnecting, disruptive, and hurtful addictive-based behaviors that would be less tolerated from other groups are deemed more acceptable. When these behaviors are alcohol induced and remain within a culturally sanctioned "functional" climate (such as career networking, recreational activities, and bonding experiences) the effects of their damage are often normalized or excused. Many of these behaviors, when placed in another context, would be deemed inappropriate, at best, and abusive, at worst.

Another issue that seems to influence a person's susceptibility to addiction is prior physical, sexual, or emotional abuse. In a study exploring violence and addiction, MacDonald & Wells (2001) found that addiction issues were seen to be more common among women who suffered violent injuries when compared to women who did not suffer violent injuries. They reported that according to emergency room studies, women admitted for a violent injury were more likely than those admitted for other causes to have positive blood alcohol concentration.

Sexual and physical abuse histories among women were more likely to result in addiction, particularly in relation to substances, prescriptions, and eating disorders. The research clearly portrays the cyclical nature of abuse and addiction. Being a victim of abuse increases the possibility of addiction. Suffering from addiction increases the risk of becoming a victim of violence (McDonald & Wells, 2001).

Adolescence also brings with it risks for addiction. The research literature on current adolescent addiction issues states that, regardless of issues of race and gender, a strong indicator of lifetime alcohol dependence is the age at onset of substance use. Grant & Dawson (1998) note that individuals who begin drinking in adolescence are far more likely to become alcohol dependent than if they had started using substances later in life. They are four times more likely to become alcoholic if they began consuming alcohol at age 14 than if they had started subsequent to reaching 20 (Grant & Dawson, 1998).

Family attitudes toward drinking bring with it increased risks. Shen, Locke-Wellman, & Hill (2001) note that family attitudes toward drinking alcohol influenced and reinforced alcoholic behaviors. For example, adolescents who shared or experienced the drinking behaviors of family members were considered high risk.

Gender specific sociocultural factors also put us at risk. For example, societal expectations still exist for males and females with regard to our expressions of feelings and relational styles. These expectations can be confusing, inhibiting, and destructive. The means by which we manage these expectations either facilitate or exacerbate existing problems and can lead to an "emotional phobia" (Greenspan, 2003, p. 63) for many, ultimately culminating in addictions.

Greenspan (2003) writes, "Ours is a dissociative culture—a culture that separates body from mind, body from spirit, feeling from thinking. The result is that for many men, the capacity for emotional sensitivity is drastically impaired, with devastating effects on men's abilities to be at home in their own emotional bodies and to connect in relationship" (p. 21). She adds, "Double-binded from the start, women are expected to carry, feel, and display the emotional vulnerability that is taboo for men; but at the same time, we are devalued for doing so" (Greenspan, p. 21).

Simply put, in our culture risks abound for substance abuse and addiction. It is no wonder that addiction is difficult to identify and has become a prevalent social and relational problem.

WHAT DOES AN ADDICT LOOK LIKE?

At this time, allow yourself to form an image when you read the word "addict." What comes to mind? What are your notions of what addicts look like? I asked students during class to write down their first, instinctual impression of an addict. Images among the group varied. However, by and large, reported images of poverty, youth, and ethnicity were dominant. Then one young man quietly noted, "My first impression of an addict is a person who is fun, economically sound, socially protected, and callous."

Certainly, there are culturally driven images of what an addict looks like. Addicts come in all shapes and forms. Some function in many areas of life and others do not function well at all. Some become addicted as youngsters and others develop addictions later in life. Certainly, privilege brings with it a cushion, in the form of education, status, power, or money that protects some addicts from the reality of their addiction, since they are in better positions to normalize them.

Behaviorally, addiction is broad based. Some addicts are seen as abusive, angry, and out of control, causing those closest to them great fear and generating in them self-protective attitudes and behaviors. Others are seen as benign, nostalgic, and sentimental, often lamenting lost loves or unlived lives. Still others are seen as charming, amusing, and entertaining characters. And others drink alone while feeling depressed, bewildered, and determined to escape the complexities of life. The bottom line is that addicts are confused and confusing and they share one common theme: Their relationships suffer.

Within a traditional understanding of addiction, "nonfunctional addicts" are unable to function well in the world because of an obsession with an addicted source, usually drugs, alcohol, sex, or gambling (Bratter & Forrest, 1985; Peele, 1985). Some addicts lie, cheat, and steal in order to maintain access to an addictive source. Their ability to maintain a job or relationship is limited. Addicts live in chaos, experiencing unpredictability in their most intimate connections. Highly dependent, not only on the source of their addiction, but also on the supportive qualities of those around them, the lives of these addicts become unmanageable. Sometimes they kill others or die. If they are lucky, they "hit bottom" without tragic consequences and begin the climb back up (Beattie, 1987; Bratter & Forrest, 1985).

Much of the work on addiction has addressed the nonfunctional addict. There is, however, another category of an addict that is less discussed in the literature. These addicts are sometimes known as "functional" addicts. As the name suggests, functional addicts function well in many areas of their lives, particularly socially and professionally. Many have the social skills and networks, academic accomplishments, professional successes, and financial security that would, by classic definition, preclude addiction. Still, they, too, cope with life's difficulties by becoming dependent on substances, circumstances, or objects, but unlike the traditional addict, do not show their vulnerabilities on first look (Bratter & Forrest, 1985). The consequences of their addictions become most clear as the addictive belief system comes into play in relation to others.

Functional addicts are especially confusing. Like their nonfunctional counterparts, they are controlled by their addictive beliefs but are additionally protected by their performance, success, financial security, and practical functionality. As such, their "bottom" is harder to hit. With respect to women, Covington (1994) notes, "How can we have a problem if we still get the kids to school on time, balance the checkbook, do all the chores, and show up for work everyday? How can life be unmanageable if it looks so orderly?" (p. 20).

In a relational context, there is little to distinguish the nonfunctional addict from the more functional one. Addiction involves a distorted belief system that impedes emotional growth and impacts a person's capacity to participate in relationships that foster growth, connection, sensitivity, and mutual care. As such, in this chapter, reference is made to both groups when addressing the concerns of the "addict." The goals for the treatment of relational issues relative to addiction for both groups is twofold: (a) to specifically address painful relational voids; and (b) to provide a context for developing the emotional, psychological, and relational tools for forging connections with others and creating meaning. Meaning, in this respect, is seen as the mutuality experienced and enjoyed in relationships with others (Nakken, 1996).

THE IMPACT OF ADDICTION ON RELATIONAL MOVEMENT

RCT (Miller, 1976; Stiver, 1990; Miller & Stiver, 1993) provides a context for conceptualizing basic relational dynamics and will serve as a framework for discussing the impact of addiction on relationships. According to RCT, we all have a yearning for connection with others. That is, we all have a desire to be genuine, authentic, and fully present in our relationships, representing our feelings and needs to the important people in our lives without fear that we will be punished or rejected. When we are not able to be fully present and to represent ourselves authentically in these relationships, we begin to hide aspects of ourselves in order to maintain the relationship in whatever form it is available. These patterns of relating are initially seen as strategies for survival. Unfortunately, in time, these strategies for survival become strategies for disconnection.

Although all relationships experience disconnections, and we each employ such strategies in our relationships at various times, this paradox *defines* the basic structure of relationships within an addictive context. For example, according to models of addiction, certain factors impact an addict's developed capacity to form relationships and connect with others (Jampolsky, 1991; Nakken, 1996; Peele, 1985). These factors influence his or her capacity to introspect, give, be accountable to, and offer care for self and others.

As individuals are able to see their own authentic needs, they can better see the needs of others. With addiction, however, their capacity to be authentic, genuine, supportive, or to provide mutuality is thwarted. When people suffer empathic failures, when they are not understood, or when they feel let down, acute disconnections can develop (Jordan, 2001). This results in what RCT refers to as "condemned isolation" (Miller & Stiver, 1997, p.72). Condemned isolation is a term used to describe an experience of "being locked out of the possibility of human connection" (Miller & Stiver, 1997, p. 72). According to Jordan (2001), experiencing this form of isolation for most people is a primary source of pain.

Although this dynamic can and does take place in a variety of contexts, such as, friend to friend, romantic partners, or coworkers, for the purposes of this chapter, this pain can best be illustrated in the experiences of children. Certainly, there are a myriad of ways in which children are hurt and violated by their parents' words or behaviors. How parents handle these hurts, however, determines the means by which children increase their relational competencies and enjoy the freedom and comfort that mutuality in relationships can bring. When parents are in the throes of addiction, these competencies are threatened.

For example, when we consider all the factors that operate to maintain an addiction, including secrecy, an illusion of health, denial, and blame, we can see that these factors make authentic responsiveness in relationship difficult. If disconnections are chronic, the relationship and the mental health of each person are threatened and both parties can move to a place of condemned isolation. At this point a downward spiral takes place and various self-destructive patterns can emerge for both parties.

Children typically attempt to form connections, even if they are constructed by being inauthentic and playing by relational rules. In order to "play by the rules," they can develop strategies of disconnection that keep their true feelings at bay. These strategies also serve to keep the addiction invisible. They help the child maintain some sort of connection; it protects the addict from knowing its impact on others; and it keeps them from directly experiencing the impact of addiction on them personally. Because people become conditioned to relate inauthentically to the addict by keeping their true feelings out of relationship, the impact remains hidden. This, of course, preserves the illusion, once again.

RCT posits that we develop relational images that reflect our beliefs about ourselves and about how we expect to be treated by others (Jordan, 2001). This theory proposes that our relational images are based on real-life experiences in our important relationships. Given that, consider the relational images formed by children who grow up with at least one parent disconnected and unavailable, in this case, because of some form of addiction. Depending on the relational images they form, what value(s) would these children place on relationships and what patterns of behavior and interpersonal dynamics could develop in relation to others?

Certainly, when parents are addicts, developing relational competency and positive relational images can be particularly difficult for all members of the family. For example, people with addictions tend to rationalize or justify their behaviors, and whether or not they were abusing substances at the time of violation, many have little memory of their actions. As such, the means by which parents acknowledge the hurt they cause their children directly impacts the relational movement of the children, particularly with regard to their learned capacity to be authentic with others. Although hurts and violations created within an addictive context can be physically, sexually, or emotionally abusive, subtler violations also injure a child's sense of worth and possibility for relational connection.

As an example, a child is hurt because his father does not attend a very important event in the child's life, a baseball tournament with an awards banquet to follow. The father had promised to attend the tournament but lost track of time while meeting friends for a drink after work. The child, looking anxiously into the crowd, is visibly upset when he cannot find his father. Even at the point of receiving his trophy, he looks around, once again, just in case his father arrived without his notice. The ceremony ends and the child returns home. Hours later, the father arrives. The child tells the father he is hurt and disappointed. The father acknowledges his regret and sorrow for having missed the event. Beyond that, the father remembers how this impacted his son and does his best to attend subsequent important events, or tells the child in advance if he is not able to do so.

As a result, the child feels that he matters to his father. He learns that he can communicate with him, regardless of the discomfort it may bring. He also learns that he is relationally effective, that what he feels and reports can actually make a difference to others, and that he is able to participate in shaping the relationship. As such, this boy grows to learn that he can represent himself honestly, and without reservation, within the context of a safe relationship.

How would this affect his relationships in later life? For one, he would recognize his impact on his relationships and would have a model for using this impact in a mutually empowering manner. Also, he would have a greater potential for staying authentic in relationship and of staying connected personally and with other people.

Now, let's look at what this situation would look like if the father was unable to affirm his son's reality; if instead of recognizing the pain he caused, he became angry at the child, even covertly minimizing the problem, rejecting the son, withdrawing in anger or hurt, and denying the consequence to the child. Jordan (2001), states:

> In such a situation, the child begins to disconnect from his or her own experience and begins to hide or twist the experience to fit what is acceptable to the more powerful adult. In this instance, the child learns, "When I am hurt or angry . . . I lose the connection with this important person. (p. 96)

The child would blame himself or herself and begin to disconnect from his/her inner experience. Ultimately, the child's understanding of reality would be altered.

As we can see, this experience leads to a downward spiral in which the child attempts to create safety by playing by the rules, but as a consequence becomes less authentic and feels less genuine, seen, and understood. Imagine how this pattern of relating would affect a person's expectations of relationship and manner of being in relationship. How helpless would we feel if the only way to negotiate a relationship with someone that mattered to us would be to remain quiet, not notice hurts, not even speak out when someone is figuratively standing on our foot? These responses, however necessary for a child in some circumstances, create "chronic disconnections which actually move people into isolation, self-blame, and immobilization, the hallmarks of what clinicians call pathology" (Jordan, 2001, p. 96).

Jordan (2001) notes that although the yearning for real connection remains, people can feel too vulnerable and unsafe to risk being authentic when their expression of feelings and needs has been minimized, denied, or even attacked. At this point, children are faced with what RCT refers to as the central relational paradox. Although the yearning to connect deepens, so, too, does the fear of being authentic and vulnerable in the relationship. She adds, "The person is then caught between the intense yearning for connection and the terror of it" (p. 96).

RCT submits that the goal of development over the life span is to increase our relational capacities (Miller & Stiver, 1987; Jordan, 2001). Given that addictions not only impede the development of our relational capacities but also thwart deepened, substantive relational movement, they keep us from building or working toward developing growth-fostering relationships. Since the model examines the impact of nonmutual, abusive, or chronic disconnections in relationships, especially with regard to marginalized individuals (Covington & Surrey, 2000), particular attention will be given to the relational consequence of experiencing such violations within an addictive context.

A RELATIONAL LOOK AT ADDICTION: STRATEGIES FOR DISCONNECTION

The literature on addiction describes common cognitive and relational patterns seen in addiction. These are commonly known as defense mechanisms and enabling qualities. RCT would call these "strategies for disconnection." In therapy, it is important to identify these patterns of relating and to interpret them in such a way that we create a context for deepening connections and developing growth-fostering relationships. If we can give a name to the issues related to relational ruptures and illuminate the purpose they serve, two things can happen. First, we can view the problem through a relational lens. Second, when we can understand the impact that the behavior, pattern, or strategy has on the relationship, we can move forward with an intervention to give direction for therapy.

Given the roles that strategies for disconnection play in a person's life and the purposes they have served, asking addicts to relinquish these by examining their responsibilities in the relational ruptures and disconnections experienced with others can be frightening. Even more frightening is the prospect of asking them to examine the impact they have had on the well-being of others. Certainly, it is uncomfortable for any of us to acknowledge when we have been hurtful. How much more uncomfortable could it be for us to acknowledge the damage we cause others when we have dedicated our lives to maneuvering escape routes from these realities?

Peabody (1994) describes some common patterns and addictive worldviews that impede a person's relationship potential, including:

> fears of losing his or her identity; fears of dependency/avoids bonding; creates rigid personality boundaries (won't let people in); is sensitive to everything that leads to bonding; seduces and withholds to avoid bonding; minimizes feelings that lead to bonding; gets nervous when things go well or bonding occurs; wants more space or has to run; can't make commitment; is indifferent to others; feels entitled to be taken care of (his or her way); won't put up with discomfort; has complete control of the schedule; says to partner "just stay put and let me come and go." (p. 129)

Relationships cannot thrive and be enjoyed freely with such inflexible and unilateral restrictions.

In describing the experience of women addicts, Covington (1994) describes how women lose contact with their core selves while using and abusing substances. As a result, they lose connection with their own values to be responsible, loving, creative, and available. Their lives then become "filled with dishonesty, rigidity, fear, and distrust" (p. 16). These experiences bring pain and suffering for all involved.

To illustrate this section, I will present an account by a man we will name Tom. Tom is a 52-year-old newly self-professed functional alcoholic. According to Tom, he began abusing alcohol and other substances in high school and college. He acknowledges that alcohol has been an integral part of his lifestyle throughout his adult life. An avid athlete, Tom enjoyed excellent health, and for the most part, has felt invincible. Three years ago, Tom suffered his first heart attack. Last year, he suffered his second. Tom has undergone surgery but he is not out of the woods. He is scared. Tom's health has slowed down—and so has his pace. Not surprisingly, he is conducting a life review.

A successful executive in a multimillion-dollar firm, Tom acknowledges working hard and playing hard. According to Tom, he has consumed life to the fullest. Still, when faced with his immortality, Tom is less comfortable.

> I have done anything I wanted to do with little to no concern for how it would affect anyone else. It has truly never occurred to me that I wasn't entitled to this life. I felt fortunate, without a doubt, but I felt this life was mine to enjoy and to do with as I chose. Now, I'm not so sure. I have an uneasy feeling about this. I have been very fortunate that most things have worked out for

me. The things that haven't, I haven't allowed to matter much. My biggest failures have come with women. I loved some of them. Hell, I don't know.

The last one was tough. She was great. Then, something as simple as seeing
her sit in a chair watching television upset me. Having her open the refrigera-
tor would annoy me. I would tease her/criticize her relentlessly. After two
months, I couldn't sleep facing her. I had to sleep with my face to the wall. I
either made snide remarks or criticized her openly. I can still see the confusion
on her face. I took the extended cross-country vacation I have always wanted
to take. I was gone two months; didn't tell her I was leaving; didn't write or
call. She was frantic. A friend finally told her where I was. When I returned,
she was gone. There's no doubt she loved me. Strangest thing is, I loved her,
too.

Part of the irony of addictive relating is that although people are able to have
loving feelings, they are not generally able to do much with the feelings in a tan-
gible, productive way. To use relational terms, they are not able to participate in
growth-fostering relationships where connection, investment, and mutuality are
possible. In an effort to feel good about who they are, some make sense out of
their experiences in a manner that preserves their self-image but that unfortunately continues to create havoc and damage in their relationships. For example,
according to Tom, throughout his relationship history, he would not remember
many, if any, of his hurtful behaviors when confronted. "I rarely remembered saying or doing these things." The literature speaks to the countless ways that individuals will deny their circumstances or interpret them in ways that justify behaviors
and rationalize their life goals. This denial creates an illusion that distorts reality.

Another common strategy for disconnection is known in the addiction literature as *rationalization*. As such, people talk themselves into believing information
that is comfortable for them. When they do not take the feelings of those around
them into account, others are frequently hurt by their rationalizations. Tom
describes a situation involving his ex-wife's daughter's marriage. "I didn't like the
idea that we were going to be paying for this thing. She has a father. Why wasn't
he doing this? In any event, I agreed."

He continues:

My wife was financially self-sufficient and a strong contributor to our
finances, so I could hardly object much. Still, when it came down to it, I didn't
like it. And I didn't want to go to the wedding. And I didn't want to deal with
discussing it. I told her to go ahead without me—I'd catch up with her. Never
did. I got on the couch, watched television, turned off the message recorder,
and took a nap. The way I saw it, I paid for the damn thing. Why would I need
to go, too? I wouldn't expect her to do anything she didn't want to do.

Tom's way of conceptualizing this situation and his reluctance to discuss his
position made negotiation of personal preferences and important life events
impossible. He describes his feelings when the relationship was dissolving. "I had
some guilt." Still, Tom was able to justify his behaviors by saying, "She knew the
score." He did not consider the impact of his behaviors on her because although

he professed to love her, he had provided due warning by letting her know how difficult relationships have been for him.

Another strategy for disconnection occurs when people deny negative qualities and behaviors in themselves and instead see them in and "project" them onto another person. Tom muses, "I did anything I felt like doing and didn't think much about whether she liked it or not. Still, at the break up, I was perplexed, thinking, 'Why does it have to be all about you [her]?'" When people project behaviors onto others, they blame them for the discomfort or pain in the relationship. Blaming is a particularly powerful strategy for disconnection and occurs when someone fails to assume responsibility for the grief they cause and instead attribute the cause to the other person.

Although his inflexible relating patterns created a high-maintenance relationship, he placed the "need" issue onto her. For example, when discussing the end of the relationship, Tom noted that he could not meet her "needs." Tom is coming to see that his focus on his own needs and comfort made mutuality in the relationship impossible because the needs and comfort of his partner(s) have been virtually ignored. As such, as many times as Tom has initiated relationships, they have historically represented burdens for him to manage. Because of his stated desire to have a comfortable, mutually based relationship, making a decision to connect emotionally and to consider another person's comfort and safety have become relational goals for him.

ENABLING IN CONTEXT: REDEFINING CODEPENDENCY

"Codependency" is a term originally coined to describe individuals who attempt to maintain relationships with addicts by overcompensating, ". . . overfunctioning . . . or by being obsessed with trying to control him or her" (Covington & Surrey, 2000, p. 8). The addictive literature since the early 1980s describes varying characteristics of individuals carrying this "syndrome." These include obsessive thoughts, controlling behaviors, entrenched denial systems, boundary concerns, anger and rage, and sexual dysfunctions, to name a few (Beattie, 1987; Wegscheider-Cruse, 1985).

Indeed, individuals who found themselves in an addictive relationship were seen to be addicts themselves. These theories question why a person would maintain such an unsatisfying relationship. Traditional writings on addiction purport that individuals involved in such a relationship, particularly if they accept these behaviors, are also addicts, or codependents. According to Stone Center writings, "Women are defining themselves as "codependent" in increasing numbers and seeking treatment in 12-Step programs or specialized treatment centers" (Covington & Surrey, 2000, p. 8).

Although the Twelve Step tradition (Alcoholics Anonymous World Services, Inc., 1976) is known for its contributions toward growth and healing, Covington & Surrey (2000) note two limitations to the idea of codependency. For one, they propose that the very nature of the definition is so inclusive and ambiguous that everyone in the culture could be seen as codependent. Another issue they raise is that the term "codependency" is gender linked. They note, "Rather than affirming and revisioning women's potential strengths in relationships, and validating their motive for connection, the codependency concept tends to pathologize their relational orientation, thus putting women in a cultural double bind" (p. 8). Their concern is that the label suggests an intrapsychic issue rather than a relational one. Given the Stone Center's mission to examine the impact of non-mutual, abusive, or disconnected relationships, this concern is a formidable one.

Covington (1994) adds that women have central themes running through their lives to which they must become conscious. Although these patterns are initially established in the service of connection, they, unfortunately, can complicate a relationship and, instead, create strategies of disconnection.

For example, in examining the addictive process, Covington (1994) notes that women may assume perfectionistic qualities that drive them to "look attractive at all times, never make mistakes, and know just what to say and do under all circumstances" (p. 72). Our culture and the media perpetuate such values (Kilbourne, 1999). Women may protect the others' egos and may also turn against themselves by denying or excusing bad behaviors. According to Covington (1994), women talk themselves into accepting unacceptable behaviors because they do not want to risk losing their relationships. Later, they blame themselves for not having known better, done it differently, or done something to affect positive change.

As an example, Lisa was in an eight-year relationship with Ben. Although she reports many happy, comforting, and exciting times with Ben throughout their time together, she also reports experiences of detachment, withdrawal, outright abusiveness, and bewildering rejection. Given the cyclical nature of her experience, Lisa's own relational images, and her staunch investment in the relationship, she has felt stuck wanting connection with someone who ultimately set unrealistic parameters around what their relationship would look like.

Lisa begins, "Towards the beginning of our time together, I remember Ben saying, 'I'm not good at relationships;' coyly adding, 'Well, that's not true. I am good at getting into them. Not as good at staying in them.' I'm embarrassed to say that I laughed. When he said it so casually, as if it was funny, I dismissed it. Later, I came to see how 'not funny' it really was." After one particularly painful breakup, Ben initiated reconciliation. Lisa reports:

> It was one of the happiest times of my life. He was loving and affectionate and appeared to sincerely want me in his life. Then, later that evening, he said, 'I hope I can do this. I've never been very good at commitments.' Ben slept the next day away. When he awoke, he seemed uncomfortable, edgy, and finally uninvited me to join him on holiday. This had become a pattern I could no longer handle. I don't want to even remember the hurt I felt. I told him that it could never work between us. He said, 'It won't if you don't want it to.' Ben placed the responsibility for the success of our relationship onto me. I was already doing that. That was a sure-lose.

In order for relationships to deepen, both persons must be capable of risking expressing their preferences, needs, and experiences. When people use work, substances, entertainment, or other distractions during painful or uncomfortable situations, they become accustomed to speaking in code. Although one or both may appreciate the short term benefit of such avoidance, an unfortunate consequence is the natural buffer to intimacy that gets set in place between them. This creates a rhythm of relational swings ranging from intense enjoyment during the good times to painful distance and awkwardness during trying times. It is on these occasions that fear replaces spontaneity. Lisa continues:

> I can't speak for Ben, but in my case, I can see now how my fears got the best of me. It was the hardest thing to love him so much and to feel scared to say so . . . to have the feelings and to have to fight saying them out loud. There was a point where Ben told me I should have given him an ultimatum. That got me thinking. Perhaps I should have. I had never considered that to be my style; nor did I expect that giving anyone an ultimatum would be productive. Still, he was right in that I had given into my fears. I simply didn't know what else to do anymore. This time, I packed up my things and went home.

A look at Lisa's family history sheds light on her relational fears. Since childhood, Lisa had felt a close association with her family. Still, she witnessed relational struggles between people she loved and valued. Perplexed and helpless, Lisa reports asking herself, "How could two wonderfully incredible, smart people who obviously love one another not be able to work this out?" Her childlike desire to protect and "make better" was unrequited. So, too, was her desire to see two people who appeared to love one another work things out in a mutually productive manner. Like many children, Lisa compensated for any strife by attempting to achieve and by accommodating to others; this was sometimes accomplished at the expense of being authentic. As such, she developed a well-sanctioned pattern of hiding difficult and painful realities from others to avoid being a burden and to escape further loss. Unfortunately, this dynamic set the stage for several relational consequences in her relationship with Ben.

For one, given Ben's pursuit of her, she assumed Ben's sense of loyalty to the relationship and investment in its success. When Ben would retreat without explanation, Lisa would become confused and would also pull back, triggering a heartfelt loss. She, in fact, enjoyed the playfulness, sweetness, and intimacy that she perceived to be characteristic of their time together; it was the very real pattern of retreat and withdrawal that she found hard to bear. During these periods of disconnection she would recoil into a place of hurt, depression, and self-blame. Months later, when Ben would return, contrite, sorry, and full of declarations of love, she would reconsider. Lisa adds:

> I'd remember the best and forget the worst. I'm dealing with the reality of the relationship now, and that's been hard. I told Ben that I thought he had built a case against me in his own mind from the start. He agreed. How strange that someone would do that and at the same time continue to pursue you. I guess

I really wanted to believe that he loved, appreciated, and valued me. I wish I could still believe that but there is so much evidence to the contrary. Still, as hard as it's been, I've come to a much deeper appreciation for what love can look like, for what it can feel like. Do I love Ben? Absolutely. At the same time, I also very deliberately choose to invest in relationships where love and mutuality are possible.

Sometimes Lisa sees herself as an addict. Other times she sees herself as recovering from naiveté. Still other times she simply sees herself as a woman who loved deeply. Whatever her conceptualization, she describes the harrowing experience of release:

The truth is, I invested too much not to care. I get the picture now but I genuinely fell in love with him before I knew what I was getting into. I think of him . . . wish it could be different. Sting's "Fields of Gold" and Dylan's "Don't Think Twice" both make me think of him. One is a bittersweet love song. The other one is just bitter. Another song that reflects this experience for me is "Tainted Love." Walking away while loving him because I was self-destructing by staying is the single hardest thing I've ever had to do.

Again, according to Relational-Cultural theory, we all yearn for connection, but paradoxically engage in disconnecting strategies that keep us out of the connection we desire. Some people use these strategies to maintain the illusion of control, freedom, and independence. In reality, when people misuse drugs, alcohol, work, the media, etc., these serve as relational distancers and short-term emotional Band-Aids. In fact, it is under these circumstances that people have the least control of their lives; their freedom is limited by their preconceived fears, inflexible relating patterns, and rich fantasy life.

HOPE FOR RELATIONAL TRANSFORMATION

In this chapter we have discussed addiction in a variety of forms. We have seen that unless treated successfully, addiction negatively and sometimes dangerously impacts our lives throughout the life span. In some cases, it can result in death. We have explored the relational consequences of addiction and have seen how strategies for disconnection impede closeness and genuine care, resulting for many in a state of condemned isolation. Now, we will briefly discuss some means for working with the addictive process to facilitate connection with self and others.

A number of treatment interventions are commonly used with addiction issues. Briefly stated, the most common treatment strategy employed in working with addicts is some derivative of the Twelve Step Program, founded by Bill Williams in 1939 (Alcoholics Anonymous World Services, Inc. 1976). Spiritually based, this program has been used by millions of individuals. These programs propose that a power greater than ourselves is ultimately in control of our lives and that we are powerless over our addictions. Ironically, it promotes a structure of empowerment by acknowledging our very powerlessness. The Twelve Steps provide a structure for personal awareness, assisting individuals to revisit their attitudes, feelings, and thoughts, and to then place these constructs into practical action.

At the same time, Covington (1994) notes that the Twelve Steps were written with male alcoholics in mind more than 55 years ago. Given that, a need has existed to provide a contemporary framework for women's experience of addiction and recovery. In her book, *A Woman's Way Through the Twelve Steps*, Covington provides just that.

In addition to treatment modalities, there are personal and relational dynamics that must be considered. Working with addiction, *in some respects,* is not unlike working with any other population. Each of us must experience safety in order to risk opening up emotionally, particularly when our self-protective patterns have been long-standing. Therefore, as therapists we must ask ourselves, "How do we best facilitate this safety?" Reflecting on our own experiences with safety is a viable first step.

At this time, let me ask you to think about a personal limitation, previous experience, or current experience in your life that simply hurts. Whatever the context, imagine what it would be like to feel stuck in this experience . . . to feel there is no way out. Now imagine that you are telling someone about this experience or limitation. What does it feel like for you to self-disclose? What response are you hoping for from the listener? What response could shut down your communication? Considering how you communicate and what people need to feel safe and connected helps to facilitate relational movement in therapy.

I will use the case of Tom, presented earlier in the chapter, to illustrate the twists and turns of relational movement in therapy when addiction is a core issue. Tom first entered my office with a swagger and a chuckle. He sat down on the couch and, looking me straight in the face with a gleam in his eye and wearing an impish grin said, "You're a very pretty lady." He followed this by asking me a personal question. I was caught off guard and recall feeling uncomfortable. Looking at him with focus and intention, I responded by asking him what he hoped to accomplish in our work together. As we reviewed the informed consent, I began to tell him a bit about me and about what therapy would look like with me. I related that my clinical work had begun with children who were dying. And I told him that I took my work seriously. That was the beginning of our therapeutic relationship.

More and more, our work has deepened and I have grown to appreciate Tom's willingness to risk; I have come to appreciate his courage. In our work together, I have kept in mind that he has charmed his way through life, maintaining a highly functional persona through the power and privilege he enjoys, masking his pain and vulnerability. My hope has been to communicate care and to acknowledge *him*, while helping him recognize and deal honestly with not only his pain, but also the pain experienced by others in relation to him. I have responded to his queries honestly and sometimes brutally. Humor and dead seriousness have both been integral parts of our therapy. It has appeared that through this very direct relational exchange we have been able to maintain and deepen our connection. Safety, for *him,* has come in knowing that his attempts at manipulation will no longer go unnoticed—or unaddressed.

Although Tom and I have deepened our connection, we have, at times, lapsed into periods of disconnection. Our therapeutic relationship has been at times difficult but always genuine. Tom has come to acknowledge his alcoholism, problems with commitment and intimacy, and loneliness. I have heard stories that rattle my sensitivities and challenge me to remember the pain that would drive some of his behaviors. As a woman, I am sometimes challenged to work with my own feelings and to respond appropriately when he reports casual abuses of power with women. As the therapist, I am aware of the power I hold in the relationship. I have become even more aware of that as time goes on. I am challenged and committed to using my power in the relationship consciously and constructively. Ultimately, Tom has been able to look at his own pain, and more recently, to look at the ways in which he has created suffering for others. Our work has not been easy or simple. It *has* been authentic.

CONCLUSION

As we have seen throughout the chapter, addiction is characterized by pain and suffering, whether or not it is acknowledged. Much of our work as therapists is to create a space where such acknowledgment is possible. This acknowledgment, of course, brings with it feelings of confusion and bewilderment when our clients begin to recognize that the freedom they have staunchly protected and fought for was never really there. It also brings feelings of frustration in knowing that although they may have been free to come and go, they have not had the freedom to stay and to do so contentedly.

Hope for transformation comes in knowing that to have genuine freedom, we must develop the capacity to connect, invest, and provide mutual care for and with the important people in our lives. Indeed, substantive relief from addiction requires that we access our vulnerabilities and connect with our authenticity. As such, we discover our courage and strength to participate in growth-fostering relationships and to tap into our sense of commitment when disconnections within these relationships arise. With that, we are ultimately positioned to more fully experience and enjoy the fruits of relational depth and the boundless possibilities for relational freedom.

REFERENCES

Alcoholics Anonymous World Services, Inc. (1976). *Alcoholics Anonymous.* New York: Alcoholics Anonymous World Services, Inc.

Beattie, M. (1987). *Codependent no more.* Center City, MN: Hazelden.

Bratter, T. & Forrest, G. (1985). *Alcoholism and substance abuse: Strategies for clinical intervention.* New York: The Free Press: A Division of Macmillan, Inc.

Comstock, personal correspondence, September, 2003.

Covington, S. (1994). *A Woman's way through the twelve steps.* Center City, MN: Hazelden.

Covington, S. & Surrey, J. (2000). The relational model of women's psychological development: Implications for substance abuse. *Work in Progress, No. 91.* Wellesley, MA: Stone Center Working Paper Series.

Grant, B. F., & Dawson, D. A. (1998). Age of onset of alcohol use and its association with DSM-IV alcohol abuse and dependence: Results from the National Longitudinal Alcohol Epidemiological Survey. *Journal of Substance Abuse, 9,* 103–110.

Greenspan, M. (2003). *Healing through the dark emotions: The wisdom of grief, fear, and despair.* Boston & London: Shambhala Publications, Inc.

Jampolsky, L. (1991). *Healing the addictive mind: Freeing yourself from addictive patterns and relationships.* Berkeley, CA: Celestial Arts.

Jordan, J. (2001). A relational-cultural model: Healing through mutual empathy. *Bulletin of the Menninger Clinic, 65*(1), 92–103.

Kilbourne, J. (1999*). Deadly persuasion: Why women and girls must fight the addictive power of advertising.* New York: Free Press, Inc.

MacDonald, S. & Wells, S. (2001). Factors related to self-reported violent and accidental injuries. *Drug and Alcohol Review, 20,* 299–307.

Matano, R. A., Wanat, S. F., Westrup, D., Koopman, C., & Whitsell, S. D. (2002). Prevalence of alcohol and drug use in a highly educated workforce. *The Journal of Behavioral Health Services & Research, 29* (1), 30–44.

Miller, J. (1976). *Toward a new psychology of women.* Boston: Beacon Press.

Miller, J. & Stiver, I. (1993). A relational approach to understanding women's lives and problems. *Psychiatric Annals, 23* (8), 424–431.

Miller, J. & Stiver, I. (1997). *The healing connection: How women form relationships in therapy and in life.* Boston: Beacon Press.

Nakken, C. (1996). *The Addictive personality: Understanding the addictive process and compulsive behavior.* Center City, MN: Hazelden.

Orford, J. (2001). Addiction as excessive appetite. *Addiction 96,* 15–31.

Peabody, S. (1994). *Addiction to love.* Berkeley, CA: First Celestial Arts Printing.

Peele, S. (1985). *The meaning of addiction: Compulsive experience and its interpretation.* Lexington, MA, and Toronto: Lexington Books, D. C. Heath and Company.

Peele, S., Brodsky, A., & Arnold, M. (1991). *The truth about addiction and recovery.* New York: Fireside/Simon and Schuster.

Stiver, I. (1990). Dysfunctional families and wounded relationships. *Work in Progress, No. 41.* Wellesley, MA: Stone Center Working Paper Series.

Shen, S., Locke-Wellman, J., & Hill, S. Y. (2001). Adolescent alcohol expectancies in offspring from families at high risk for developing alcoholism. *Journal of Studies on Alcohol, 62*(6), 763–772.

Stroebe, M., Gergen, M. M., Gergen, K. J., and Stroebe, W. (1993). Hearts and bonds: Resisting classification and closure. *American Psychologist, 48,* 991–992.

Valliant, G. E. (1985). Loss as a metaphor for attachment. *American Journal of Psychoanalysis, 45:1,* 50–67.

Wegscheider-Cruse, S. (1985). *Choicemaking.* Pampano Beach, FL: Health Communications, Inc.

15

Spiritual Development

By Melanie Munk, M. A.

REFLECTION QUESTIONS

1. *What does spirituality mean to you?*
2. *How is religion different from spirituality? How is religion related to spirituality? Are they even different or related in your belief/experience?*
3. *If you have a "spiritual aspect," how did that develop for you? If you don't have a "spiritual aspect," how did that develop for you?*
4. *How has your spirituality changed or developed over your lifetime and what caused it to change?*
5. *Have you ever experienced difficulties with your spirituality? What was that like for you?*

INTRODUCTION

Spirituality has long been a passion of mine as it has had such a defining role in the meaning of my life thus far. Shortly after being asked to write this, I came excited to my "Gender and Ethnicity Throughout the Lifecycle" class, as the student presenters that night were to discuss gender and spirituality. I was prepared to take copious notes, find many more resources, and participate in what I thought would be a lively discussion. The presenters instead chose to break us into small discussion groups and presented us with a questionnaire they felt would elicit information about our views of our spirituality and our gender. I was dismayed and angry after reading just the first question, "What is your conceptualization of the supreme being?" The question implied to me first that I *should* have a concept of a supreme being, second that there is in fact a supreme being, and third that there is only one supreme being. In other words, the questions implied a Judeo-Christian religious worldview as being the only possible representation of spirituality.

When I objected to the questions and refused to answer after explaining their implications and my rejection of them as a thinly veiled litmus test of my level of Christian indoctrination, rather than an exploration of our personal spiritualities, I was confronted. I repeatedly refused to answer whether or not I was a Christian, and just what were my beliefs. I was offended by the questionnaire. I felt marginalized and threatened but left that night with an increased understanding. There is confusion even among professionals as to what defines religion versus what defines spirituality.

My purpose here is to attempt to define spirituality and examine its roles in the context of development within different populations. There exists a paucity of research in the areas of spirituality and spiritual development across the life span, as well as in contexts such as within cultures, sexualities, genders, etc. Therefore, this chapter will take an exploratory route through some of the research that is available. It is my strong belief, as well as that of others, that the area of spirituality is an area of cultural competence that counselors need to explore as part of their education program to avoid such marginalization of clients (Burke & Miranti, 1995; Pate & Bondi, 1992). I will refrain from discussing specific religious beliefs and doctrines except where it has a demonstrable impact on certain populations, as it does within the bisexual, gay, lesbian, and transgendered/transsexual (BGLT) community. A better question therefore to begin a discussion of spirituality would be: What is spirituality to you?

For me, spirituality is the connection I feel beyond myself to other people and the forces of nature. It includes the following examples. It's Sunday afternoon and I've been working feverishly on a project for school. For the past couple of weeks, I have been experiencing intense flashbacks that have made it very difficult for me to concentrate. My friends have been noticing that I have been withdrawing and generally reacting negatively to things. And though I've been seeing my therapist and talking often with my friends, I am experiencing extreme difficulty maintaining a sense of routine and balance. I put down what I am reading, and take a moment to just breathe. In that moment I experience a connection to something outside myself. I write the following down on my reading, "It occurs to me that I must live more by my spiritual values and less by my old tapes, by unquestioned cultural values, and by the rest of those things that come automatically. Mindfulness is the key."

It's Friday afternoon after work, and I felt particularly ineffective today. In my work with homeless women and men, and with women living at the battered women's shelter, I know that support from my colleagues is critical because of the level of trauma we deal with on a daily basis. Unfortunately for about the past month or so we haven't had much time to support one another and our staff meeting was especially brief today. I walked outside to my vehicle not wanting to be alone, but everyone else has already gone their separate ways for the weekend. I know the reaction that I'm having right now is a combination of both what I'm going through personally as well as the challenges of doing the work that we do. Today though I'm really feeling an intense need to be connected with and validated by my coworkers, and I am upset that I did not get a chance to talk about what was going on with me and my needs (our staff meetings usually include this component). Before getting into my vehicle, I take a moment to ground myself and to get centered. In those moments I reexperience my connection to my family at work as well as my connection to everything else. I receive the validation that I needed, and a sense of calm and peacefulness surrounds me. I am comforted in the knowledge that even though we can't always be present for one another we are always there for one another in spirit.

I begin this discussion of the definition of spirituality with two examples from my own life. Spirituality is an elusive term to define within the literature. Within the general public however most people seem to be able to give examples of spirituality in their own lives. What I have defined as spiritual for myself above, someone else might not see as spiritual at all. Instead, my first experience could be named "cognitive insight," and my second experience could be some type of "emotional reframing" or a "self-soothing technique." In this context, spirituality appears to be a subjective experience.

DEFINING SPIRITUALITY

Spirituality has only recently become an area of interest for therapists, being seen in the past as belonging to religious leaders rather than a part of an integrated human growth and development theory. As such, professionals still debate what constitutes spirituality versus what constitutes religion. This debate has a tendency to mirror our respective internal processes, and thus the reader may leave this section feeling confused. There can be a tendency to limit one's definition of spirituality to one's own stage of spiritual development and views on organized religion, creating remarkably different definitions. Differentiating spirituality from other components of individual identity development is therefore difficult.

Haldeman (1996) stated that "the spiritual components of identity development have scarcely been mentioned in the mental health literature" (p. 882). Davidson (2000) discusses "spiritual and psychological identity as two parallel and overlapping processes" and thus therapists can experience "difficulty in separating psychological therapeutic processes from spiritual development" (pp. 420–421). Benner (2002) argues that the difference between spirituality and psychology is in orientation. Spirituality orients a person toward the transcendent while psychology orients the person toward groundedness (p. 359). Ingersoll (1995) states that spirituality is "an organismic construct endemic to human beings," meaning that everyone possesses a spiritual dimension regardless of whether the individual exercises that dimension or not (p. 12). Faiver et al. (2001) build on Ingersoll's definition of spirituality and state that "we believe spirituality is an innate human quality . . . the experience of spirituality is greater than ourselves and helps us transcend and embrace life situations" (p. 2).

Spirituality is defined many different ways throughout the literature. In attempting to define and construct a humanistic-phenomenological spirituality separated from religious components and applicable across religious differences, Elkins et al. (1988) define spirituality as "a way of being and experiencing that comes about through awareness of a transcendent dimension and is characterized by certain identifiable values in regard to self, others, nature, life, and whatever one considers to be the Ultimate" (p. 10). The nine components of spirituality synthesized by these researchers are:

1. A transcendent dimension
2. Meaning and purpose in life
3. Mission in life
4. Sacredness of life
5. Material values (meaning that material things are valued but not sought as the end of spiritual pursuits)
6. Altruism
7. Idealism
8. Awareness of the tragic
9. Fruits of spirituality (pp. 10-12)

Pearlman and Saakvitne (1995) view spirituality as a frame of reference through which individuals interpret experiences. The authors list four components of spirituality: "orientation to the future and sense of meaning in life, awareness of all aspects of life, relation to the nonmaterial aspects of existence, and sense of connection with something beyond oneself" (p. 63). They state that spirituality is "the meeting place of identity and world view" (p. 63). Ingersoll (1995) lists seven dimensions of spirituality:

1. One's conception of the divine, absolute, or "force" greater than one's self.
2. One's sense of meaning or what is beautiful, worthwhile.
3. One's relationship with Divinity and other beings.
4. One's tolerance or negative capability for mystery.
5. Peak and ordinary experiences engaged to enhance spirituality (may include rituals or spiritual disciplines).
6. Spirituality as play.
7. Spirituality as a systemic force that acts to integrate all the dimensions of one's life. (p. 11)

Buchanan et al. (2001) discuss two types of orientations regarding religion and spirituality: intrinsic and extrinsic. An intrinsic orientation is a "reliance on an internal authority, meaning that the expert is the individual, truth is derived from individual experience, and great value is placed on personal insight" and is "identified more closely with spirituality" (p. 436–437). Spirituality is defined as a "personal belief system" (pp. 436–437). Religion is identified with an extrinsic orientation as "religion is usually associated with institutionalized beliefs and attitudes" (p. 436). Extrinsic orientation is seen in religion "where truth is maintained by religious leaders, scriptures, and institutions" (p. 437).

The commonalities in these various definitions and conceptualizations of spirituality inform a view of spirituality as an intrinsic organismic potential that provides the individual with a frame of reference from which to integrate the various dimensions of one's life as well as a way of being and experiencing. It usually includes some type of transcendent aspect (although what some experience as transcendent, others may experience as immanent); creates a sense of meaning in one's experiences and one's chosen path; and includes some type of sacredness, appreciation, or mystery. In addition, spirituality informs how one relates to the self, to others, and to something beyond oneself whether that is seen as some god(dess), god(desse)s, nature, the universe, the higher self, the interconnectedness of everything, the ground of being, or some other conceptualization. When seen as such a broad aspect of meaning-making and defining ways of being and experiencing, it becomes vital for counselors to explore and understand their own and their clients' spirituality and spiritual development.

SPIRITUAL DEVELOPMENT

The actual process of spiritual development across the life span has rarely been studied. Within the research literature there exist few qualitative studies and even fewer longitudinal studies that address the issue of spiritual development in adulthood. When spiritual development has been studied, it has generally been in relation to issues of gerontology and dealing with the end of life such as religious coping mechanisms. Research that addresses spiritual development as a change over time or that takes into account the sociocultural context of spirituality is a virtually unexplored area of adult development (Tisdell, 2002). The model most often cited as a basis for spiritual development research is James Fowler's stages of faith model.

Briefly, Fowler's (1981) model of faith development is a stage model consisting of six separate, consecutive phases. Similar to Kohlberg's moral development model, Fowler does not believe that all persons progress through all stages of growth, with development generally arresting at the third stage for many adults. When building his faith development model, Fowler drew upon both Kohlberg's moral development model and Piaget's cognitive development model. Fowler's six stages of faith development are as follows:

1. *Intuitive-projective faith.* This stage represents the faith of a young child up to early verbal stages. The child responds to the emotional environment of the family and congregation, learns cultural "rights" and "wrongs," and develops imagination. An example of this stage of faith is a child's belief in Santa Claus.
2. *Mythic-literal faith.* This stage represents the verbal young child up to later childhood. Here the child is fascinated by the myths of a faith and responds to these myths literally without any abstract meaning attached. The child may also respond to mythic symbols of the faith such as crucifixes, statues, etc. An example of this stage of faith is a child's belief in a parental image of god as one who punishes the bad and rewards the good.
3. *Synthetic-conventional faith.* This is the stage of older childhood. Faith is seen as that which brings people together and provides a unifying concept and sense

of belonging for family, congregation, and society. For many, this is the terminal stage of development. In this stage, individuals do not acknowledge differences in faith practices of others and view their faith as the "one right, true, only way." An example of this stage can be seen in adolescents who form groups based on fitting in: if you wear these clothes, listen to this type of music, like these people, etc., then you are part of the group. At this stage, any image of deity is seen as a companion and ally. Faith is rule bound and hierarchical with no questioning of the group's norms and beliefs.

4. *Individuating-reflective faith*. This stage may begin in adolescence or later. In this phase, the individual senses the need to develop a personal faith within oneself for the purpose of creating meaning of one's experiences and one's life in relation to others and society. The individual turns away from external sources of spiritual authority and begins to focus more on what is personally meaningful, developing an internal spiritual orientation. The focus of faith moves away from being viewed as the unifying concept of the group and more as making sense of the individual. This stage may occur in those who stay within organized religious practice, as well as in those who leave. An example of someone in this stage of faith is a person who begins to critically examine their beliefs and reflect on values, ideals, and meaning.

5. *Conjunctive faith*. At this stage, the individual acknowledges the paradoxes inherent within the faith journey. Faith is tested by these paradoxes, and faith is deepened along with empathy for others on their own faith journeys. The individual gains a keener awareness of the complexities of faith and is more comfortable with ambiguities. The individual becomes more open to religious and spiritual traditions different from one's own. An example of someone at this stage is a person willing to respect the validity of another's "truth" even when it contradicts one's own, while simultaneously being able to communicate one's own authentic "truth."

6. *Universalizing faith*. Viewed more as a category of individuals whose examples of faith have a profound impact on others' faith journeys. These persons have given up their own lives to actualize their spiritual values in the world. Recent examples include Mother Teresa, Martin Luther King Jr., and Gandhi.

A major drawback to Fowler's model of faith development is that it was constructed from a sample consisting of 97 percent White Judeo-Christians. Its generalizability is therefore limited.

SPIRITUAL DEVELOPMENT IN LATER ADULTHOOD

One model of spiritual development in later adulthood views spiritual growth "as the positive outcome of the maturation process" (Wink & Dillon, 2002, p. 80).

This general theory of adult spiritual growth proposed by Jung and others views middle to later adulthood as the most important time for spiritual growth. Prior to this time, the individual is seen as being engaged in a number of external tasks such as pursuing a career and forming a family, which removed attention from internal reflection. A second model of spiritual development in later adulthood "suggests that ageism and age discrimination [push] many older adults to spirituality, presumably by fostering disengagement and curtailing life choices" (p. 80). Both of these models propose that spirituality advances more in middle to late adulthood due to increased time to be introspective on the part of the individual. What they do not take into consideration is spiritual development that happens earlier in life despite a purported focus of the individual on external tasks.

Atchley (1997) constructed a theory of spiritual development in later adulthood based on an Eastern perspective. Atchley states that "spirituality is not an activity or an idea, but instead is an *experience*" (p.127). In his view, the "contemplative mystical experience is the psychological dynamic that accounts for spiritual development in later life" (p.126). Atchley believes that while aging in and of itself does not bring about spiritual development, "aging does alter the conditions of life in ways that could heighten awareness of spiritual needs and that can stimulate spiritual development" (p. 129). When people are slowed by age, there's more time available for practices that encourage mystical experiences such as meditation and contemplation. These mystical experiences in turn encourage further spiritual practice which will lead to further mystical experiences, thereby setting up a type of feedback loop in spiritual development.

Stokes (1990) stated that "changes in faith tend to occur more during periods of transition, change and crisis than during times of relative stability" (p. 177). This certainly characterizes middle to late adulthood when parents are sending children out into the world for the first time, friends and loved ones are passing away more frequently, and people are beginning to retire. One longitudinal study of spiritual development across the adult life cycle found "a significant increase in spirituality from late middle (mid-50s/late 60s) to older adulthood (late 60s/mid-70s)" (Wink & Dillon, 2002, p. 91). These authors found that spiritual development occurs most often in individuals "who are psychologically minded, invested in the world of ideas, and tended to experience adversity in their lives" (p. 93). It is important to note that all but two of the 290 participants in this study were White, and all participants professed a Judeo-Christian belief system, limiting generalizability of this study.

Within traditional psychological theory, Erik Erikson addresses spirituality only obliquely in his final stage: Integrity versus Despair. During this final stage, it is believed that the individual's task is to review one's life as having been meaningful and satisfying. While these theories do not give a definitive model for examining spirituality in the context of older adulthood, they do highlight the importance of spirituality in these persons' lives. As spirituality is a significant meaning-making process in many people's lives, it is important to understand spirituality in the context of coming to terms with reviewing one's life, and the end of one's life.

SPIRITUAL DEVELOPMENT IN THE CONTEXT OF IDENTITY DEVELOPMENT

Identity development and developmental crises are often the impetus that drives individuals to consider their own spirituality. These crises may be a onetime predictable crisis such as moving away from home for the first time or having a baby. The crisis could be a onetime traumatic crisis such as the loss of a loved one, a violent attack or rape, or a natural disaster. Lastly, the developmental crisis may be a chronic trauma or adversity such as child abuse, spouse abuse, chronic illness, racism, sexism, heterosexism, and addiction. The remainder of this chapter will examine spiritual development within the context of culture, gender, sexual orientation, and chronic trauma through the use of my personal narrative. In a discussion of chronic adversity and chronic trauma, it is important to understand the role of shame.

In Relational-Cultural theory (RCT), shame is a major source of disconnection and isolation. Shame can be defined as a deep sense of unlovability within the individual, as though one has an essential defect (Jordan & Miller, 2002). When in shame, the individual does not trust that she can bring her entire self into relationship. The individual does not bring the shamed parts of herself into relationship, so they don't grow and change. Shaming can be an aspect of sociopolitical force used to isolate and silence people such as that which occurs in racism, sexism, and heterosexism. The result of shame on the individual is a feeling of condemned isolation, feeling locked out of the possibility of human connection. This feeling of isolation leads to self-blaming—feeling like the individual is the problem. Shame can also lead to victim blaming. Lastly, I believe that shame and the awareness of the shamed parts of our selves are a large part of what can initially create a "loss of faith" or a "crisis of faith."

Individual Context

My own identity development involved a dramatic shift and healing of shame within my relationship with the divine. I grew up in a White, working class, Catholic family that rarely attended services except as affected the chances of my sister and me receiving the sacraments. In addition to these obvious aspects of my family there was the family secret: the daily physical and emotional abuse that my sister and I endured. As a young person, I always felt a particular pull to that which was transcendent but which I also felt was immanent within me and especially within nature. This spiritual connection was what kept me sane growing up in my home and what gave me hope that I would one day be able to leave.

When I moved away from home, I threw myself into Catholicism and became a deeply devoted person. My relationships were primarily centered around people I knew in church, around the adolescents in my youth group, and especially with sisters in various religious orders. After my faith was firmly established, I began seriously questioning my sexual orientation. Based on my relationship with the divine as I understood it, I came to the conclusion that even though I knew I was a lesbian, it would be wrong for me to live my life as such. Therefore I began forming relationships with various religious orders with the intention of becoming a nun. Having been starved of connection as a child, I craved the openness, support, and deep connections that I saw within the religious orders I visited. I had no idea how to form these relationships in my own life.

As the years went on, and the tension within me grew between who I felt I was and with whom I felt my relationships, including my relationship with the divine, said I "had to be," my relationships gradually began to shift. I moved away from where I had been living at the time shortly after I "came out." The consequences of coming out for me were that I lost all the relationships within my church community, which at the time comprised all of my primary relationships (I had been in the army, where there is strict taboo against homosexuality). I was left only with my relationship to the divine, and I felt very disconnected in that relationship as I could not reconcile who I felt myself to be and the shame I was taught in my church that was attached to my identity. I struggled a long time with feeling as though I had been persecuted, or I was being tested, wondering if the universe just didn't truly care about me. I already felt unlovable by my own parents and now by the people who had been in my life, now I began to feel unlovable by the divine. My shame led me to a loss of faith.

Gradually I left the church, and began to heal my relationship with the divine through examining other possible spiritual relationships. I searched for a long time, and experienced much doubt and guilt along the way because I was afraid of what I had been taught, that there was truly only one way. I knew, however, that I could not exist in this position of shame and that I needed to feel connected to the divine. As I gradually changed my spiritual identity, my relationship with the divine began to heal, and other relationships in my life began creating further growth opportunities. Ultimately it was the shame I felt because of my religious upbringing that initiated my crisis of faith and the healing of that shame that continued my spiritual development.

Identity Development

Within the current literature there are numerous different models of identity development. I would like to discuss here a realist theory of identity, which I believe can help illuminate the process of spiritual development. According to Moya (1997), "[T]he advantage of a realist theory of identity is that it allows for an acknowledgement of how the social facts of race, class, gender, and sexuality function in individual lives without *reducing* individuals to those social determinants" (p. 136). A realist theory of identity is grounded in the individual's social location and is relevant to this discussion of spiritual development as such.

There are six claims that Moya makes in constructing a realist theory of identity. The first claim "is that the different social facts (such as gender, race, class, and sexuality) that mutually constitute an individual's social location are causally relevant for the experiences she will have" (p. 137). In my own identity development and spiritual development, the interaction of gender (my femaleness), culture (White, working class, Catholic), and sexuality (lesbian) were relevant.

The second claim of a realist theory of identity "is that an individual's experiences will influence, but not entirely determine, the formation of her cultural identity" (p. 137). In my spiritual development, my acceptance of a lesbian identity allowed me to move away from my religious upbringing as a Catholic. The resolution of the shame I felt enabled me to move through my loss of faith to develop a deeper, more personally meaningful spirituality. Another in a similar position may not make the same choices that I did. One may choose instead to remain in the Church and work for social justice within the structure, or one may choose to accept her lesbian identity but not act on her sexuality. One may also choose to abandon entirely a spiritual identity; the permutations are endless and subjective.

Moya's third claim is "there is a cognitive component to identity which allows for the possibility of error and of accuracy in interpreting experience" (p. 138). For example, in my own coming out I had never experienced discrimination on the basis of my sexuality. Rather than considering that my experience with others was based on heterosexism, I interpreted my experience as a confirmation of the abuse I had been through in the past, that there was actually something fundamentally unlovable about myself. Gradually as I met others with different viewpoints, I was able to reinterpret my experience as heterosexism. As I explored other spiritual paths that did not condemn homosexuality, my connection with the divine was further strengthened.

The fourth claim Moya makes is that "some identities, because they can more adequately account for the social facts constituting an individual's social location, have greater epistemic value than some others that same individual might claim" (p. 138). Epistemic value "refers to a special advantage with respect to possessing or acquiring knowledge about how fundamental aspects of our society (such as race, class, gender, and sexuality) operate to sustain matrices of power" (p. 136). In my own experience, my identities as a lesbian and as a Wiccan illuminate best for me the systems of power within our society.

The fifth claim of a realist theory of identity "is that our ability to understand fundamental aspects of our world will depend on our ability to acknowledge and understand the social, political, economic, and epistemic consequences of our own social location" (p. 139). For example, when I was first coming out, I did not understand the ramifications of that action in terms of my own social location. My whiteness had allowed me the privilege of "not seeing" and thus I attributed my new social location to myself rather than understanding the ramifications of heterosexism.

The last of Moya's claims "is that oppositional struggle is fundamental to our ability to understand the world more accurately" (p. 140). It was only after I began to reject the shame I felt at being a lesbian that I began to understand the concepts of power and privilege. I began this chapter with an example of a class I attended where the presenters assumed a Christian value system in a discussion of spirituality. Part of my frustration in that class was a blindness of others to understand what my objections were to what was to them a harmless questionnaire. My own "oppositional struggle" which I have achieved through my spiritual development has helped me to understand better at times and to see the true importance of individual context.

Cultural Context

In a qualitative study of "women of various race and class backgrounds guided by feminist and antiracist educational perspectives, who have also had to renegotiate their spirituality in light of having been raised in patriarchal religious traditions" Tisdell (2002) found four relevant themes of spiritual development in women adult educators for social change (p. 128). The four common themes that Tisdell found in her research were:

1. Moving away-spiraling back.
2. A deepening awareness and honoring of the Lifeforce.
3. The development of an authentic identity.
4. The requirement of social action.

Moving away and spiraling back is associated with "a re-evaluation process and the reworking of such childhood symbols and traditions and reshaping them to be more relevant to an adult spirituality" (p. 133). This moving away from childhood spiritual systems was attributed by participants to "underlying sexism, heterosexism, racism, or other hypocrisy within those traditions" (p. 133). A deepening awareness and honoring of the Lifeforce was associated with mystical experiences in which participants experienced mystery and "the interconnectedness of all things" (p. 134). The development of an authentic identity occurred in the context of their spiritual connection despite teachings of participants' current or previous religions against those identities. In my own spiritual development, it was this formation of an authentic identity that caused a crisis of faith and a moving away from my previous religion.

The requirement of social action was partially a function of the selection criteria for this study, but may also have been influenced by cohort effects as "most of these women were strongly affected by the civil rights movements and the women's movement of the 1960's and 1970's" (p. 137). It seems that the outcome of the participants' deep questioning may have led to social justice efforts or resulted from their social action involvement. The author further notes that "spiritual development emerges in the overall context of lives" (p. 132).

The Context of Gender

As stated before, one aspect of spirituality and spiritual development is that the individual develops a relationship with her conceptualization of the divine and uses her spirituality to further inform other relationships in her life. Within this context, RCT becomes very informative for women's spiritual development. Feminist spirituality is a form of spirituality that is informed by RCT and has grown out of the feminist movement. Characteristics of feminist spirituality are:

"being rooted in women's experiences, placing women at the center, reverencing the earth and all creation, valuing women's body and bodily functions, seeking an interconnectedness with all living things, and placing an emphasis on ritual" (Neu, 1995, p. 189).

From the standpoint of RCT, Warwick (2001) recently undertook to examine women's religious identity in therapy. In her work, Warwick defines spirituality as "the expression of connection to the forces of life and death, as well as the practice of personal rituals, myths, and activities designed to develop or enhance that connection" (2001, p. 124). Warwick notes that just as the quality of our relationships with others may vary (for example, growth-fostering or destructive) so may our relationship with the divine vary from person to person and range from healthy to unhealthy, open or closed. Warwick states that "sometimes women need to change religious identities as they redefine themselves" (p.127). This can be especially true when women begin to heal the relationships in their lives, and also through the points in their lives when their understanding of their identity shifts. In conclusion, Warwick (2001) illustrates the point that I feel is necessary to reiterate here: "We are well trained to explore relationships with clients, and . . . a client's relationship with her gods and/or her religious community is fair territory to explore" (p. 130).

SPIRITUALITY AND SEXUAL ORIENTATION

As previously examined, spiritual development is impacted by early religious upbringing. In order to examine spirituality in the context of sexual orientation, it is important to understand various religions' views of sexual orientation. This section will examine religion and sexual orientation as well as spirituality and sexual orientation.

Religion and Views of Sexual Orientation

Institutionalized religion, particularly the three major monotheistic religions of the West—Judaism, Christianity, and Islam—have traditionally perpetuated intolerance and oppression of openly bisexual, gay, lesbian, and transgendered/transsexual (BGLT) persons. Davidson (2000) states that "many experts have articulated a belief that because of the oppression of heterosexist society, gay, bisexual, and lesbian persons are more in need of, and more open to, spiritual nourishment than others" (p. 409). Examining major religions' views on BGLT persons is essential to understanding both the intersection of sexuality and spirituality, and in understanding the oppression of BGLT persons as well as their spiritual needs.

Judaism traditionally has prohibited sexual acts between same-sex partners. Davidson states that "the union between male and female is viewed as a fundamental basis for Jewish family life" (p. 412). Some movements have been made toward tolerance of gays and lesbians in Reform and Reconstructionist branches in recent times. The predominance of Judaism, however, believes that homosexuality is a sin, and "for Orthodox Jews, sexual acts by same-sex partners remains a violation of nature" (pp. 412–413).

The Roman Catholic Church offers the clearest stance on homosexuality. In a 1986 letter to the Bishops of the Catholic Church on the pastoral care of homosexual persons, Cardinal Ratzinger states that "although the particular inclination of the homosexual person is not a sin, it is a more or less strong tendency ordered toward an intrinsic moral evil; and thus the inclination itself must be seen as an objective disorder" (1994, p. 40). Cardinal Ratzinger's letter leaves little doubt to the "faithful" on the Catholic Church's stance against homosexuality and the Church's intolerance of the BGLT person. Within the Church, a suborganization known as Dignity has arisen to address the needs of homosexuals in the Church, although the movement is not widespread especially in rural areas, and is not sanctioned by Rome. It is perhaps the perfect irony in view of the Church's stance on homosexuality that "gay men represent a disproportionately large percentage of Catholic clergy" (Davidson, 2000, p. 415).

Most recently, the Congregation for the Doctrine of the Faith on June 3, 2003, published "Considerations Regarding Proposals to Give Legal Recognition to Unions between Homosexual Persons." Within this document, the Catholic Church further states that "homosexuality is a troubling moral and social phenomenon" (Ratzinger & Amato, June 2003). Furthermore, Ratzinger and Amato call on Christians to protest homosexual unions by:

> unmasking the way in which such tolerance might be exploited or used in the service of ideology; stating clearly the immoral nature of these unions; reminding the government of the need to contain the phenomenon within certain limits so as to safeguard public morality and, above all, to avoid exposing young people to erroneous ideas about sexuality and marriage that would deprive them of their necessary defences [sic] and contribute to the spread of the phenomenon. Those who would move from tolerance to the legitimization of specific rights for cohabiting homosexual persons need to be reminded that the approval or legalization of evil is something far different from the toleration of evil. (2003)

Ratzinger and Amato further state:

> the absence of sexual complementarity in these unions creates obstacles in the normal development of children who would be placed in the care of such persons. They would be deprived of the experience of either fatherhood or motherhood. Allowing children to be adopted by persons living in such unions would actually mean doing violence to these children, in the sense that their condition of dependency would be used to place them in an environment that is not conducive to their full human development. This is gravely immoral. (2003)

In this most recent document, the Catholic Church continues to propagate false stereotypes against BGLT persons despite overwhelming evidence to the contrary, especially in regard to the impact on the development of children for which there is no demonstrable impact (for example, see Chapter 8, "Coming Out and Living Out Across the Life Span," by Stacee Reicherzer).

Protestant denominations, lacking one single authoritative head, range in views on homosexuality from acceptance to tolerance or qualified acceptance to the outright condemnations of biblical fundamentalists. Protestant religions that openly welcome gays and lesbians include the Unitarian Universalists and the Quakers. According to Davidson, "qualified acceptance often means adopting a position of hating the sin (sexual acts between same-sex partners) but loving the sinner (LGB persons)" (p. 413). This position is often not helpful to BGLT persons as it sets up what Davidson terms a double-bind where "affirmation by the church community is available only through celibacy or (false) heterosexual marriage" (p. 413). Both options ask the BGLT person to be other than what she or he is. The Metropolitan Community Church (MCC) is one specifically gay and lesbian sect that has formed in response to the needs of Protestant homosexuals. Most recently, on August 5, 2003, the Episcopal Church confirmed the nomination of the Church's first openly gay bishop, the Reverend Gene Robinson, which has created much controversy within that denomination.

Islamic doctrine views "the individual's role in heterosexual marriage and family . . . as a primary and compelling obligation to God and to society" (p. 414). While Islam does not encourage celibacy as does the Catholic faith, it views heterosexual marriage as the only acceptable norm. Davidson states that "sex outside of marriage is a sin—theoretically punishable by death" (p. 412).

As a result of the hostility that BGLT persons have faced from organized religion, many have turned to alternative practices and beliefs and many have turned away from spirituality altogether. Traditional Native American beliefs view two-spirited people (bisexual persons, lesbians, and gay men) as sacred. These beliefs have led to two-spirited people being the traditional shamans and healers of a tribe. Other alternative spiritual practices that BGLT persons have turned to include neo-paganism, witchcraft, New Age practices, and Buddhism (pp. 415–416). Starhawk (1982), herself a bisexual woman and a leader of Reclaiming (a modern witchcraft movement), believes that lesbians are threatening to traditional patriarchical institutions and religions "because they are women refusing to serve as ground for the male self" (p. 141). Because of the oppression BGLT persons experience within traditional organized religion, it is no surprise that many seek refuge in alternative spiritualities or abandon spirituality altogether.

Views of BGLT Spirituality

According to Ritter and O'Neill (1989), "for a religious experience to be viable it must be philosophically reasonable, morally helpful, spiritually illuminating, and communally supportive" (p. 9). In accordance with these principles, the authors explain how the traditional Judeo-Christian heritage of the United States is none of these to BGLT persons. Judeo-Christian beliefs are not philosophically reasonable in that they "often [fragment] individuals by seeming to demand that individuals change or be forgiven for their *very* nature" (p. 10). It is not morally helpful in that the moral advice given to BGLT persons is "posited in the negative and focuses directly on the prohibition of intimate and sexual relationships between same-gender couples" (p. 10). The Judeo-Christian heritage is not spiritually illuminating to BGLT persons because:

> Most mainline churches have not provided models of sacred gay lives and same-gender relationships nor identified the gay or lesbian person's central role in a variety of sacred traditions nor presented a mythology broad enough to encompass the life questions of lesbian and gay individuals, and thus have not offered them a pathway to holiness. (p. 10)

In addition, Judeo-Christians are typically not communally supportive and "openly gay men and lesbians rarely find much communal support from traditional organized religion" (p. 10).

Buchanan et al. (2001) believe that the "extent of the struggle with sexual and spiritual identities depends on whether a person has an intrinsic or extrinsic orientation to religion" (p. 438). A person with an intrinsic orientation is believed to be more likely to accept internal messages regarding one's sexual identity, whereas "this process is made more difficult for those who have had a religious upbringing" (p. 438). The authors further state that "traditional extrinsic orientations to religion increase the level of internalized homophobia, delay the development of a homosexual identity, and add conflict to an already difficult path that gays and lesbians are taking" (p. 439). According to these authors, "an extrinsic orientation would lead a person to 'use' his/her religion to achieve acceptance, societal status, or personal desires for security" (p. 437). The authors define fundamentalism as an extreme form of extrinsic orientation and state that "fundamentalism can form an extreme challenge to forming sexual identity, if that identity does not adhere to the requisite religious beliefs" (p. 437).

In their article, Buchanan et al. list two possible outcomes to the struggle lesbians and gay men face when dealing with their sexual identity and their spiritual/religious identity. The first is choosing between two worlds. For example, within the Catholic Church, a lesbian can "either follow church doctrine and reject the inclinations toward a . . . lesbian identity, or she . . . can reject [her] affiliation with the church and accept the emerging sexual identity" (p. 440). In addition, many other mainline churches preach that the BGLT person must give up her or his sexual identity or forever live in sin against God's will. The other option available with this outcome is the "rejection of religious tenets and the church as an institution" (p. 440). A study cited by the authors found that 69 percent of gay men drawn from a community sample have rejected their religion (p. 440).

The second possible outcome of this struggle that Buchanan et al. note is the integrating of two worlds. The authors state that "rejecting a part of the self, be it either the spiritual or sexual identity, may have a negative effect on a gay or lesbian's mental health" (p. 440). They cite such organizations as Dignity and MCC as well as alternative spiritual paths such as Buddhism, Yoga, meditation, Wicca, neo-paganism, and Native American traditions as being options that exist for lesbians to explore their spirituality rather than rejecting it altogether.

CONCLUSION

Through my own individual context I have attempted to illuminate facets of spiritual development. There are many more contexts that exist and that can influence an individual's overall identity development as well as his or her spiritual development. One's spirituality does not develop in a vacuum, but rather is subjective and constructed within the overall social location of the individual. My own social location as a White, working class, female, Catholic-now-Wiccan, lesbian provides the context in which my own spirituality develops. It is my hope that my conversation will encourage you to examine your own contexts and their influences on spiritual development.

REFERENCES

Atchley, R. C. (1997). Everyday mysticism: Spiritual development in later adulthood. *Journal of adult development, 4*(2), 123–134.

Barret, R., & Barzan, R. (1996, October). Spiritual experiences of gay men and lesbians. *Counseling & Values, 41*(1), 4–15.

Benner, D. G. (2002). Nurturing spiritual growth. *Journal of psychology and theology, 30*(4), 355–361.

Buchanan, M., Dzelme, K., Harris, D., & Hecker, L. (2001). Challenges of being simultaneously gay or lesbian and spiritual and/or religious: A narrative perspective. *The American journal of family therapy, 29*, 435–449.

Burke, M. T. & Miranti, J. G. (Eds.) (1995). *Counseling: The spiritual dimension.* Alexandria, VA: American Counseling Association.

Davidson, M. G. (2000). Religion and spirituality. In R. M. Perez, K. A. DeBord, & K. J. Bieschke (Eds.), *Handbook of counseling and psychotherapy with lesbian, gay, and bisexual clients* (pp. 409–433). Washington, DC: American Psychological Association.

Elkins, D. N., Hedstrom, L. J., Hughes, L. L., Leaf, J. A., & Saunders, C. (1988, Fall). Toward a humanistic-phenomenological spirituality: Definition, description, and measurement. *Journal of humanistic psychology, 28*(4), 5–18.

Faiver, C., Ingersoll, R. E., O'Brien, E., & Menally, C. (2001). *Explorations in counseling and spirituality: Philosophical, practical, and personal reflections.* Belmont, CA: Brooks/Cole Thomson Learning.

Fowler, J. (1981). *Stages of faith.* New York: Harper & Row.

Gall, M. D., Gall, J. P., & Borg, W. R. (2003). *Educational research: An introduction (7th ed.).* Boston: Pearson Education, Inc.

Haldeman, D. C. (1996). Spirituality and religion in the lives of lesbians and gay men. In R. P. Cabaj & T. S. Stein (Eds.), *Textbook of homosexuality and mental health* (pp. 881–896). Washington, DC: American Psychiatric Press.

Harding, S. (1991). *Whose science? Whose Knowledge? Thinking from women's lives.* Ithaca, NY: Cornell University Press.

Heyward, C. (1989). Coming out and relational empowerment: A lesbian feminist theological perspective. *Work in Progress, No. 38.* Wellesley, MA: Stone Center Working Paper Series.

Heyward, C. (1989). *Touching our strength: The erotic as power and the love of god.* San Francisco: HarperSanFrancisco.

Hunt, M. E. (1995). Psychological implications of women's spiritual health. *Women & therapy, 16*(2&3), 21–32.

Ingersoll, R. E. (1995). Spirituality, religion, and counseling: Dimensions and relationships. In M. T. Burke & J. G. Miranti (Eds.), *Counseling: The spiritual dimension* (pp. 5–18). Alexandria, VA: American Counseling Association.

Jordan, J. V., & Miller, J. B. (2002, October). Introductory theoretical concepts. Fall (level 1) intensive training: Founding concepts and recent developments. Seminar by Jean Baker Miller Training Institute conducted at St. Stephen Priory, Dover, Massachusetts.

Moya, P. M. L. (1997). Postmodernism, "realism," and the politics of identity: Cherrie Moraga and Chicana feminism. In M. J. Alexander & C. T. Hohany (Eds.) (1997). Feminist genealogies, colonial legacies, democratic futures (pp. 125–150). New York: Routledge.

Neu, D. L. (1995). Women's empowerment through feminist rituals. *Women & therapy, 16*(2&3). 185–200.

Pate, C. H. & Bondi, A. M. (1992). Religious beliefs and practice: An integral aspect of multicultural awareness. *Counselor educations and supervision, 32,* 108–115.

Pearlman, L. A. & Saakvitne, K. W. (1995). Constructivist self development theory and trauma therapy. *Trauma and the therapist: Countertransference and vicarious traumatization in psychotherapy with incest survivors* (pp. 55–74). New York: W. W. Norton & Company.

Ratzinger, J. (1994). Letter to the bishops of the Catholic Church on the pastoral care of homosexual persons. In J. S. Siker (Ed.), *Homosexuality and the church: Both sides of the debate* (pp. 39-48). Louisville, KY: Westminster John Knox Press.

Ratzinger, J., and Amato, A. (2003, June). Considerations regarding proposals to give legal recognition to unions between homosexual persons. Congregation for the Doctrine of Faith. Retrieved September 6, 2003, from http://www.vatican.va/roman_curia/congregations/cfaith/documents/rc_con_cfaith_doc_20030731_homosexual-unions_en.html.

Reinharz, S. (1992). *Feminist methods in social research.* New York Oxford University Press.

Reynolds, A. L. & Hanjorgiris, W. F. (2000). Coming out: Lesbian, gay, and bisexual identity development. In R. M. Perez, K. A. DeBord, & K. J. Bieschke (Eds.), *Handbook of counseling and psychotherapy with lesbian, gay, and bisexual clients* (pp. 35–55). Washington, DC: American Psychological Association.

Ritter, K. Y., & O'Neill, C. W. (1989, September/October). Moving through loss: The spiritual journey of gay men and lesbian women. *Journal of counseling & development, 68,* 9–15.

Starhawk. (1982). *Dreaming the dark: Magic, sex, and politics.* Boston: Beacon Press.

Stokes, K. (1990). Faith development in the adult life cycle. *Journal of religious gerontology, 7*(1&2), 167–184.

Tisdell, E. J. (2002, April). Spiritual development and cultural context in the lives of women adult educators for social change. *Journal of adult development, 9*(2), 127–140.

Warwick, L. L. (2001). Self-in-Relation theory and women's religious identity in therapy. *Women & Therapy, 24*(3&4), 121–131.

Wink, P. & Dillon, M. (2002, January). Spiritual development across the adult life course: Findings from a longitudinal study. *Journal of adult development, 9*(1), 79–94.

16

Fostering Resilience
Throughout Our Lives:
New Relational Possibilities

Linda M. Hartling, Ph.D.

REFLECTION QUESTIONS

1. *Think about an individual who contributed to your ability to be resilient during a difficult time. Specifically, what did this person do that made the difference?*
2. *How did this person contribute to your sense of worth?*
3. *Think about a time that you felt you contributed to the growth or resilience of another person who was going through a difficult time or experience (such as a child, a sibling, a family member, a friend, a colleague, etc.). Specifically, what did you do that seemed to make a difference? What did you do that contributed to that person's ability to become more resilient?*
4. *How did you feel about participating in the growth of another?*

INTRODUCTION

After the September 11 terrorist attacks on New York City and Washington, DC, millions of people in the United States and others around the world were faced with the impact of a devastating human tragedy followed by a period of national economic insecurity and the threat of international conflict. In response, many individuals, families, and communities joined with others to form an unprecedented sense of collective courage with which to begin the process of addressing these shared disasters. This collective courage is not an anomalous reaction to one unimaginable tragedy. Rather, it is likely derived from a reservoir of interpersonal courage that individuals, families, and communities must call upon to overcome difficulties and hardships that are a part of daily living. Previous chapters of this book have explored and described effective responses to specific difficulties, including troubled and disrupted relationships, chronic illness, addictive behavior, grief, death, and social discrimination. Building on these discussions, this chapter will focus on understanding the factors that contribute to the general ability of people to overcome profound difficulties and extreme hardships, in other words, the study of *resilience.*

Much research reinforces the notion of resilience as a unique, internal trait possessed by a few fortunate people. This view asserts that resilient individuals exhibit *special, internal* qualities—such as a good-natured temperament, higher intelligence, higher self-esteem, and internal locus of control—that allow them to achieve positive outcomes despite encounters with potentially overwhelming obstacles. The preference for conceptualizing resilience as an individual, internal phenomenon may have emerged from Western European theories of psychological development that tend to focus on individual experience, self-development, individual achievement, and self-sufficiency (Cushman, 1995). These theories posit that healthy development involves progressive separation from relationships throughout one's life, in order to become independent and autonomous (Jordan & Hartling, 2002). Based on this view, a resilient individual is someone who defies disaster and learns to "stand on one's own two feet."

In this chapter, we will explore a new view of resilience that moves beyond an emphasis on individual, internal traits to a view that places relationships at the center of our thinking. Judith Jordan introduced this perspective in her 1992 paper, in which she urged clinicians and researchers to enlarge and expand their narrowly defined understandings of resilience:

> [W]e can no longer look only at factors within the individual which facilitate adjustment; we must examine the relational dynamics that encourage the capacity for connection. (p. 1) . . . Few studies have delineated the complex factors involved in those relationships which not only protect us from stress but promote positive and creative growth. (p. 3)

Jordan encourages us to explore the relational factors that shelter people from extreme difficulties and facilitate their progress toward healthy development. Incorporating over 20 years of theory-building, research, and practice that began at the Stone Center, and is continuing through the work of the Jean Baker Miller Training Institute at Wellesley College, this chapter will examine research on resilience from the vantage point of Relational-Cultural theory (RCT).

RCT posits that *growth-fostering relationships* are a central human necessity throughout our lives and the lack of these relationships impairs our ability to be resilient and results in psychological problems (Jordan, Kaplan, Miller, Stiver, & Surrey, 1991; Jordan, 1997; Miller & Stiver, 1997). Furthermore, RCT emphasizes that the cultural context highly defines people's relationships and influences their ability to engage in these relationships. In particular, sociocultural conditions can facilitate or impede the development of growth-fostering relationships, which are necessary for resilience. Growth-fostering relationships are characterized by movement toward mutuality, mutual empathy, and mutual empowerment. All too often, the research on resilience has framed relationships as incidental or obstructing factors; in contrast, RCT proposes that *growth-fostering relationships are the essential ingredient of lifelong resilience.*

This chapter explores some of the individual characteristics commonly identified in the research on resilience and offers a relational reframing of these characteristics guided by RCT. In addition, this chapter will address the following questions:

1. What is resilience and what is relational resilience?
2. What are some of the individual characteristics commonly associated with resilience and what are the relational aspects of these characteristics?
3. What are some relational ways to strengthen resilience throughout one's life?

This chapter will also include a case example demonstrating how a growth-fostering relationship can help a client begin to transform adversity and experience the rewards of resilience.

HOW CAN RCT ENLARGE OUR UNDERSTANDING OF RESILIENCE?

By moving beyond an individualistic understanding of resilience, RCT can help clinicians and researchers think about the complex, multilayered interpersonal and cultural dynamics that affect one's ability to be resilient. In particular, RCT emphasizes the importance of examining the impact of privilege (McIntosh, 1988, 1989), power (Miller, 1976; Miller, 2002; Walker, 2002b), social stratification (Walker, 1999, 2002a), and biased perspectives (sexism, racism, classism, heterosexism, homophobia, and the like) that can skew all clinical research and practice, including the research on resilience. Many years ago, Jean Baker Miller (1976) observed that "the close study of an oppressed group reveals that a dominant group inevitably describes the subordinate group falsely in terms derived from its own systems of thought" (p. xix). This raises serious questions about the assumption that research is politically neutral and objective. Miller's statement cautions us to be vigilant when examining research designed, developed, conducted, and interpreted primarily by members of a dominant, privileged group. Privileged groups will tend to define themselves and their values as the norm or ideal, concomitantly implying that others are deficient or abnormal.

With regard to the study of resilience, researchers who subscribe to or support the values of the dominant group in a society may conduct research that is more attentive to the characteristics exhibited by this group and describe these characteristics as the ideal for all people (Minnich, 1990). When this happens, the experiences of subordinate groups, such as women and marginalized men, may be miscon-strued or completely overlooked. RCT summons us to pay particular attention to the impact of dominant and subordinate relationships that can shape how research and clinical practice are developed and conducted. As practitioners of this theory, we must engage in an ongoing analysis of the cultural contexts in which people are expected to grow and develop. In many contexts, certain individuals in a soci-ety receive beneficial support that goes unnoted because these advantages are automatically accrued to members of a privileged group, while people belonging to marginalized groups may be chronically deprived of resources that could help them overcome hardships (Walker, 2002a).

For example, Peggy McIntosh (1988) has described the advantages taken for granted by those with White privilege in America, such as increased access to health care, educational opportunities, and employment opportunities. RCT guides clinicians and researchers to pay attention to these issues. It also encourages practitioners to attend to the strengths of people living in oppressive, alienating sociocultural contexts. Often these people have had to develop and practice extraordinary resilience to accomplish ordinary tasks (Genero, 1995).

The following example of early research exploring resilience to stress illus-trates the limitations of focusing on individual, internal characteristics while neglecting the Relational-Cultural context in which these characteristics develop. In the 1970s, Kobasa (1979; Kobasa & Puccetti, 1983) identified an individual, internal quality associated with resistence to stress that she called "hardiness." According to this research, "hardy" individuals exhibit three characteristics: 1) *Commitment*: being able to easily commit to what one is doing; 2) *Control*: a general belief that events are within one's control; and 3) *Challenge*: perceiving change as a challenge rather than a threat (Kobasa, 1979). This conceptualization of resilience was well received in academic and clinical communities. Over the years, hardiness has been held up as a standard of stress resilience and applied across diverse populations of men, women, and children.

However, today, we are conscious of the limitations of the initial research that led to the conceptualization of hardiness. In the past, the demographic character-istics of a research sample were often not deemed significant. Today, investigators realize that understanding this information is crucial. Kobasa was examining the experience of a narrowly defined group, specifically White, male, middle- to upper-level business executives. Knowing this helps us understand that the inter-nal qualities of commitment, control, and challenge (i.e., hardiness) may have been useful for describing stress resilience of the subjects in the circumscribed research. Nevertheless, the researcher's conclusions may have also triggered "faulty generalizations" (Minnich,1990) about the stress resilience of women and others not represented in the study. Today, we expect researchers to pay attention to soci-ocultural contexts of their investigations and we are more aware of the limitations of early research. Indeed, we know that business executives in the 1970s were the beneficiaries of a silent system of extensive support comprised of secretaries, wives, mothers, and undervalued service providers (experts in providing relational support) who likely made it possible for these privileged professionals to be "hardy."

If the researchers had investigated a more diverse population in this early research, they might have identified many other characteristics associated with stress resilience. For example, Elizabeth Sparks (1999) described relational practices, rather than the individual internal qualities, that strengthened the resilience of a sample of African American mothers on welfare. This marginalized group of women engaged in connection, collaboration, and community action to overcome the destructive impact of poverty, racism, and social stigmatization. By broadening the research population and taking a relational perspective—rather than restricting the population and focusing on individual, internal qualities or traits—researchers can make visible some of the relational, collaborative characteristics that contribute to the resilience of many populations. As this example illustrates, a Relational-Cultural theoretical perspective can enrich, enlarge, and deepen our understanding of resilience.

WHAT IS RESILIENCE AND WHAT IS RELATIONAL RESILIENCE?

The research literature typically describes resilience in three ways: 1) good outcomes defined as the absence of deviant or antisocial behavior after experiencing adverse conditions, 2) maintaining competence under conditions of threat, and 3) recovery from traumatic experiences (Jordan & Hartling, 2002; Masten, Best, & Garmezy, 1990). These definitions, combined with traditional thinking about psychological development, tend to perpetuate a focus on individual personality characteristics, such as temperament, intelligence, self-esteem, optimism, and internal locus of control. Yet, even researchers conducting some of the earliest studies noted that *relationships* contributed to their subjects' ability to be resilient. Based on a groundbreaking 40-year study of 700 multi- and mixed-racial children living in adverse conditions on the Hawaiian Island of Kauai, Emma Werner (1993) observed that: "The resilient youngsters in our study all had at least one person in their lives who accepted them unconditionally, regardless of temperamental idiosyncrasies, physical attractiveness, or intelligence" (p. 512).

Rather than continuing to emphasize internal personality traits, RCT helps us to encompass and explore the Relational-Cultural geography of people's lives. In particular, it proposes that growth-fostering personal and social relationships allow people to withstand tremendous hardships and trauma. From this perspective it is useful to examine how people establish, engage in, and sustain growth-promoting relationships throughout their lives. This activity is the process Jordan (1992) calls building *relational resilience*. Ultimately, resilience may be an outcome of one's ability and opportunities to connect, reconnect, and resist disconnection.

With this broader view in mind, the next section of this chapter will reexamine the research on resilience and explore some of the relational factors that contribute to one's ability to be resilient. It will review some of the individual characteristics commonly associated with resilience and explore the relational aspects of these characteristics.

RELATIONALLY RETHINKING
INDIVIDUAL RESILIENCE

The literature identifies many individual characteristics associated with resilience. Here, we will review a sample of the research describing characteristics that are signs of individual resilience and delve into the relational factors that influence the development of these individual characteristics. Specifically, we will discuss the individual and relational aspects of temperament, intelligence, self-esteem, internal locus of control, mastery, and social support.

Temperament and Relationships

Temperament is an internal, relatively stable, individual characteristic often associated with resilience (Rutter, 1978; Werner & Smith, 1982). Focusing on individual differences, a study by Werner and Smith (1982) suggested that "good-natured" boys and "cuddly" girls were more resilient than other children. In later research, Rutter (1989) offered a relational understanding of temperament when he observed that children with adverse temperaments were twice as likely to be the targets of parental criticism. His work suggests that a child's temperament has a significant impact on the *parent-child relationship*, either protecting the child or putting the child at risk. In other words, temperament affects one's ability to engage in relationships, either enhancing or inhibiting one's possibility of experiencing a growth-fostering, resilience-promoting relationship.

Based on Rutter's observations, one might conclude that children with difficult temperaments would be less resilient. However, RCT encourages us to take a broader view and examine how the cultural context intersects with a child's temperament and his or her relational opportunities. For example, a study of East African Masai children found that children with more difficult temperaments were more likely to survive in extreme drought conditions (de Vries, 1984). Because the Masai culture values assertiveness, the researchers theorized that the difficult (assertive) children were more able to effectively access the relational resources they needed to survive. This is one example of how a Relational-Cultural view can expand our understanding of temperament and its influence on a child's relational opportunities.

Connected Intellectual Development

How does intelligence, another individual trait cited in the research, contribute to resilience? Ann Masten and her colleagues (Masten, Best, & Garmezy, 1990; Masten,1994, 2001) offer a few explanations. Individuals with greater intelligence may be more able to discern danger and find escape routes, or they may have more educational advantages when compared to others, or they may have more capable parents. Exploring a key aspect of intellectual development, Daniel Siegle (1999) emphasizes that interpersonal relationships are the central source of experience that influences how the brain develops. Relationships provide children with experiential opportunities that activate certain neural pathways in the brain, either "strengthening existing connections or creating new connections" (Siegle, p. 13).

In other words, "Human connections create neuronal connections" (Siegle, p. 85). Again, this suggests that relationships significantly contribute to the development of individual characteristics associated with resilience. Siegle's observations should encourage researchers and clinicians to conscientiously investigate the qualities of relationships that foster intellectual development and strengthen resilience throughout our lives.

Rethinking Self-Esteem

Self-esteem has long been viewed as an important personality characteristic associated with resilience (Dumont & Provost, 1999). Yet, RCT raises some important questions about this particular characteristic. As Judith Jordan (1994) observes, Western European culture has tended to construct self-esteem based on the values of individual achievement and self-sufficiency, rather than collaboration and connection. As a result, individuals may work to establish "healthy" self-esteem by engaging in hierarchical comparisons in which their self-esteem depends upon feeling superior to others in some way. Within this context, the process of developing "healthy" self-esteem can become an endless competitive exercise. In addition, those who subscribe to collaborative models of achievement—rather than competitive models—may be perceived as lacking a healthy sense of self-esteem (Kohn, 1986). Yet even in competitive business settings, individuals can gain a greater sense of self-esteem by incorporating a relational belief system that promotes interdependence and mutually empowering work practices (Fletcher, 1999).

For a long time, it was thought that Black children had lower self-esteem than White children. Bernadette Gray-Little (2000) has noted that this assumption grew out of a 1947 study in which Black children were asked to choose between two dolls that were identical except for skin color. In this study, the Black children chose the White doll, a choice researchers interpreted as a sign of Black children's low self-esteem. Gray-Little challenged this and other similar interpretations of research data after examining over 261 studies of over half a million children. This careful investigation of multiple studies indicated that the self-esteem of Black children is at least as high as White children and, in some cases, higher. She realized that research inspired by the 1947 study using White and Black dolls may offer only an indication of how a racial group is valued in society, but not an indication of Black children's self-esteem. In other words, the Black children in the study may have selected White dolls because they knew that White people were more valued in society. In addition to questioning the conclusions of earlier research, Gray-Little challenges the assumption that self-esteem grows out of comparing and accumulating individual achievements. In a *Washington Post* article describing her investigations, she stated, "Self-esteem is determined by our interactions with people significant to us personally" (Fletcher, 2000). Her relational observation is supported by other research that shows that self-esteem correlates with a child's closeness to his or her mother (Burnett & Demnar, 1996).

Taking a cultural approach to rethink self-esteem, Yvonne Jenkins (1993) proposes that individualistic constructions of self-esteem may have limited relevance to people of color. Rather than self-esteem, Jenkins describes a group-centered, relational understanding of self-esteem, which she calls *social esteem*. Social esteem is developed and strengthened through a group-related identity that values "interdependence, affiliation, and collaterality" (p. 55). Jenkins proposes that *"for collective societies, group esteem is practically synonymous with the anglocentric conceptualizations of self-esteem"* (p. 55). Social esteem may be an essential part of healthy psychosocial development for populations in which the unit of operation is the family, the group, or the collective society. Within these collective contexts, social esteem may enhance one's ability to cope with socially imposed adversities such as racism, discrimination, and other forms of humiliation. Social esteem may help members of marginalized and oppressed groups find the strength to survive and defy injustice.

The insights of Jenkins and Gray-Little remind us that we need to expand our understanding of the relational factors that promote one's sense of esteem. Or, perhaps, we need to move to another concept altogether. Rather than self-esteem, Jean Baker Miller (1986) suggests that people develop a *sense of worth* through participating in mutually empathic, mutually empowering, growth-fostering relationships. Just as a parent can assure a child that she is loved and valued despite inevitable encounters with failure, it may be that relationships cultivate a sense of worth that fortifies our ability to be resilient through other difficulties.

From Internal Control to Relational Empowerment

Internal locus of control (ILOC) is another individual characteristic often associated with resilience (Masten, Best, & Garmezy, 1990; Werner & Smith, 1982). To define this characteristic, Roediger and his colleagues state, "Children who take responsibility for their own successes and failures are said to have an internal locus of control" (1991, p. 352). This definition describes individual experience, but it neglects the impact of racism, sexism, heterosexism, or other forms of discrimination that can influence one's ability to take responsibility for successes or failures. For example, it may be much easier for individuals to take responsibility for their experience if they are members of the social group that is viewed as the norm or ideal group in a society, that is, the dominant group (Miller, 1976). Members of dominant groups benefit from unearned advantages that facilitate their successes and mitigate their failures. Therefore, they have an easier life experience. Unearned advantages make it easier for members of this group to assume they have internal control over their experience.

Furthermore, the dominant group might find it advantageous to persuade subordinate and marginalized groups that they *should* have an internal locus of control—that they *should* feel responsible for their lack of success as well as their failures. If this is accomplished, subordinate groups may be less likely to resist or challenge the insidious social practices that obstruct their progress and achievement. For example, historically women have been made to feel responsible for being the victims of sexual assault. In direct and indirect ways, women were often told they were responsible for their experience because they wore the wrong clothes, they were in the wrong place, they acted in the wrong way. Attributing sexual assault to something within women's individual control (a.k.a., blaming the victim) can paralyze women with shame, immobilizing their energies that could be used to challenge social practices that increase the risk of women being assaulted. Furthermore, focusing on individual responsibility and internal control conveniently allows dominant members of society to absolve themselves of any responsibility, even though they may be practicing attitudes and behaviors that precipitate sexual assault. Today we know that no one deserves to be sexually assaulted and society must change social practices and attitudes that propagate assaults on women. Unfortunately, many patriarchal societies continue to attribute sexual assault to women's lack of internal, individual control (Firmo-Fonta, 2003).

A recent study helps us begin to rethink our understanding of ILOC (Magnus, Cowen, Wyman, Fagen, & Work, 1999). This study compared stress resilient (SR) White and Black children to stress affected (SA) White and Black children. The results showed that SR White children had a greater sense of internal locus of control than SA White children, but no significant difference was found between the SR and the SA Black children. In response to these findings, these researchers proposed that, compared to White families, Black families may not emphasize internal control because it might encourage a false belief that one can or should be able to control pervasive, socially constructed adversities such as racism and other forms of discrimination. Many researchers devised instruments to accurately assess internal-external locus of control (e.g., Levenson, 1981; Nowicki & Duke, 1974; Rotter, 1966). However, Paul Kline (1993), in his *Handbook of Psychological Testing*, summarizes the problematic aspects and psychometric deficiencies of a number of these scales. Alternatively, investigators at the Stone Center at Wellesley College formulated *The Mutual Psychological Development Questionnaire* (Genero, Surrey, & Miller, 1992) and *The Relational Health Indices* (Liang, Taylor, Williams, Tracy, Jordan, & Miller, 1999) that may help researchers assess the construct of mutual empowerment as opposed to the locus of control. It is possible that people are more resilient when they feel they have the ability to influence their experience through relationships that empower them to take action on behalf of themselves and others.

From Mastery to Relational Competence

Mastery is a widely used term to describe how people develop individual skills that contribute to their ability to be resilient. Nevertheless, a Relational-Cultural perspective reveals the very subtle and questionable connotations of this term. Judith Jordan (1999) notes that "'to master' is to reduce to subjection, to get the better of, to break, to tame" (pp. 1–2). As a result, "[M]astery implicit in most models of competence creates enormous conflict for many people, especially women and other marginalized groups, people who have not traditionally been 'the masters'" (p. 2). Because of this, Jordan suggests that "competence" may be a more useful, less tainted term for describing the development of skills that contribute to one's ability to be resilient (Jordan, 1999). With this in mind, how does one develop a sense of competence? Ann Masten and her colleagues (1990) identified three specific, relational ways parents can foster their children's sense of competence:

1. Parents (and other adults) can model effective action for their children.
2. Parents can provide their children with opportunities to experience competence.
3. Parents can verbally affirm the competence of their children and affirm their children's ability to develop new skills and utilize these skills effectively.

Not surprisingly, the competence associated with resilience appears to grow through participation in supportive and encouraging relationships.

Social Support and Movement Toward Connection

The relationship between social support and resilience has been well documented in psychological and health research (Atkins, Kaplan, & Toshima, 1991; Belle, 1987; Ganellen & Blaney, 1984; Ornish, 1997). When taken at face value, social support appears to be a relational construct, especially when it is compared to the constructs discussed thus far. However, social support is most often described in the research as a *one-way, unidirectional* form of relating, something that one gets from others (Fiore, Becker, & Coppel, 1983). In contrast, RCT has focused on the *two-way, bi-directional* nature of relationships; that is, the two-way, growth-promoting quality of relating known as *connection* (Jordan, 1992). Connection is fostered through mutual empathy, mutual empowerment, and movement toward mutuality in relationships (Jordan, 1986). Though the two-way impact of relationships has been a central tenet of RCT throughout its development, only recently have researchers begun to note the importance of understanding the bi-directional nature of relationships that foster resilience (Masten, 2001; Masten, Hubbard, Gest, Tellegen, Garmezy, & Ramirez, 1999).

Moving beyond the one-way notion of social support toward the RCT concept of connection, Renée Spencer (2000) reviewed a number of studies that suggest that children who have at least one supportive relationship with an adult can achieve good outcomes despite encounters with adverse conditions, such as parental mental illness (Rutter, 1979), separation from a parent (Rutter, 1971), marital discord (Rutter, 1971), divorcing parents (Wallerstein & Kelly, 1980), poverty (Garmezy, 1991), maltreatment (Cicchetti, 1989), and multifaceted or a combination of risk factors (Seifer et al., 1996). A study of over 12,000 adolescents by Michael Resnick and his colleagues (1997) found that sense of *connection*—to parents, family members, or other adults—is the most important factor associated with a reduced risk of substance abuse, violence, depression, suicidal behavior, and early sexual activity regardless of race, ethnicity, socioeconomic status, or family structure. This finding draws into question many theories that have suggested that healthy development requires progressive separation from relationships. By "cutting those apron strings" and separating from our children we may be increasing the risk that they will develop psychological or behavioral problems. Rather than separation, perhaps parents and other adults need to find healthy ways to stay connected to their children as they grow and change. This is indeed one of the most difficult challenges for parents, perhaps because there has been little emphasis on cultivating our connections over the life span.

Connection appears to be particularly important for children in school. A 2002 report based on data from the *National Longitudinal Study of Adolescent Health* emphasizes the benefits that occur when young people feel connected within their school settings (Blum, McNeely, & Rinehart, 2002). This report described the results of a survey of over 90,000 adolescents from 80 different communities during one academic year. Students who felt connected were less likely to use cigarettes, alcohol, or drugs; less likely to engage in early sexual activity, violence, or become pregnant; and less likely to experience emotional distress. According to the authors, "When students feel they are a part of school, say they are treated fairly by teachers, and feel close to people at school, they are healthier and are more likely to succeed" (p. 2). Unfortunately, while the majority of students who responded to this survey said they experienced a sense of connection in their school settings, the data also showed that 31 percent of students do not feel connected at school. This information provides additional evidence of the urgent need to promote healthy development through connection throughout children's lives.

Beyond childhood, a sense of connection is important at any point in people's lives. Berkman and Syme (1979) demonstrated that men and women who were married, who had contact with close friends and relatives, or who had informal or formal group associations had "lower mortality rates than respondents lacking such connections" (p. 188). In his national analysis of social connectedness, Harvard professor Robert Putnam (2000) noted that studies "have established beyond reasonable doubt that social connectedness is one of the most powerful determinants of our well-being" (p. 326). Furthermore, "[H]appiness is best predicted by the breadth and depth of one's social connections" (p. 332). Clearly, this shows that relationships—with family members, friends, and others—are central to our ability to be happy and resilient throughout our lives.

Connection may be especially important for promoting women's resilience. For many years researchers assumed that the two key responses to stress were *fight or flight*. Taylor and her colleagues (2000) have challenged this assumption and proposed that women may utilize a *tend and befriend* response to stress. Taylor et al. take the position that the fight/flight response in women may be inhibited by brain chemistry that reduces their feelings of fear, decreases their sympathetic nervous system activity, and simultaneously triggers their caretaking and affiliative behavior. As a result, women may engage in caretaking activities or the creation of a network of associations to protect themselves and others (such as children) from a threat. In support of this analysis, other studies have determined that women are more likely to mobilize social support in times of stress, maintain close relationships with female friends, and engage in social networks more often than men (Belle, 1987). It is also possible that Taylor's analysis might apply to some aspects of men's behavior, but this research has yet to be conducted.

Recently, a few researchers have begun to question the value of some forms of social support. Some have suggested that social support can trigger commiseration in misery, causing additional stress to individuals participating in support groups (Belle, 1987). These real concerns are why it is important to explore and articulate a more sophisticated, complex understanding of relationships. Researchers and clinicians must learn to distinguish the intricate characteristics of relationships, as well as explore the behaviors practiced *within* relationships that impede or promote growth and resilience.

The following case illustrates how a more complex understanding of the intersection between relationships and resilience can enrich our perceptions of a client's ability to be resilient and lead us to new ways of supporting a client's efforts to overcome adversity.

THE CASE OF "MARTA"

Questions to Consider:

1. What are some of strategies of disconnection (a.k.a., strategies of survival) Marta used that impeded her ability to engage in relationships?
2. What are some the new skills that Marta developed that enhanced her ability to be more relationally resilient?
3. How did the therapist practice resilience in the therapy relationship?

Graduating from high school is a major milestone in the lives of most young people, yet, this event was not the most significant of Marta's lifetime accomplishments. To reach graduation, Marta had to surmount a childhood marked by neglect, emotional and physical abuse, abandonment, homelessness, foster care, depression, disordered eating, and substance abuse. At one point, U.S. marshals were pursuing Marta's family because of her parents' international custody dispute. While earning her high school diploma was something she could celebrate publicly, privately I knew her life experience suggested that she deserved much more, perhaps a doctoral degree in courage and resilience.

Three years before, after months of conflict with her mother, Marta was essentially abandoned by her mother at a local shelter. Following this, social workers arranged for Marta's placement in foster care and referred her to a community mental health agency for therapy. I was assigned her case. At that time, one of Marta's foster parents described her as extremely difficult, highly sensitive to criticism, adversarial, suspicious, reactive, volatile, demanding, and aggressive. Her relationships in her foster home and at school were perpetually strained and unstable. Based on these problems, her social workers described her as "borderline" (having a borderline personality). The foster parents and the social workers hoped that therapy would assist Marta in resolving her traumatic history, prevent her from being thrown out of her foster home, and help her develop the skills she needed to live independently. As her history suggests, Marta's case was extremely complex, requiring treatment for trauma, substance abuse, eating difficulties, depression, destructive family dynamics, and other issues. For this discussion, I am going to examine one aspect of Marta's case: the process of how she became more resilient during our three years of working together.

In the beginning, I noted that Marta would have to overcome many obstacles working against her ability to be resilient. She was not able to identify a single relationship in her life that had offered her a sense of consistent love or connection; she was known for her volatile temperament and behavior; she had consistently felt she had little influence over her life experiences (lack of internal locus of control); she couldn't identify any significant skills she had to offer (lack of a sense of competence); she questioned her value as a human being (low self-esteem); and she felt she had little or no social support, not from friends, from teachers, or from her foster parents.

Given the desolate relational landscape of Marta's life, her sensitive, volatile temperament began to make sense. Perhaps the behaviors that were attributed to her difficult temperament were primarily the outgrowth of her desperate efforts to secure what she needed to survive in the neglectful, dangerous living conditions she described to me. By being difficult, she was able to gain access to things she needed as a child and simultaneously keep potentially threatening people at a distance. This seemed to be her "strategy of survival" (i.e., strategy of disconnection; Miller & Stiver, 1994). Marta would describe the repeated conflicts she would have with her foster parents, other foster children in the home, her peers, and her teachers. Aggressive behavior appeared to be her way of eliciting connection to the only relationships that were available to her and it simultaneously provided her some protection from being hurt. Furthermore, her pugnacious behavior would consistently ensure that she would gain attention from one or more of her service providers. However, this strategy also kept her from revealing the vulnerable, compassionate aspects of her personality and from forming more growth-fostering relationships that might have helped her become more resilient in response to the adversities she had experienced. With this in mind, our work together began with exploring new ways for her to engage in relationships.

One of the essential steps in this process was helping Marta recognize that she could influence relationships in her life without being aggressive, manipulative, or demanding. We both recognized that this behavior had served her well to get her needs met in the context of unresponsive and hurtful relationships that were available to her in the past (Miller & Stiver, 1994). Nevertheless, this way of engaging others in relationships came at a high price, distancing her from opportunities for more positive and supportive relationships. In therapy, we began to explore how she could get her needs met and stay connected without falling back on her strategies of survival/disconnection.

Over time, Marta began to discover that she could ask for what she wanted utilizing other skills, such as being vulnerable, direct, open, sharing her feelings, and being willing to negotiate. At one point, she surprised me when she told me she had joined an environmental club that worked to increase recycling at her high school. While in the club, she was able to practice negotiation skills with school officials she had formerly viewed as unresponsive adversaries in her life. Her new skills didn't mean she always got what she wanted, but she eventually learned her views would be heard, valued, and sincerely considered. She also learned that she had the possibility to get what she wanted without having to engage in harsh intimidation or aggressive behavior. Knowing that she and her wishes mattered to others allowed her begin to believe that she could have an impact on her relationships. This appeared to reduce her need to always "win" arguments with her teachers, her friends, and her foster parents.

Eventually our therapy relationship grew large enough to navigate challenging disagreements and still sustain connection, which was a sign of relational resilience (Jordan, 1992). Of course, this process involved moving through periods of difficult disconnection, particularly early on. During these periods, I had to assume responsibility for what Irene Stiver (1992) described as "holding the potential for connection," or having faith that the relationship can endure. Occasionally, I had to call on external relational resources to sustain my resilience in response to a difficult disconnection with Marta, such as consulting colleagues, supervisors, and other service providers. These occasions made it clear to me that we were *both* engaged in a process of strengthening our skills of being resilient in relationship, that we were building resilience together. And, our efforts paid off. By moving through the obstacles to relationship we were able to begin to address many of the other important issues and difficulties that were disrupting Marta's life: past trauma, depression, disordered eating, substance abuse, and a seriously troubled family history that resulted in her being placed in foster care and all of which had a tremendous impact on her relational development.

As Marta expanded her skills for engaging in our therapy relationship, she was also expanding her skills for engaging in other relationships. Gradually, the relationships in her life that had been characterized by mutual hostility and strain began to diminish and be replaced by more constructive relationships.

(continues)

Throughout this process, her understanding of relationships became more sophisticated. Rather than framing relationships in simplistic, binary terms (good/bad), she gained greater relational flexibility and competence, effectively coping with the complex ups and downs, connections and disconnections of relationships without automatically slipping into her old, aggressive ways of behaving. She became more empathic with others, her relationships, and herself. As she gained greater relational confidence and competence, she also became more proficient at making constructive choices and more involved in her school activities (particularly art, writing, and her ecology club). She was able to identify people whom she felt supported her growth (teachers, friends). She experienced less turmoil in her foster home, even though the home was far from an ideal setting (sadly, one of her foster parents had hidden his alcohol dependence). And, she was able to gain and sustain a job as a waitress, where she had to regularly exercise her relational skills with many different, occasionally difficult, customers.

All of her expanding skills eventually led to her successful graduation from high school. In a session prior to her graduation, Marta and I reviewed the resilience she had developed in response to our relationship and in response to her relationships with others. To my surprise and delight, she shared with me that she had been keeping an account of her successes in what she called her "Happy Book." In this book she noted the growing number of good experiences that began occurring in her life, and she gave herself credit for achievements that would never show up on her academic record. With every entry, she gradually transformed the tragedies of her life into tiny, deeply touching triumphs. She documented her history of resourcefulness in the face of extreme adversities, her emerging successes at school, the accolades she received from public recognition of her creative talents, her progress with new relationships, and her dreams for the future. For me, this book was the real "report card" on her achievements in a harsh school of life, convincing me she deserved a doctoral degree in resilience.

Although many traditional psychological theories suggest that adolescents need to learn to be more independent as they grow up, Marta didn't need to learn how to be more independent. Marta needed to learn how to participate in relationships, in particular, growth-fostering relationships. Through connection, first in therapy and then in her life, she was able to develop a sense of resilience despite the profound hardships and trauma she encountered in her early childhood. Her story helps us understand that *growth-fostering relationships are the essential ingredient for building resilience* and are ultimately at the core of our mental health and well-being.

RELATIONAL WAYS FOR STRENGTHENING CLIENTS' RESILIENCE

While Western European theories of psychological development have tended to point toward independence and autonomy as the ultimate outcomes of healthy development, the research on resilience seems to guide us in a more relational direction. Perhaps resilience is not a unique, internal quality that some people have and others do not, but, rather, it is an interpersonal process, a process that develops through engagement in one or more sufficiently growth-fostering relationships in one's life. If relationships are the key to resilience, then all of us can contribute to the resilience of others and, in the process, grow ourselves. All of us need and can find ways to participate in relationships that support and promote mutual empathy, mutual empowerment, and movement toward mutuality. Clinicians have a special opportunity to offer their clients resilience-building relationships. Below are a few ways therapists might begin to strengthen their clients' resilience through relationships:

1. Assess the client's access to relationships that support his or her resilience, relationships that are responsive to his or her unique individual (e.g., temperament, intellectual, etc.) characteristics.
2. Help clients identify and expand relationships that create conditions for growth and resilience.
3. Strengthen clients' ability to create growth-fostering relationships.
4. Help clients to not just feel good about themselves but also to believe that they have something to offer to others (Jordan, 1994).
5. Help clients feel that that they can have an impact on their relationships (Miller, 2002) and discern when a relationship is no longer moving toward mutuality and mutual growth.
6. Strengthen clients' ability to take positive action on behalf of themselves, others, and their relationships.
7. Help clients find models of effective action (e.g., mentors, role models, teachers).
8. Help clients find opportunities to experience competence and help them recognize their abilities to be effective.
9. Help clients move toward mutual empathy in relationships where this is possible.
10. Help clients create good connections rather than exercising strategies of disconnection or power over others.

These suggestions can be summarized as finding more and more ways to expand our clients' experiences of growth-fostering relationships, relationships characterized by mutual empathy, mutual empowerment, mutuality, zest, clarity, increasing sense of worth, and a desire for more connection (Miller & Stiver, 1997).

CONCLUSION

In or out of therapy, this chapter proposes that relationships are at the core of well-being, growth, and resilience. Earlier chapters offered pathways to healing and hope in response to specific psychological, social, and behavioral problems. Most often, these pathways are forged through engagement in effective therapeutic relationships that guide injured individuals beyond their experiences of pain and suffering toward greater health and well-being. This chapter explores the generalized power of relationships to foster resilience throughout people's lives. It proposes that resilience is both a relational process as well as an outcome of participating in growth-fostering relationships.

By fully appreciating the relational sources of resilience and healing, we are presented with new possibilities for strengthening the well-being of all people. It becomes clear that *everyone* can encourage the healing and resilience of others—clinicians, parents, grandparents, brothers, sisters, teachers, coaches, employers, etc. We, as counselors and therapists, are often on the forefront of these efforts. Hence, we are granted special opportunities, as well as responsibilities, to help people rebuild the vital relationships they need to heal, grow, and be resilient throughout their lives.

REFERENCES

Atkins, C. J., Kaplan, R. M., & Toshima, M. T. (1991). Close relationships in the epidemiology of cardiovascular disease. *Advances in Personal Relationships, 3,* 207–231.

Barnard, C. P. (1994). Resiliency: A shift in our perception? *American Journal of Family Therapy, 22*(2), 135–144.

Belle, D. (1987). Gender differences in the social moderators of stress. In D. Belle (Ed.), *Gender and stress* (pp. 257–277). New York: Free Press.

Berkman, L. F., & Syme, S. L. (1979). Social networks, host resistance, and morality: A nine-year follow-up study of Alameda County residents. *American Journal of Epidemiology, 109*(2), 186–204.

Blum, R. W., McNeely, C. A., & Rinehart, P. M. (2002). *Improving the odds: The untapped power of schools to improve the health of teens* (Monograph). Minneapolis: University of Minnesota Center for Adolescent Health and Development.

Burnett, P. C., Demnar, W. J. (1996). The relationship between closeness to significant others and self-esteem. *Journal of Family Studies, 2*(2), 121–129.

Cicchetti, D. (1989). How research on child maltreatment has informed the study of child development: Perspectives from developmental psychopathology. In D. Cicchetti & V. Carlson (Eds.), *Child maltreatment: Theory and research on the cause and consequences of child abuse and neglect.* Cambridge, England: Cambridge University Press.

Cushman, P. (1995). *Constructing the self, constructing America: A cultural history of psychotherapy.* Reading, MA: Addison-Wesley/Addison Wesley Longman.

de Vries, M. W. (1984). Temperament and infant mortality among the Masai of East Africa. *American Journal of Psychiatry, 141,* 1189–1194.

Dumont, M. & Provost, M. A. (1999). Resilience in adolescents: Protective role of social support, coping strategies, self-esteem, and social activities on experience of stress and depression. *Journal of Youth and Adolescents, 28*(3), 343–363.

Firmo-Fonta, V. (2003, September). The dialectics of humiliation: Polarization between occupier and occupied in Post-Saddam Iraq. Paper presented at the First Annual Meeting of Human Dignity and Humiliation Studies, Paris, France.

Fletcher, J. (1999). *Disappearing acts: Gender, power, and relational practice at work.* Boston: MIT Press.

Fletcher, M. A. (2000, March 26). Studies challenge belief that black students' esteem enhances achievement: Putting value on self-worth. *The Washington Post,* p. A03.

Fiore, J., Becker, J., & Coppel, D. B. (1983). Social network interactions: A buffer or a stress. *American Journal of Community Psychology, 11*(4), 423–439.

Ganellen, R. J., & Blaney, R. H. (1984). Hardiness and social support as moderators of the effects of stress. *Journal of Personality and Social Psychology, 47*(1), 156–163.

Garmezy, N. (1991). Resiliency and vulnerability to adverse developmental outcomes associated with poverty. *American Behavioral Scientist, 34*(4), 416–430.

Genero, N. (1995). Culture, resiliency, and mutual psychological development. In H. I. McCubbin, E. A. Thompson, A. I. Thompson, & J. A. Futrell (Eds.), *Resiliency in ethnic minority families: African-American families* (pp. 1–18). Madison: University of Wisconsin.

Genero, N., Surrey, J., & Miller, J. B. (1992). The Mutual Psychological Development Questionnaire. *Project Report, No. 1.* Wellesley, MA: Stone Center Working Papers Series.

Gray-Little, B. (2000). Factors influencing racial comparisons of self-esteem: A quantitative review. *Psychological Bulletin, 126,* 26–54.

Hartling, L. M. (2002). Strengthening our resilience in a risky world: It is all about relationships. *Wellesley Centers for Women Research and Action Report, 24*(1), 4–7.

Hartling, L. M., & Ly, J. (2000). Relational references: A selected bibliography of theory, research, and applications. *Project Report, No. 7.* Wellesley, MA: Stone Center Working Papers Series.

Jenkins, Y. M. (1993). Diversity and social esteem. In J. L. Chin, V. De La Cancela, & Y. M. Jenkins (Eds.), *Diversity in psycho-*

therapy: The politics of race, ethnicity, and gender (pp. 45–63). Westport, CT: Praeger.

Jordan, J. V. (1992). Relational resilience. *Work in Progress, No. 57.* Wellesley, MA: Stone Center Working Papers Series.

Jordan, J. V. (1994). A relational perspective on self-esteem. *Work in Progress, No. 70.* Wellesley, MA: Stone Center Working Papers Series.

Jordan, J. V. (Ed.). (1997b). *Women's Growth in Diversity: More writings from the Stone Center.* New York: The Guilford Press.

Jordan, J. V. (1999). Toward connection and competence. *Work in Progress, No. 83.* Wellesley, MA: Stone Center Working Papers Series.

Jordan, J. V., & Hartling, L. M. (2002). New developments in Relational-Cultural Theory. In M. Ballou & L. Brown (Eds.), *Rethinking mental health and disorder.* New York: The Guilford Press.

Jordan, J. V., Kaplan, A. G., Miller, J. B., Stiver, I. P., & Surrey, J. L. (1991). *Women's growth in connection: Writings from the Stone Center.* New York: The Guilford Press.

Kline, P. (1993). The *handbook of psychological testing.* New York: Routledge.

Kobasa, S. C. (1979). Stressful life events, personality and health: An inquiry into hardiness. *Journal of Personality and Social Psychology, 37,* 1–11.

Kobasa, S. C., & Puccetti, M. C. (1983). Personality and social resources in stress resistance. *Journal of Personality and Social Psychology, 45*(4), 839–850.

Kohn, A. (1986). *No contest: A Case against competition.* Boston: Houghton Mifflin.

Liang, B., Taylor, C., Williams, L. M., Tracy, A., Jordan, J. V., & Miller, J. B. (1999). The Relational Health Indices: An exploratory Study. *Work in Progress, No. 293.* Wellesley, MA: Center for Research on Women.

Levenson, H. (1981). Differentiating among internality, powerful others, and chance. In H. M. Lefcourt (Ed.), *Research in locus of control construct, Vol. 1* (pp. 15–63). New York: Academic Press.

Magnus, K. B., Cowen, E. L., Wyman, P. A., Fagen, D. B., & Work, W. C. (1999). Correlates of resilient outcomes among highly stressed African-American and white urban children. *Journal of Community Psychology , 27*(4), 473–488.

Masten, A. S. (1994). Resilience in individual development: Successful adaptation despite risk. In M. C. Wang & E. W. Gordon (Eds.), *Educational resilience in inner-city America: Challenges and prospects* (pp. 3–25). Hillsdale, NJ: Lawrence Erlbaum.

Masten, A. S. (2001). Ordinary magic: Resilience processes in development. *American Psychologist, 56*(3), 227–238.

Masten, A. S., Best, K. M., & Garmezy, N. (1990). Resilience and development: Contributions from the study of children who overcome adversity. *Development and Psychopathology, 2,* 425–444.

Masten, A. S., Hubbard, J. J., Gest, S. D., Tellegen, A., Garmezy, N., & Ramirez, M. (1999). Competence in the context of adversity: Pathways to resilience and maladaptation from childhood to late adolescence. *Development and Psychopathology, 11,* 143–169.

McIntosh, P. (1988). White privilege and male privilege: A personal account of coming to see correspondences through work in women's studies. *Working Paper, No. 189.* Wellesley, MA: Center for Research on Women.

McIntosh, P. (1989). White privilege: Unpacking the invisible knapsack. *Peace and Freedom,* July/August, 10–12.

Miller, J. B. (1976). *Toward a new psychology of women.* Boston: Beacon Press.

Miller, J. B. (1986). What do we mean by relationships? *Work in Progress, No. 70.* Wellesley, MA: Stone Center Working Papers Series.

Miller, J. B. (1988). Connections, disconnections, and violations. *Work in Progress, No. 33.* Wellesley, MA: Stone Center Working Papers Series.

Miller, J. B. (2002). How change happens: Controlling images, mutuality, and power. *Work in Progress, No. 96.* Wellesley, MA: Stone Center Working Papers Series.

Miller, J. B., Jordan, J. V., Stiver, I. P., Walker, M., Surrey, J., & Eldridge, N. (1999). Therapists' authenticity. *Work in Progress, No. 82.* Wellesley, MA: Stone Center Working Papers Series.

Miller, J. B., & Stiver, I. P. (1997). *The healing connection: How women form relationships in therapy and in life.* Boston: Beacon Press.

Miller, J. B., & Stiver, I. P. (1994). Movement in therapy: Honoring the "strategies of disconnection." *Work in Progress, No. 65.* Wellesley, MA: Stone Center Working Papers Series.

Minnich, E. (1990). *Transforming Knowledge.* Philadelphia: Temple University Press.

Nowicki, S., & Duke, M. P. (1974). The locus of control scale for college as well as non-college adults. *Journal of Personality Assessment, 38,* 36–137.

Ornish, D. (1997). *Love and survival: The scientific basis of the healing power of intimacy.* New York: HarperCollins.

Putnam, R. (2000). *Bowling Alone: The collapse and revival of American community.* New York: Simon & Schuster.

Resnick, M., Bearman, P., Blum, R., Bauman, K., Harris, K. Jones, J., Tabor, J., Beuhring, R., Sieving, R., Shew, M., Ireland, M., Bearinger, L., & Ury, J. R. (1997). Protecting adolescents from harm: Findings from the National Longitudinal Study on Adolescent Health. *Journal of the American Medical Association, 278*(10), 823–832.

Roediger, H. L., Capaldi, E. D., Paris, S. G., & Polivy, J. (1991). *Psychology.* New York: Harper Collins.

Rotter, J. B. (1966). Generalized expectancies for internal versus external control of reinforcement. *Psychological Monographs, 80,* No. 609.

Rutter, M. (1971). Parent-child separation: Psychological effects on the children. *Journal of Child Psychology and Psychiatry, 12,* 233–260.

Rutter, M. (1978). Early sources of security and competence. In J. Bruner & A. Garton (Eds.), *Human growth and development* (pp. 33–61). Oxford, England: Clarendon Press.

Rutter, M. (1979). Protective factors in children's responses to stress and disadvantage. In M. W. Kent & J. E. Rolf (Eds.), *Primary prevention of psychopathology: Vol. 3 Social competence in children* (pp. 49–74). Hanover, NH: University Press of New England.

Rutter, M. (1989). Temperament: Conceptual issues and clinical implications. In G. A. Kohnstamm, J. E. Bates, & M. K. Rothbart (Eds.), *Temperament in childhood* (pp. 463–479). New York: Wiley.

Seifer, R., Sameroff, A. J., Dickstein, S., Keitner, G., Miller, I., Rasmussen, S., &

Hayden, L. C. (1996). Parental psychopathology, multiple contextual risks, and one-year outcomes in children. *Journal of Clinical Child Psychiatry, 25*(4), 423–435.

Seigle, D. J. (1999). *The developing mind: Toward a neurobiology of interpersonal experience.* New York: The Guilford Press.

Sparks, E. (1999). Against the odds: Resistance and resilience in African American welfare mothers. *Work in Progress, No. 81.* Wellesley, MA: Stone Center Working Papers Series.

Spencer, R. (2000). A comparison of relational psychologies. *Project Report, No. 6.* Wellesley, MA: Stone Center Working Papers Series.

Stiver, I. P. (1992). A relational approach to therapeutic impasses. *Work in Progress, No. 58.* Wellesley, MA: Stone Center Working Papers Series.

Taylor, S. E., Klein, L. C., Lewis, B. P., Gruenewald, T. L., Gurung, R. A. R., & Updegraff, J. A. (2000). Biobehavioral responses to stress in females: Tend-and-befriend, not fight-or-flight. *Psychological Review, 107*(3), 411–429.

Walker, M. (1999). Race, self, and society: Relational challenges in a culture of disconnection. *Work in Progress, No. 85.* Wellesley, MA: Stone Center Working Papers Series.

Walker, M. (2001). When racism gets personal. *Work in Progress, No. 93.* Wellesley, MA: Stone Center Working Papers Series.

Walker, M. (2002a). How therapy helps when the culture hurts. *Work in Progress, No. 96.* Wellesley, MA: Stone Center Working Papers Series.

Walker, M. (2002b). Power and effectiveness: Envisioning an alternative paradigm. *Work in Progress, No. 94.* Wellesley, MA: Stone Center Working Papers Series.

Walker, M., & Miller, J. B. (2001). Racial images and relational possibilities. *Talking Paper, No. 2.* Wellesley, MA: Stone Center Working Papers Series.

Wallerstein, J. S., & Kelly, J. B. (1980). *Surviving the breakup: How children and parents cope with divorce.* New York: Basic Books.

Werner, E. E., & Smith, R. S. (1982). *Vulnerable but invisible: A study of resilient children.* New York: McGraw-Hill.

Index

TO THE OWNER OF THIS BOOK:

I hope that you have found *Diversity and Development: Critical Contexts That Shape Our Lives and Relationships*, useful. So that this book can be improved in a future edition, would you take the time to complete this sheet and return it? Thank you.

School and address:_____

Department:_____

Instructor's name:_____

1. What I like most about this book is:_____

_____

_____

2. What I like least about this book is:

_____

_____

3. My general reaction to this book is:

_____

_____

4. The name of the course in which I used this book is:

_____

5. Were all of the chapters of the book assigned for you to read?_____

    If not, which ones weren't?_____

6. In the space below, or on a separate sheet of paper, please write specific suggestions for improving this book and anything else you'd care to share about your experience in using this book.

_____

_____

_____

THOMSON
BROOKS/COLE

BUSINESS REPLY MAIL
FIRST-CLASS MAIL   PERMIT NO. 34   BELMONT CA

POSTAGE WILL BE PAID BY ADDRESSEE

Attn: Social Work/Amy Lam

BrooksCole/Thomson Learning
10 Davis Drive
Belmont, CA   94002-9801

OPTIONAL:

Your name:_____ Date: _____

May we quote you, either in promotion for *Diversity and Development: Critical Contexts That Shape Our Lives and Relationships*, or in future publishing ventures?

Yes: _____ No: _____

Sincerely yours,

Dana Comstock